Computer Methods for Science and Engineering

JAMES N. HAAG, Ph.D.
Professor of Computer Science and Physics
University of San Francisco

Consulting Editor, Computer Sciences

Computer Methods for Science and Engineering

Robert L. LaFara

Research Mathematician, Consultant Staff
Applied Research Department, Naval Avionics Facility
Indianapolis, Indiana

HAYDEN BOOK COMPANY, INC.
Rochelle Park, New Jersey

In memory of my mother, Pearl Jane LaFara

ISBN 0-8104-5766-0
Library of Congress Catalog Card Number 72-94292

Printed in the United States of America

1 2 3 4 5 6 7 8 9 PRINTING

73 74 75 76 77 78 YEAR

Preface

The purpose of this text is to provide a working engineer or scientist with basic skills in numerical methods. It may also be used in an undergraduate course, provided that the students have the necessary prerequisites, namely, mathematics through calculus and a programming course, preferably FORTRAN. It would be desirable that the student have as a prerequisite or corequisite a course in differential equations. Although some advanced concepts are included, the text is not intended to be used in a graduate course. It has been designed primarily for the graduate engineer or scientist whose undergraduate education did not include this subject. However, it is assumed that the reader has had a programming course and/or programming experience.

Emphasis in this book is on application rather than theory. *Application* is meant in the sense of incorporation of these methods into useful computer programs. Theory is not neglected, however. Enough theory is given to satisfy the reader of the theoretical basis of the methods discussed. Where proofs and derivations would be long and tedious, they are omitted.

The material is presented in a logical order so that if the book is used as a course text, the instructor can present the subject matter in the same order as given in the text. Variations in the order of presentation are also possible, since each chapter is somewhat independent of other chapters. As an additional aid to the instructor, all equations and formulas have been numbered, even though not all are referenced by the text.

Derivations, presentations, and proofs of numerical methods alone are not sufficient for a thorough understanding and appreciation of these methods. Experience is also necessary, and this book contains numerous exercises in order to provide such experience. It is strongly recommended that whenever possible, the exercises be done by computer. One reason for this is that many of the exercises, once programmed, may be used with little or no modification by the student later in his own professional practice. Also, even though a person may thoroughly understand a numerical method *and* computer programming, he may find it difficult to organize a problem for computer solution. The pitfalls encountered may not be in the numerical methods or in the programming but rather in the organization of both into a single entity. Problem organization is difficult to teach. However, through the use of examples in the text and through experience with the exercises, it is hoped that the student will be able to bridge this gap. Numerous flowcharts, based on current, standardized American National Standards Institute shapes for symbols, have been included to aid the student in problem organization.

Last, but not least, this volume may be used as a reference book. The methods and examples are each assigned their own section or subsection so that they may be quickly located without sifting through extraneous material.

A note of acknowledgment is due those persons who helped, advised, and inspired me during the preparation of this book. In particular, I owe special thanks to Dr. Martha Cox, who carefully and critically reviewed the classroom notes that formed the basis for this book. The students who used those notes—working engineers and scientists of the Naval Avionics Facility in Indianapolis—also contributed significantly through their comments and questions. Finally, I must express my appreciation to my family for their tolerance and quiet during the many evening hours I spent on the manuscript.

ROBERT L. LAFARA

Contents

1

History

1.1 THE ANCIENT ERA

The use of computing methods is as old as recorded history. The oldest known written document is a set of Sumerian business records that are over 5000 years old. The ancients were well versed in computing. Many clay tablets and papyri exist that have been identified as mathematical texts. The Sumerians, Babylonians, and Egyptians had complex civilizations requiring a knowledge of numerical methods for the division of lands, building construction, tax assessment, etc. Although none of the above examples seem very complex today, it is difficult to imagine how such calculations were carried out with the clumsy number systems then in use. Most of us are familiar with Roman numerals, which are particularly inconvenient for computing. Some systems predating the Roman's system were somewhat more convenient. The Sumerians used a sexagesimal system (based on the number 60) that even included a method for representing fractions. Basically the system was similar to our hours, minutes, seconds scheme and is probably responsible for it. That number system was used in various forms from 2500 B.C. until as late as 800 A.D.

It is interesting to note the antiquity of some concepts.

1. The method for calculating the area of a rectangle was known as early as 3500 B.C.
2. The Babylonians were solving quadratic equations about 2000 B.C.
3. The Egyptians surveyed 700 miles of the Nile River for an irrigation system about 1800 B.C.
4. The Pythagorean theorem was known at least 1000 years before the time of Pythagoras.
5. The value of π was known to some precision by the Egyptians and others. The pyramids are constructed such that a circle whose radius is equal to the height of a pyramid has approximately the same perimeter as the base of the pyramid.

The most important development for computing was the use of place value in numeric notation (i.e., the value of a symbol, or group of symbols, depends upon the position of that symbol, or group of symbols, within a string of symbols). This development has been compared in importance to the development of alphabetic writing as opposed to pictograph writing. The use of place value was even present in the ancient sexagesimal system. That system originally did not include a symbol for zero, although some works contain blanks to indicate zero. At least by the Hellenistic age (300 B.C. to 150 A.D.), the Greeks were using a symbol for zero.

1.2 THE MEDIEVAL ERA

Before this period there was a renaissance in astronomy culminating in the publication of Ptolemy's *Almagest* about 140 A.D. This book contained tables for plane trigonometry based on chords of circles and a theory (although incorrect) of planetary motions. The Ptolemaic theory became the standard in astronomy for many centuries and brought mathematical precision to astronomical computations.

The history of the decimal number system is somewhat obscure. It is believed that the system was derived from a Hindu system of about 800 A.D. known as Devangari numerals. We know these same numerals (essentially) as Arabic numerals. The reason for this seeming paradox is that an Islamic mathematician Abu 'Abdallah Mohammad ibn Musa al-Khwarizmi popularized this system by writing a book describing the system in 825 A.D. Although no copies of this book are in existence, we know of it through a 12th-century translation. The advent of the decimal number system simplified computing. The next major event in the simplification of computing was the invention of logarithms by John Napier (1550-1617).

The previous discussion gives a sketchy picture of the history of numbers up to the 17th century. The history of astronomy which has had a profound effect on computing methods can be divided into three eras:

1. Ancient (prior to Ptolemy),
2. Medieval (Ptolemy to Galileo),
3. Modern (after Galileo).

The history of mathematics does not divide this clearly. We cannot clearly define the Ancient from the Medieval period. The development was more gradual and during some periods even regressive. The significant thing to note is that during the 17th century, the Modern age of mathematics as well as astronomy arrived.

1.3 THE MODERN ERA

The challenge of the Ptolemaic theory by Nikolaus Copernicus (1473-1543), the deduction of the laws of planetary motion by Johannes Kepler (1571-1630), and the telescopic verification of these by Galileo Galilei (1564-1642) set the stage for the new era. The new era was entered when Sir Isaac Newton (1642-1727) formulated the Law of Gravity and invented the Calculus.

The development of mathematics and other sciences since Newton is well documented and will not be treated here. However, the development of computing methods and computers will be discussed. Newton's contributions led to other developments. The need for computing and computing methods likewise increased. It is surprising to learn that, although computers are relatively new, many of the computing methods (or algorithms) that will be covered in subsequent chapters were developed in the 19th century. One of the greatest mathematicians was Karl Freidrich Gauss (1777-1855). In later chapters, numerical methods developed by Gauss will be discussed.

I would like to interject an interesting note at this point before continuing. The word *algorithm*, which means a computing method or scheme, is believed to be a corruption of the name of the Arabic arithmetician al-Khwarizmi. The similar sounding word *logarithm* is derived from the Greek words *logos* (meaning word or proportion) and *arithmos* (meaning number).

1.4 THE COMPUTING ERA

Although the development of useful computing methods was relatively swift, the development (until recently) of computing equipment was painfully slow. Blaise Pascal (1623-1662) of France built the first successful adding machine in 1642. The Englishman Charles Babbage (1792-1871) in the early and middle 19th century designed and built, in part, various calculating machines. His concepts were very advanced; in fact, too advanced for his time. Neither the need, the understanding, nor the technology was available to implement his ideas successfully.

The most significant development from a practical point of view was the invention of punched card tabulating equipment by Herman Hollerith (1860-1929). His work on his "Census Machine" started in 1882. He applied for a patent in 1884 which was issued in 1889. Although his machine was developed primarily for the U. S. Bureau of the Census, it was done at his own expense and he had to compete with two other systems to be selected for use in the 1890 census. Hollerith eventually sold his company, Tabulating Machines Company, which later merged with other companies to become International Business Machines (IBM).

The use of punched card tabulating equipment for business data processing grew from this beginning. The introduction of automatic calculators is less clearly defined. Many claims have been made about who developed the first computer. At least part of the confusion is because of semantics. More will be said on that subject later.

It is impractical in this space to attempt to discuss the development and evolution of computing equipment in detail. Like many other technological breakthroughs, many persons, working independently, were thinking about or building computing equipment in the 1930s. The only meaningful developments came as the result of large expenditures of time and money. One such series of developments was carried out at Bell Laboratories by George R. Stibitz and others. The first machine, designed to perform the four basic mathematical operations on complex numbers, was put into operational use at Bell Laboratories, in New York City, in January 1940. Communication with this machine was via a teletypewriter. This was a fortunate choice as it allowed Dr. Stibitz to give live demonstrations when he presented a paper before the American Mathematical Society, at Dartmouth College, in September 1940. This first demonstration of remote use of a computer proved to be an interesting precursor of things to come. Other Bell Lab machines included such features as programming and data entry via punched tape, and floating-point arithmetic.

Meanwhile, Dr. Howard Aiken of Harvard, working under a government contract, designed the Mark I calculator which was built by IBM and completed in 1944. It, like the Bell machine, was a relay calculator whose instructions were read as needed from punched tape. J. Presper Eckert and Dr. John M. Mauchly, working at the University of Pennsylvania, designed and built the ENIAC (Electronic Numerical Integrator And Calculator). This machine was built for U.S. Army Ordnance and was completed in 1946. It was the first electronic calculator, but it was program controlled by punched tape like the Mark I.

Also, in 1946, a historic paper was published by John von Neumann *et al.* describing the logical design of an electronic computing instrument. The most notable concept that was included was the idea that the device has a memory that can store both computed values and instructions. It is this concept that differentiates a computer from a calculator. That is, a computer is a calculator that can store its own instructions and can alter its sequence of instructions based on values of computed quantities. This paper led to the design and construction of EDVAC, the first electronic computer. This machine was also built for Army Ordnance by the University of Pennsylvania.

The first production model computer was UNIVAC I designed by Eckert and Mauchly and built in 1951 by the Remington-Rand Corporation. It is significant to note that serial number one was sold to the Bureau of the Census.

The year 1951 clearly marks the beginning of the Computer Age. The introduction of commercially available computers has made this possible. Since that time, we have seen vacuum-tube machines give way to solid-state machines, and those in turn are being replaced by integrated-circuit machines. Although the hardware developments have been remarkable, the development of software (programming systems, aids, etc.) has been even more revolutionary. The introduction of FORTRAN (FORmula TRANslation) in 1956 and COBOL (COmmon Business Oriented Language) in 1961 were major milestones in software development. Another important milestone was the introduction of conversational timesharing systems. Although work on such systems dates back to the late 1950s, the most important step, from a practical viewpoint, was the development of a system at Dartmouth in 1964. This system was developed under a National Science Foundation grant by Profs. John G. Kemeny and Thomas E. Kurtz. Their system has done for the timesharing computer industry what the Model T did for the automotive industry. As was noted earlier, the first remote computer entry demonstration was made at Dartmouth in 1940. At that time, Dr. Stibitz, who presented the paper, thought that the newspaper reporters were being extravagant when they predicted that such devices would someday be used by students to do their homework. The fact that students are using computers to do their homework and that engineers and scientists use computers daily in their work makes it imperative that they learn numerical methods in order to capitalize on the power of the computer.

2

Computers and Computing

2.1 INTRODUCTION

It is assumed that the user of this text has had some previous training and experience with some programming language such as FORTRAN. Therefore, no chapters on coding are included. However, since the presentation of methods is strongly computer oriented, it is appropriate that some discussion of computers and computing in general be included. The topics Computer Organization, Problem Organization, and Flowcharting are given in later sections of this chapter. The section on flowcharting is especially important, since the concepts and standards presented will be used throughout the remainder of the text.

The term *programming* is often used ambiguously and will probably continue to be used ambiguously. Therefore, an understanding of how this term is used is more important than a precise definition. In the course of preparing a problem for computer solution, three activities are involved: systems analysis, programming, and coding. Systems analysis consists of defining the problem and outlining, in general, the method of solution. The second activity, programming, is the preparation of the detailed plan of the method of solution. This detailed plan is usually a flowchart or a series of flowcharts giving all the steps of the procedure. The third activity is the writing of the computer program(s) in a language (or languages) that can be used by the computer to solve the problem.

It is seldom possible to separate these three activities. If the same person performs all three steps (typically the case in scientific programming), as analyst–programmer–coder he usually is constantly switching roles during the preparation of the job. In business data processing, the roles are usually delegated to different teams; but even then the roles are not clearly defined. Typically there are just two teams. One of these may consist of analysts and/or analyst–programmers, and the other team may consist of programmer–coders and/or coders. The term *programming* is used to describe all of the preceding functions, combinations of the preceding functions, or any one of these functions.

The previous paragraphs, although they do not resolve the "programming" question, do aid in defining a purpose of this text. This purpose is to teach, by example, systems analysis and programming. There are few generalizations that can be made about these subjects. The methods of solution for different problems may be quite different, and even one problem may have alternate methods of solution. The primary purpose of this text is to present a variety of computing methods that will be useful to scientists and engineers. A secondary purpose is to demonstrate through the organization of these methods how to organize larger problems for solution.

2.2 COMPUTER ORGANIZATION

Computer models differ from one another in many of their details. However, all computers have the same general organization, which is shown schematically in Fig. 2.1. There are five major elements:

1. *Memory*—A computer's memory is a device for storing data and instructions that are to be used by the computer. Although a computer usually has only one main memory, it can also have auxiliary memories. These memories may be of different types. A typical computer will have a main memory made of magnetic cores and one or more auxiliary memories of the magnetic disc or magnetic drum type. For simplicity, Fig. 2.1 shows only the main memory. The solid arrows indicate the normal flow of information between the memory and the other elements.
2. *Control Unit*—The computer's control unit is a device that calls instructions from the memory, decodes them, and controls the other units of the computer as specified by the instructions. The dashed arrows in Fig. 2.1 indicate the flow of control signals.
3. *Arithmetic Unit*—The arithmetic unit of a computer is a device that performs the fundamental operations (addition, multiplication, comparison, data manipulation, etc.) on the data.
4. *Input Unit(s)*—An input unit is a device that translates information that is stored on an external medium (cards, tape, etc.) into a form that can be entered into the computer's memory. A typical computer can have several input devices that can be of different types. (For eyample, a card reader that senses holes in cards, an input keyboard that senses switch closures, a punched paper tape reader that senses holes in tape, a magnetic tape unit that senses magnetized spots on an iron oxide coated tape, etc. Each of these devices then translates the detected information into a form that can be stored in the memory.)
5. *Output Unit(s)*—An output unit is a device that translates information from the memory into a form that can be stored on an external medium. A typical computer can have several output devices that can be of different types. (For example, a card punch that punches holes in cards, a tape punch that punches holes in tape, a magnetic tape unit that magnetizes spots on an iron oxide coated tape, a printer that prints symbols on paper, etc. Each of these devices recodes the information as it comes from the memory into the code required for the external medium. Note that a magnetic tape unit can be used for both input and output.)

A programmer working in a high-level programming language, such as FORTRAN, need not know many of the details of the computer. However, the basic computer organization, as outlined here, is useful in organizing a problem for computer solution.

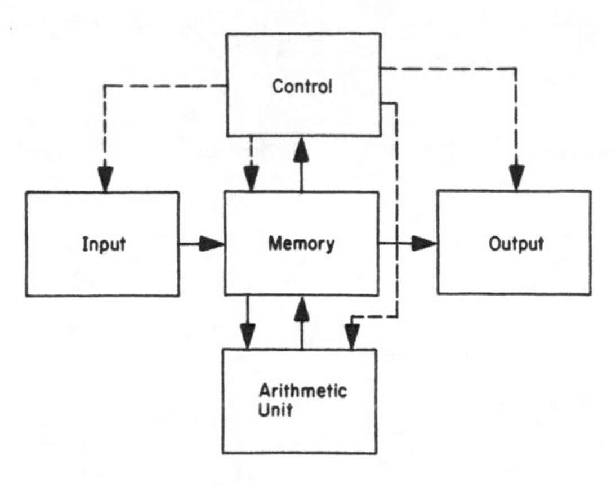

Fig. 2.1. Computer organization.

2.3 PROBLEM ORGANIZATION

Problem organization is a very large subject area. It is larger than the subject area of this book, and, in fact, is larger than an entire college curriculum. Problems range in size from "how to do simple arithmetic" to "how to land men on the Moon." It is difficult to teach methods of solving large problems. What we typically do is teach how to solve a variety of small problems and depend on experience and ingenuity in using that knowledge, to solve larger problems.

The subject matter of this text is very useful in helping a person solve large problems. It is helpful in that the methods covered may be needed to solve large problems; but it is also helpful because the experience gained in organizing numerical methods will be applicable to organizing large problems.

The basic elements in organizing problems are similar for large or small problems. They are:

1. We must first define the problem. That is, we must determine what our objectives are.
2. We must determine what parameters are fixed and what parameters can be varied.
3. We must hypothesize a method of solution.
4. We must select parameter values, proceed with the solution, and examine the results.
5. If the results are not satisfactory, we must select different parameter values, change the method of solution, or change our objectives.

These basic elements of problem organization are all essentially human activities. However, the method of solution will be wholly or partially automated. Furthermore, the method of solution selected (element 3) will be directly or indirectly involved in all the other elements. The method of solution is called an *algorithm.* It should be a step-by-step procedure that leads to unambiguous results in a finite number of operations. The algorithm should include steps for aborting the job (or a portion thereof) if it is not

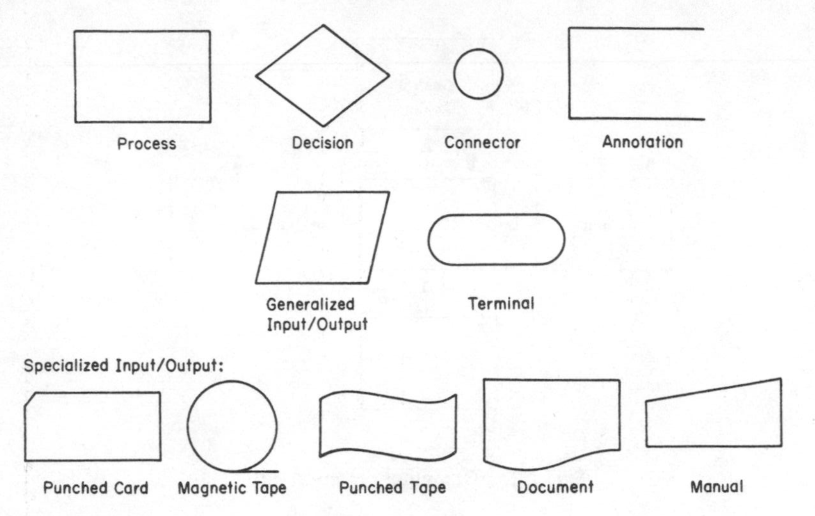

Fig. 2.2. Standard flowchart symbols. See Appendix for relationships of these symbols to standard FORTRAN statements.

possible to obtain unambiguous results in a finite number of operations. In such a case, the algorithm should include methods for identifying the source(s) of difficulty.

The method of solving a problem may include several algorithms combined into one large algorithm. In this text, algorithms will be discussed. Usually such algorithms will only be parts of larger algorithms. Therefore, the algorithms given will not necessarily provide a complete method of solution for a problem. In order to use these methods either in the exercises or in real problems, the user must add other steps to obtain a complete algorithm.

An algorithm may be presented either verbally or pictorially. A recipe is an example of a verbal algorithm. In this text, the methods discussed will usually be given both verbally and pictorially.

2.4 FLOWCHARTING

The method of presenting an algorithm pictorially is by use of a flowchart. *A flowchart is a diagram that outlines a step-by-step procedure for solving a problem.* Such a diagram usually consists of a set of boxes that are connected by arrows. The boxes contain sets of instructions for the various steps and the arrows define which sets of instructions may follow other sets of instructions.

The American National Standards Institute (ANSI) has published† a standard for flowchart symbols. Adherence to such standards is desirable, expecially when flowcharts are used for communication and documentation of computer problems. Some of the most frequently used symbols are shown in Fig. 2.2.

† *AMERICAN NATIONAL STANDARD: Flowchart Symbols and their Usage in Information Processing,* ANSI X3.5–1970, American National Standards Institute, 1430 Broadway, New York, N.Y.

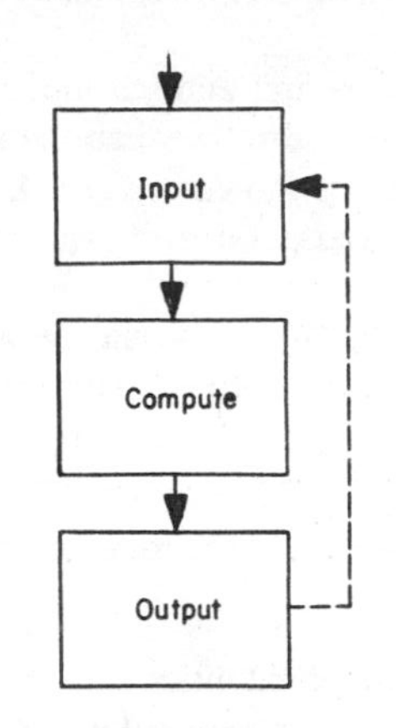

Fig. 2.3. General simplified flowchart.

A simplified flowchart of a problem to be solved by computer is shown in Fig. 2.3. When organizing a problem for computer solution, it is helpful to make several flowcharts of the same problem, each showing more detail than the previous one. In this manner, the final flowchart emerges through an evolutionary process that allows the programmer to visualize the overall program before getting too involved in details.

Relating the basic elements of problem organization, given in Sec. 2.3, to the simplified flowchart of Fig. 2.3 will aid in developing a more detailed flowchart:

1. Once the problem is defined and the objectives have been determined, that portion of the flowchart labeled OUTPUT may be given in more detail. The quantities whose values are to be output, the format of this output, and the output media may be selected.
2. Selection of those parameters whose values we wish to vary determines the detail for the INPUT portion of the flowchart.
3. Selection of an algorithm or a set of algorithms will supply detail for the COMPUTE section of the flowchart. Note that the COMPUTE section can involve portions of the INPUT and OUTPUT.
4. After the preceding steps have been completed, values for the input parameters can be selected, and the problem coded and executed.
5. Frequently, after an initial trial run, it is necessary to modify the procedure and re-execute. Sometimes it is possible to include such modifications as part of the procedure.

A flowchart is such an important tool in developing and documenting a problem for computer solution that at least one should be made for each problem. A flowchart that does not conform to ANSI standards is better than no flowchart at all.

If a problem consists simply of evaluating a set of equations once, a flowchart may not be necessary. However, this is not usually the case. Either a problem is to be executed repeatedly with various input values or it may contain various branch points that allow alternate computing steps. Usually both of the above conditions and also other conditions apply which require a complex diagram to depict the whole job. A classical example that is both easy to understand and yet illustrates the various complexities that arise is the solution of a simple quadratic equation of the form $Ax^2 + Bx + C = 0$. A simple flowchart for this problem is shown in Fig. 2.4. In this flowchart, we have the basic required elements. We will read parameter values from a card, find the required results, print them, and then repeat with new values. For a human computer with prior knowledge of quadratic equations, Fig. 2.4 would be adequate because he can deal with peculiarities as they arise. In order to prepare this problem for solution by an automatic computer, more detail is needed. The program planner must anticipate *all* of the peculiarities that may arise. In this problem we must consider the possibilities that

1. any one or any combination of input parameters may be zero,
2. the roots may be complex.

Figure 2.5 has provisions for these possibilities and includes other details not shown in Fig. 2.4. Figure 2.5 diagrams the steps required if

1. both coefficients A and B are zero,
2. the equation is only linear and therefore only one root exists,
3. the roots are real, and
4. the roots are complex.

For illustrative purposes, extra variables and steps are shown. In an actual problem, certain improvements can be made so that the actual computer program is more efficient.

Even more detail can be shown than that shown in Fig. 2.5. For instance, the symbol $\sqrt{R}$ is shown to indicate finding a square root. Since a square root function is normally available as a computer library routine, no detail is usually shown to indicate how the square root is to be found. If no such routine is available (or some unusual method is to be used), then the details of the square-root method should be shown either as part of the main flowchart or on a separate flowchart.

The primary value of flowcharts is to show loops and decision points within a procedure. Figure 2.5 illustrates both features. A *loop* is a portion of a procedure that is repeated. A loop is aptly named because it is any closed path on a flowchart that can be traced in the direction of flow. Figure 2.5 actually contains four loops, but they have some common portions. A *decision point* is a point in a procedure where some condition (usually the value of an expression) is interrogated to determine which one of multiple paths to follow in continuing the procedure. In a procedure, a decision point is usually used to determine when to terminate the execution of a loop. However, in Fig. 2.5, the execution of the procedure would continue as long as new parameter values were available.

To illustrate the usual case, Fig. 2.6 is a flowchart of Newton's iteration method for finding a square root. This procedure is based on the fact that the formula

$$r_{n+1} = 0.5\left(\frac{N}{r_n} + r_n\right)$$

yields a sequence $(r_0, r_1, r_2, \ldots)$ of approximations that approach the square root of N. This will be discussed in detail in Chap. 6. In Fig. 2.6, it is assumed that a number N exists whose square root is to be found. Initially, some first approximation is selected; then a loop is entered to repeatedly find improved approximations until two consecutive values agree within some acceptable error ε. When this condition is satisfied, execution of the loop is ended and the result R' can be used in subsequent computations.

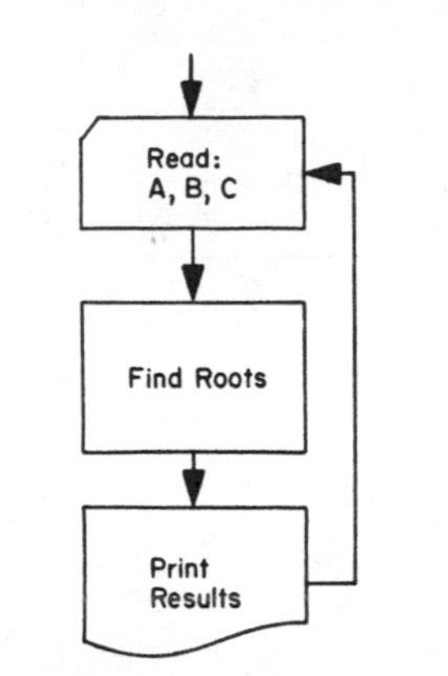

Fig. 2.4. Simplified flowchart for solution of a quadratic

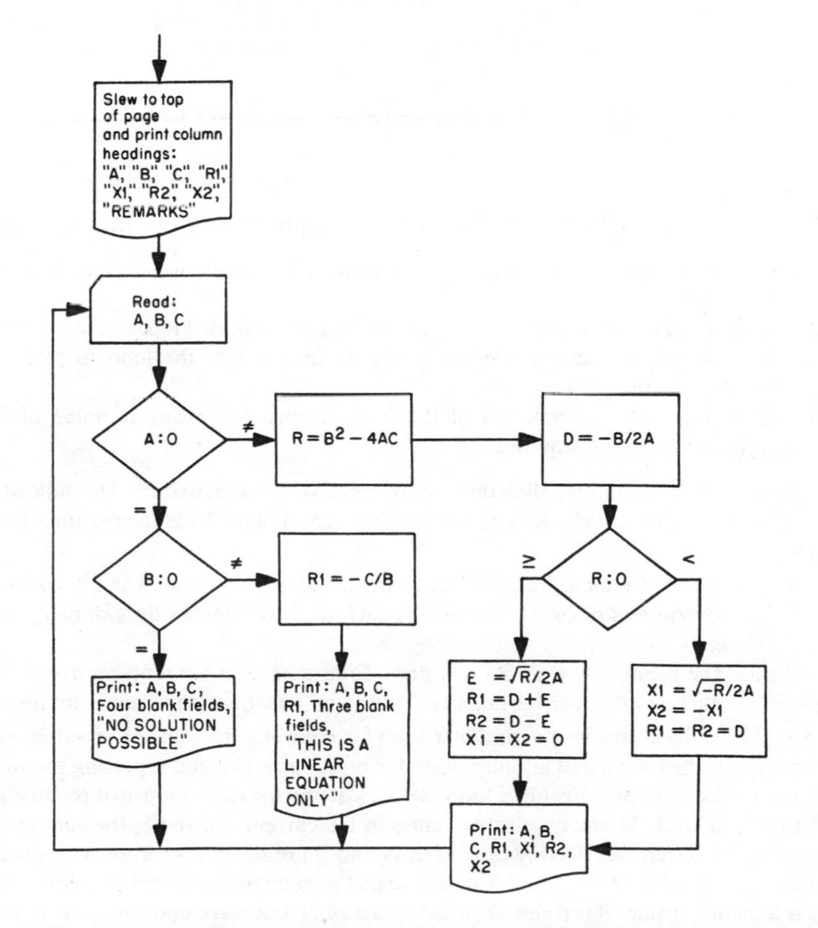

Fig. 2.5. Flowchart for solution of a quadratic equation.

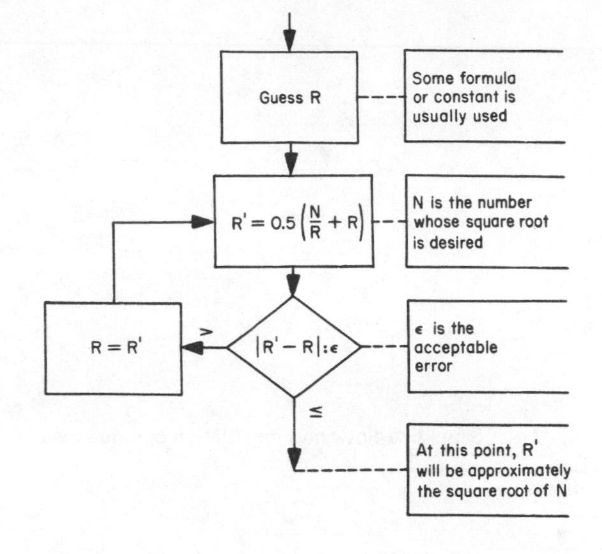

Fig. 2.6. Flowchart for finding a square root.

The following activities are usually required in a procedure that contains a loop:

1. *Initializing*—Before entering a loop, some parameters usually need to be initialized.
2. *Looping*—The actual computations of the loop are executed repeatedly.
3. *Terminating*—A decision process is usually included in the loop to stop the repetition of the loop.
4. *Finalizing*—After completion of the loop, further processing of some of the parameters may be required.

The two previous examples illustrated only some of these activities. The following example illustrates all of the activities in one problem. Figure 2.7 is the required flowchart.

Given: A file of punched cards. Each card contains one value. The file contains several sets of cards. The end of each set is designated by a dummy data card containing a value of 99999.

Find: The average of each set of values. Print a set number (consecutive values starting with one), the number of values in the set, and the average for the set.

In Fig. 2.7, we have two loops: an inner loop for counting the number of values and accumulating their sum, and an outer loop for printing results and repeating for additional sets. Consider only the inner loop. Before starting the inner loop, two parameters must be initialized: N, the number of values in the current set, and S, the sum of the values in the current set. In this case, both N and S are set to zero. After initializing, the loop is entered. A value is read from a card and is compared with 99999 to determine if it is a dummy value; if it is not, then the values of N and S are updated. (Please note that when a statement such as $S = S + X$ appears on a flowchart, it is *not* an algebraic

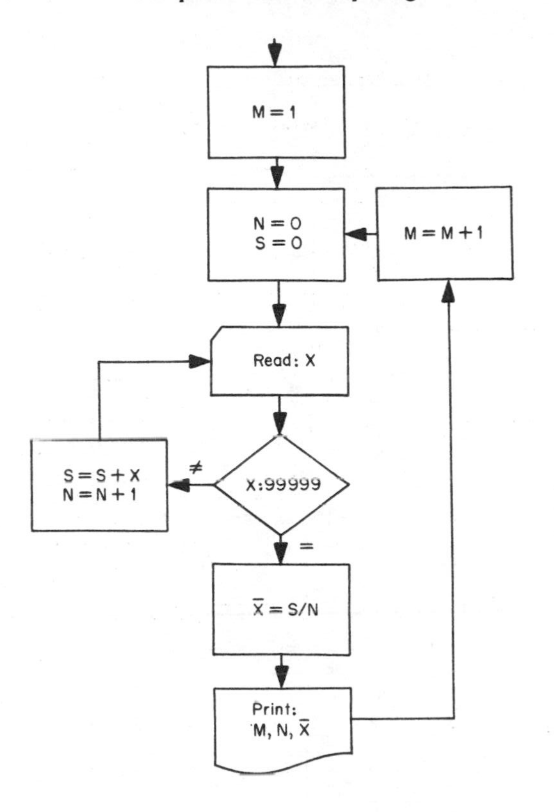

Fig. 2.7. Flowchart for determining an average.

equation, but a statement that means replace the value of the variable on the left side of the "=" with the value of the expression on the right side of the "=". In this case, the statement means replace the old value of S with the sum of the old value of S and the value of X.) When the value of X is found to be 99999, then execution of the loop is terminated and the current values of S and N are used in a final computation to determine the average.

In this example, the loops are called *nested* loops because one loop is entirely within the other. When loops are nested, care should be taken to insure that the innermost loops contain no more steps than are absolutely necessary. The reason for this is that the innermost loops are executed more times than the outer loops. Any step that can be moved from an inner loop to an outer loop should be moved, since that would improve the efficiency of the whole procedure.

For example, suppose we must find

$$S = \sum_{i=1}^{100} 3.6x_i$$

To find the individual products of $3.6x_i$ is wasteful. It is more efficient to first find the sum of the x_i's and then multiply the sum by the constant 3.6. This example is fairly

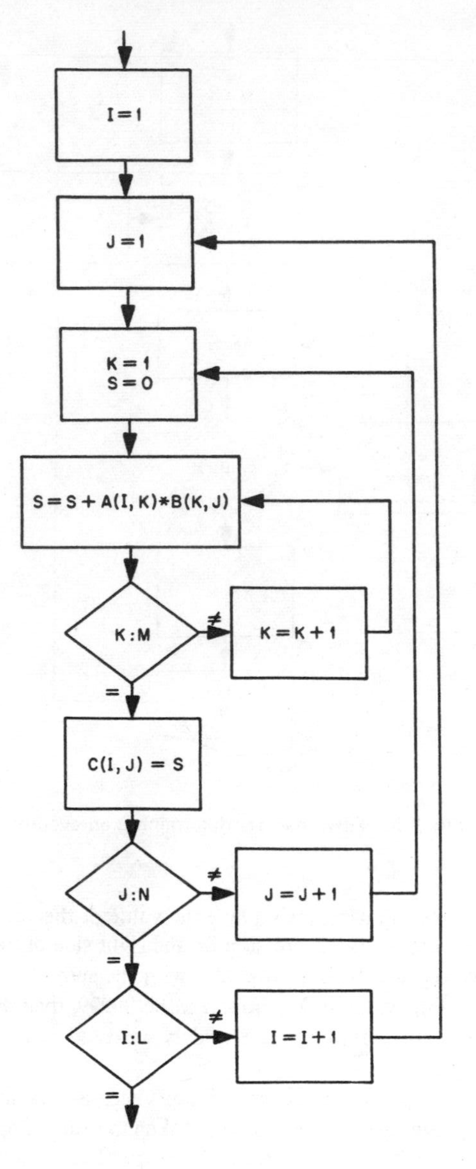

Fig. 2.8. Flowchart for matrix multiplication.

simple and the action to be taken is obvious. However, in many cases the best method of computing is not as obvious and may even require a knowledge of how the particular computer and/or programming language being used executes some operations.

Figure 2.8 is a flowchart for a matrix multiplication. The L by M matrix A is to be post multiplied by the M by N matrix B to yield the L by N matrix C. In this example,

```
      DO 1 I = 1, L
      DO 1 J = 1, N
      S = 0.
      DO 2 K = 1, M
    2 S = S + A(I,K)*B(K,J)
    1 C(I,J) = S
```

Fig. 2.9. FORTRAN coding of matrix multiplication.

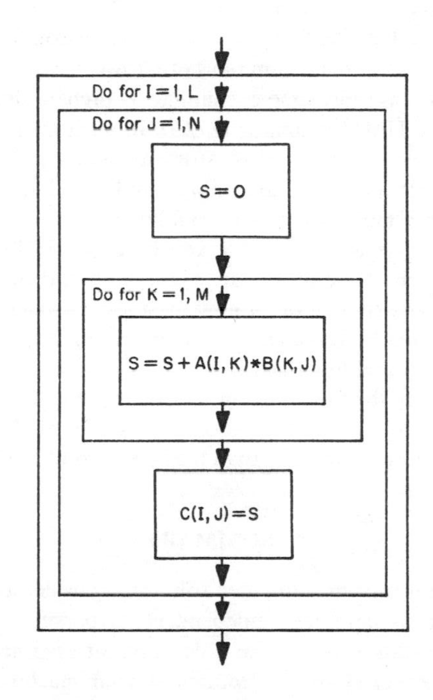

Fig. 2.10. Abbreviated flowchart for matrix multiplication.

there are three nested loops. The innermost loop is used to accumulate a sum of products. Each accumulated sum becomes an element of the result matrix C. Actually, these sums can be accumulated directly in the answer elements. The expression in the innermost loop would read

$$C(I,J) = C(I,J) + A(I,K) * B(K,J)$$

and each C(I,J) would have to be initialized to zero before entering the loop. However, the method used in Fig. 2.8 is more efficient because in most programming systems more time is required to operate on a multiply subscripted variable than to operate on an unsubscripted or singly subscripted variable. Although the amount of time saved on an individual operation may be small, the total savings may be significant. In this particular example, the innermost loop is executed M times for every one time of the next loop; this next loop is executed N times for every one time of the outermost loop; and the outermost loop is executed L times for each time the entire program is executed. Therefore, the innermost loop is executed M*N*L times for each execution of the entire program. Suppose for example that M, N, and L are each 100. In a problem this large, the innermost loop would be executed one million times. Therefore, each microsecond saved in the innermost loop will shave nearly a full second off the total execution time. Not all of this time saving will be realized, since additional operations in other loops may be necessary. However, these additional operations only absorb a small percentage of the time saved.

As stated earlier in this chapter, it is better to have a flowchart that does not conform to ANSI standards than to have no flowchart at all. The reason for this statement is to encourage the problem solver to make some sort of a flowchart. In the example of matrix multiplication just given, it is more time-consuming to prepare the flowchart (Fig. 2.8) than to prepare the FORTRAN coding (Fig. 2.9). In such an instance, the problem solver is disinclined to flowchart his problem. However, a better alternative would be to document the problem with a nonstandard abbreviated flowchart. A suggested method for flowcharting loops in a more compact form follows.

To flowchart a loop (especially those that can be coded in FORTRAN using the DO statement), first flowchart the inner part of the loop and then enclose this small flowchart in a box with an explanation of how the small flowchart is to be used. This procedure can be illustrated by flowcharting the matrix multiplication example given earlier. Fig. 2.10 is the required flowchart.

In addition to illustrating the flowcharting of one loop, Fig. 2.10 also illustrates that this method can be extended to include loops within loops. In some respects, the flowchart of Fig. 2.10 is more meaningful than the flowchart of Fig. 2.8.

2.5 SUMMARY

When preparing a problem for computer solution, consideration should be given to making the program as computer independent as possible. However, there are often extenuating circumstances that make it desirable or even necessary to capitalize on the presence (or absence) of special features of your machine (hardware) and/or programming systems (software). (For example, if your computer has no card reader, then punched card input is not possible.) Therefore, a person who is preparing a problem for computer solution should have some knowledge of the hardware and software to be used. Furthermore, a knowledge of how the hardware and software to be used differs from other hardware and software would be helpful.

The problem analyst must plan the entire problem. He must select appropriate computing methods for its various parts and must anticipate inconsistencies and ambiguities that may arise. Finally, he must integrate these steps into a complete, detailed plan and document it with a flowchart.

The flowchart serves several important functions:

1. It diagrams the problem so that the overall plan of solution can be visualized.
2. It provides a convenient "road map" to follow when coding the problem.
3. It serves as an invaluable aid when testing and correcting (i.e., debugging) the program.
4. It documents the job.

EXERCISES

1. By means of a flowchart describe how to change a flat tire.
2. Draw a flowchart of how you prepare your favorite food.
3. Given X, Y, and Z as inputs, draw a flowchart that will print out X, Y, and Z and also A, B, and C, where A is the largest value of the inputs, C is the smallest value of the inputs, and B is the remaining value of the input set. Provide for multiple sets of inputs.
4. Prepare a flowchart for calculating N factorial (N!) given N as input. N is a non-negative integer. Note that 0! = 1. Provide for printing N and N!.
5. Flowchart the process for finding the number of combinations of n things taken p at a time using the formula

$$C(n,p) = \frac{(n-p+1)(n-p+2)(\cdots)(n-1)(n)}{(1)(2)(3)(\cdots)(p)}$$

Assume that the quantities n and p are inputs that are positive integers with $n \geqq p$, and that n, p, and C are to be printed. Allow for multiple input sets.

6. Draw a flowchart for finding X^n by repeated multiplication given X and n as inputs. Assume that n is a positive integer.
7. Draw a flowchart for finding X^n by repeated multiplication given X and n as inputs. Assume that n is *any* integer, including zero.
8. The general form of a quadratic equation is

$$y = Ax^2 + Bx + C$$

Draw a flowchart that will evaluate y for values of x at intervals of Δx for $x_s \leqq x < x_f$. Assume that A, B, C, x_s, x_f, and Δx are inputs and that A, B, and C are to be printed once followed by values of x and y. Allow for multiple input sets. Start a new output page with headings for each set of inputs.

9. The x, y coordinates of a vacuum trajectory of a point mass are given by

$$x = \dot{x}_0 t \qquad \text{and} \qquad y = \dot{y}_0 t - gt^2/2$$

Prepare a flowchart for calculating the coordinates of a point mass at intervals of Δt in the variable t. Assume that t is initially zero and that $\dot{x}_0$, $\dot{y}_0$, g, and Δt are given as inputs. To provide for multiple inputs, start a new page with headings for each set of inputs. Terminate the computations after printing the first value of y that is negative.

10. The nth root X of a quantity Y can be found by repeatedly applying the recursion relation

$$X' = X + \frac{1}{n}\left(\frac{Y}{X^{n-1}} - X\right)$$

where X′ calculated at one step is used as X for the following step. Draw a flowchart of this process assuming that n and Y are inputs and the initial guess for X is 1. Stop the process when the absolute value of the difference between X′ and X is less than 0.00001. Allow for printing of the result and for multiple inputs.

11. Draw a flowchart of a recursive method for calculating the first five terms of the sine series.
12. Assume that a subroutine is available for finding the principle value of an angle given its tangent. Draw a flowchart of a process that will find any angle in the range $-\pi$ to π given both its sine and cosine.
13. Draw a flowchart for finding the sine of *any* angle. Assume that subroutines are available for finding the sine and cosine of angles not exceeding $\pi/4$ in absolute value.
14. Assume that a subroutine is available for calculating the sine of any angle. Prepare a flowchart of a routine that will print out a table of sines for arguments starting at x_s and

continuing through x_f at intervals of Δx. Assume that x_s, x_f, and Δx are available as inputs. Print the value of x and its corresponding sine, one set per line. Provide for headings at the top of each page and advance to a new page every 50 values.

15. Prepare a flowchart of a process for finding the unknown parts of a plane triangle given the following inputs: Case 1 — a, b, c (three sides), Case 2 — a, C, b (two sides and the included angle), and Case 3 — A, c, B (two angles and the included side). Print out all six triangle parts. Allow for multiple input sets.

3

Errors and Error Studies

3.1 INTRODUCTION

The solution of a numerical problem is usually not exact. Because of this, it is desirable to know either how to improve the accuracy of a solution or to determine the amount of the inaccuracy. The subjects of this chapter are *Errors* and *Error Studies.* Note that in the context of this book, the term *error* does not mean a mistake, but the difference between a quantity's exact and approximate values.

Excluding accidental errors, errors can be divided into three general classes: computing, measurement, and modeling. *Computing errors* result from the limitations of the computing devices or computing methods used. *Measurement errors* result from using approximate data. *Modeling errors* result from the use of approximate models. These classes are not distinct. Some measurement errors can be classified as computing errors while other measurement errors can be classified as modeling errors.

Errors of each of the three classes can be expressed in either one of two mathematical forms. These mathematical forms will be identified as error types and will be defined and discussed in Sec. 3.2. The source and propagation of computing and measurement errors will be presented in Sec. 3.3 along with a discussion of significant figures and examples of how to decrease error propagation. In Sec. 3.4, methods for evaluating the effect of measurement and modeling errors will be given.

3.2 TYPES OF ERRORS

There are two types of errors: absolute and relative. We will see in later sections that in some analyses it is convenient to use absolute errors and in others, relative errors.

Let x_e represent the exact value of the quantity x and x_c represent the calculated value of x. Then the absolute error ε_a is given by

$$\varepsilon_a = x_e - x_c \tag{3.1}$$ †

† In the physical sciences, absolute errors are usually expressed as $\varepsilon_a = x_c - x_e$, rather than as shown in Eq. (3.1). Which of these conventions is employed is not important, but consistency in the use of one or the other of these is important. The definition of Eq. (3.1) is given here to be consistent with later error analyses. It is common practice to express the inherent error of a numerical method or procedure such that the error when added to the approximate value gives the exact value.

The relative error ε_r is given by the ratio of the absolute error to the exact value:

$$\varepsilon_r = \frac{\varepsilon_a}{x_e} \tag{3.2}$$

Usually we do not know the exact value of a variable. If we did know it, there would be no need for error analysis. Therefore, in order to determine the absolute error, it may be necessary to substitute for the exact value a calculated value that is known to be or believed to be a more accurate value than the calculated value for which we are determining the error. In determining the relative error, we may need to substitute the calculated value for the exact value.

$$\varepsilon_r \doteq \frac{\varepsilon_a}{x_c} \tag{3.3}$$

3.3 COMPUTATIONAL AND MEASUREMENT ERRORS

It is important to emphasize that numbers do not exist in reality, but are only a concept. This concept is very useful; but in order to use it, some physical representation of numbers must be made. We may represent numbers with pebbles or sticks, or we may represent them with symbols (numerals), or, more recently, with holes in cards, magnetized particles, etc. The point is that we cannot represent all numbers exactly. Although the symbol π represents a certain transcendental number exactly, when we need to use it in a computation we must use an approximation, say 3.1415926535 in decimal notation. It is possible to devise a number system in which π would be an exact quantity, but then such simple numbers as unity could not be represented exactly.

In addition to the errors introduced by our number system representations, we may also have errors due to inaccurate measurements. The accuracy of measurements in the physical sciences is limited and therefore the accuracy of any computations based on these measurements is also limited.

3.3.1 Rounding and Truncation

Rounding and truncation errors result from our number system representations. *Truncation* is the act of discarding unwanted figures from a number, whereas *rounding* is truncation plus some adjustment of the last retained figure based on the value of the discarded figures. Truncation also can be the discarding of unwanted or unneeded terms of a series. An error resulting from this type of truncation may also be classed as a modeling error. Such an example will be illustrated later in this chapter.

Truncation errors may also occur in converting from one number system to another. The decimal fraction 0.1 becomes a repeating fraction in binary (0.000110011001100...) or in octal (0.063146314...) notations and must be terminated at some point.

3.3.2 Error Propagation in Elementary Computations

It is possible to analyze mathematically how errors in factors affect the computation results. In this discussion we will only consider the elementary operations of addition, subtraction, multiplication, division, and involution (raising a number to a power).

If two approximate quantities x_1 and x_2 with errors Δx_1 and Δx_2 are to be added (or subtracted), the approximate result is

$$y = x_1 \pm x_2 \tag{3.4}$$

If the exact values $x_1 + \Delta x_1$ and $x_2 + \Delta x_2$ are used, the result is

$$y + \Delta y = x_1 + \Delta x_1 \pm (x_2 + \Delta x_2) \tag{3.5}$$

Therefore the error in the result is

$$\Delta y = \Delta x_1 \pm \Delta x_2 \tag{3.6}$$

Since we do not necessarily know the signs of the errors, we can write

$$|\Delta y| \leqq |\Delta x_1| + |\Delta x_2| \tag{3.7}$$

Therefore, the rule for addition and subtraction is

The magnitude of the absolute error of a sum (or a difference) is equal to or less than the sum of the magnitudes of the absolute errors of the operands.

For multiplication, the approximate product of two factors is

$$y = x_1 x_2 \tag{3.8}$$

The exact product is

$$y + \Delta y = (x_1 + \Delta x_1)(x_2 + \Delta x_2) \tag{3.9}$$

Neglecting the term $\Delta x_1 \Delta x_2$, the absolute error in the product is approximately

$$\Delta y \doteq x_2 \Delta x_1 + x_1 \Delta x_2 \tag{3.10}$$

Since the relative error may be approximated by $\varepsilon_r \doteq \Delta y / y$, then

$$\varepsilon_r \doteq \frac{\Delta y}{y} \doteq \frac{x_2 \Delta x_1 + x_1 \Delta x_2}{x_1 x_2} = \frac{\Delta x_1}{x_1} + \frac{\Delta x_2}{x_2} \tag{3.11}$$

For division, the approximate quotient is given by

$$y \doteq \frac{x_1}{x_2} \tag{3.12}$$

and the exact quotient is

$$y + \Delta y = \frac{x_1 + \Delta x_1}{x_2 + \Delta x_2} \tag{3.13}$$

Multiplying the numerator and denominator of the right-hand side of Eq. (3.12) by $x_2 - \Delta x_2$ and neglecting the terms $\Delta x_1 \Delta x_2$ and Δx_2^2, we can simplify (3.12) to

$$y + \Delta y \doteq \frac{x_1 x_2 + x_2 \Delta x_1 - x_1 \Delta x_2}{{x_2}^2} \tag{3.14}$$

The absolute error in the quotient is approximately

$$\Delta y \doteq \frac{x_2 \Delta x_1 - x_1 \Delta x_2}{x_2^2} \tag{3.15}$$

and the relative error is

$$\varepsilon_r \doteq \frac{\Delta y}{y} \doteq \frac{(x_2 \Delta x_1 - x_1 \Delta x_2)/x_2^2}{x_1/x_2} = \frac{\Delta x_1}{x_1} - \frac{\Delta x_2}{x_2} \tag{3.16}$$

Equations (3.11) and (3.16) are very similar, and since we do not know the sign of the errors, we can write the following expression, which is applicable for either multiplication or division:

$$|\varepsilon_r| \leqq \left|\frac{\Delta x_1}{x_1}\right| + \left|\frac{\Delta x_2}{x_2}\right| \tag{3.17}$$

Therefore, the rule for multiplication and division is

The magnitude of the relative error of a product or a quotient is equal to or less than the sum of the magnitudes of the relative errors of the operands.

The preceding rules can also be derived by use of differential calculus. Furthermore, the use of calculus is particularly convenient when deriving rules for error propagation in more complex mathematical operations. For example, when raising a number to a power, say

$$y = x^n \tag{3.18}$$

we can find the approximate error by taking the differential of both sides:

$$\Delta y = nx^{n-1}\Delta x \tag{3.19}$$

Then, dividing Eq. (3.19) by (3.18), we arrive at the expression for the relative error:

$$\varepsilon_r = \Delta y/y = n\,\Delta x/x \tag{3.20}$$

Therefore, the relative error in y is just n times the relative error in x.

Mathematical expressions for the relative error in addition and subtraction can be derived, but we will see in the discussion of significant figures that the absolute error is more useful in analyzing these operations. Although Eqs. (3.10), (3.15), and (3.19) express the absolute errors for multiplication, division, and involution, respectively, the expressions for the relative errors are simpler and more useful in the subsequent analysis.

3.3.3 Significant Figures

Although the foregoing rules are not very difficult to apply, if used at every step, they add considerably to the complexity of the computations. These rules can be simplified, but at some cost in accuracy. The simplified rules can be used to determine, approximately, the number of figures that can be retained in a result based on the number of figures known accurately in the operands. Such figures are called *significant figures.* The validity of a significant figure analysis is somewhat dependent on the number system used. This is primarily because discarding a single digit in one number base can be equivalent to discarding more than one digit in some other number base. Also, two numbers expressed in one number base can have different numbers of significant figures, but when expressed in some other number base, both may have the same number of significant figures. Because of the possible loss of significance due to the number base used, it is wise to carry at least one uncertain figure in all computations. Some of the effects of the number base on significant figure analysis will be illustrated by an example in Sec. 3.3.4.

The significant figure rules for addition, subtraction, multiplication, and division can be derived from Eqs. (3.7) and (3.17). From Eq. (3.7) we can see that the error in the result of an addition or subtraction is essentially controlled by the error in the operand having the larger absolute error. At worst, if the errors in the operands are equal in magnitude, the error in the result may be twice the error in one operand. Normally we assume that the error in a number is no greater, in magnitude, than ½ in the least significant figure given. Therefore, the controlling operand is the one with fewer significant figures to the right of the decimal point. If neither operand has a fractional part, they should be scaled by the same scale factor so that both do have fractional parts. This will clear up the uncertainty of whether the zeros in a number like 33000 are significant. The rule for determining the significant figures in a sum or difference may be stated as follows:

> *The number of figures retained in the fractional part of the result of an addition or subtraction should be no greater than the number of figures given in the fractional part of the operand having the fewer number of figures in its fractional part.*

The application of this rule may be illustrated by the following example. If $x_1 = 1.312$ and $x_2 = 0.0327$, since x_1 has only three decimal places in its fractional part, then x_2 should be rounded to three decimal places before adding. Therefore, the sum is $y = 1.312 + 0.033 = 1.345$. To verify that this answer is reasonable, consider that $|\Delta x_1| \leqq 0.0005$ and $|\Delta x_2| \leqq 0.00005$, therefore $|\Delta y| \leqq 0.00055$. Adding and subtracting this error from the sum of x_1 and x_2 without rounding we may state that the result satisfies the inequalities $1.34415 \leqq y \leqq 1.34525$. The above single result, 1.345, appears to be a reasonable compromise.

By Eq. (3.17) we see that the magnitude of the relative error in a product or quotient is essentially controlled by the operand having the larger relative error. In order to obtain an approximation to the relative error in terms of significant figures, note that any number x can be written as a fraction f and a scale factor r^p. Therefore,

$$x = f * r^p \tag{3.21}$$

Also, assume that the fraction f has no leading zeros. If r is the base of the number system used, then $r^{-1} \leqq |f| < 1$. If the number x has q significant figures, then f will also have q significant figures. Therefore, the magnitude of the absolute error in f will be no greater than $\frac{1}{2} * r^{-q}$ and hence we can say

$$|\Delta x| \leqq \frac{1}{2} * r^{p-q} \tag{3.22}$$

Since $|f| \geqq r^{-1}$, then $|x| \geqq r^{p-1}$ and therefore,

$$\left|\frac{\Delta x}{x}\right| \leqq \frac{1}{2} * r^{1-q} \tag{3.23}$$

Thus, subject to the foregoing assumptions, the magnitude of the relative error of an operand depends only on the number of significant figures in that operand. The rule for determining the significant figures in a product or a quotient is

The number of figures retained in the result of a multiplication or division should be no greater than the number of significant figures in the operand having the smaller number of significant figures.

As an example of this rule, suppose the numbers $x_1 = 1.312$ and $x_2 = 0.0327$ are to be multiplied. By the rule, x_2 has the fewer significant figures so the product should be rounded to that same number of figures. Hence, $y = 0.0429$, which may be checked by using Eq. (3.17). Since $|\Delta x_1| \leqq 0.0005$, then $|\Delta x_1/x_1| \leqq 0.0004$, and since $|\Delta x_2| \leqq 0.00005$, then $|\Delta x_2/x_2| \leqq 0.002$. Therefore, $|\Delta y/y| \leqq 0.0024 \doteq 0.002$ and then $|\Delta y| = |\Delta y/y| * |y| \doteq 0.00008$, which says the error in y is less than one unit in the last figure retained in the above answer.

3.3.4 Examples

It was stated in the first paragraph of Sec. 3.3.3 that the number of significant figures that can be justified in a computational result is somewhat dependent on the number base used. This can be illustrated as follows. Suppose that two quantities whose values $98. \pm 0.5$ and $107. \pm 0.5$, which have been obtained empirically, are to be multiplied. The true product P then will satisfy the relations $10383.75 \leqq P \leqq 10588.75$. If the given significant figure rule for multiplication is strictly adhered to, we can only justify two figures in the result. Since $98 * 107 = 10486$, then to two places, $P = 10 * 10^3$ which implies the range $9500 \leqq P \leqq 10500$. Although this range includes most of the range given previously, it is unduly pessimistic and poorly centered. If one extra figure is carried in the result, the answer is $P = 105 * 10^2$ which implies the range $10450 \leqq P \leqq 10550$. This result, although slightly optimistic, is preferable to the two-figure result.

In another number base, base-8, three significant figures can be justified on the basis of the operands, since $(98)_{10} = (142)_8$ and $(107)_{10} = (153)_8$. Then, $(142)_8 * (153)_8 = (24366)_8$. Therefore, the three-figure result is $P = (244 * 10^2)_8$ which implies $(24340 \leqq P \leqq 24440)_8$. In decimal form, this may be expressed as $10464 \leqq P \leqq 10528$. This result is even more optimistic than the previous result, and it was reached without violating the given significant figure rule.

This example not only illustrates the dependence of the significant figure rule for multiplication on number base, but also illustrates the inexactness of this significant figure rule.

The given significant figure rules are easy to apply when performing computations manually. At each step of the computations, a value judgment can be made to determine how many figures should be retained. However, when the computations are to be performed automatically, as with a computer, the program must provide means for maintaining the most significant figures possible. This is usually done by carrying more figures than are significant and then rounding the final results. In many cases this is adequate, but in some cases significant figures can be lost even though the word length of the computer used is greater than or equal to the number of figures required in the final answer. This is because the computer must sometimes discard figures that are significant. As an example, suppose that the sum of 100 numbers is to be subtracted from the sum of another 100 numbers.

$$D = \sum_{i=1}^{100} x_i - \sum_{i=1}^{100} y_i \tag{3.24}$$

Further, suppose that all x's and y's are positive numbers greater than 0.5 but less than 1. and that each is known to seven decimal places. This means that each sum will have a total somewhere between 50 and 100. If the computer can only carry seven decimal digits, the fractional part of each sum as well as the final result will only be known to five decimal places. If, instead, the computation is performed by taking differences in pairs,

$$D = \sum_{i=1}^{100} (x_i - y_i) \tag{3.25}$$

then there would be less likelihood of dropping two digits.

The rules given are pessimistic. In most computations, the errors incurred are less than these maximums. However, care should be taken to avoid losing significant figures. This can often be facilitated by altering the method of computation as illustrated in the preceding example, where only meaningful figures were being retained. Although in some cases the extra figures carried cannot be classified as significant, they may not be entirely meaningless. Let us consider an example. Suppose that the expression $AB - CD$ is to be evaluated for $A = 1.001200$, $B = 1.002100$, $C = 1.001100$, and $D = 1.002200$. If the computer can only carry seven figures, the individual rounded products will be $AB = 1.003303$ and $CD = 1.003302$; therefore, $AB - CD = 1. * 10^{-6}$. However, if one more figure could be carried in the computations, the result would be $AB - CD = 1. * 10^{-7}$. A complete error analysis based on the fact that the individual errors are no greater than $0.5 * 10^{-6}$ will lead to the conclusion that the error in $AB - CD$ could be as large as $2. * 10^{-6}$. Since both results are less than the maximum possible error, then it appears that the first answer is as good as the second answer. The latter result should be used as it is more meaningful to say the answer is $1. * 10^{-7} \pm 2. * 10^{-6}$ rather than $1. * 10^{-6} \pm 2. * 10^{-6}$. This is more apparent if the results are written.

Eight-figure answer:

$$-19 * 10^{-7} \leq AB - CD \leq 21 * 10^{-7}$$

Seven-figure answer:

$$-10 * 10^{-7} \leq AB - CD \leq 30 * 10^{-7}$$

The important thing to remember is that there is a range of answers and this range should be centered on what is believed to be the best result.

In the event that extended precision arithmetic is not available or not convenient to use, it is often possible to extend precision by means of changing the order of computations as illustrated earlier or by altering the way the problem is expressed algebraically. By use of the substitutions $B = A + b$, $C = A + c$, $D = A + d$, the preceding problem may be written

$$AB - CD = A(b - c - d) - cd \tag{3.26}$$

Evaluating the expression on the right with the values given earlier, the result $1. * 10^{-7}$ is obtained without extended precision computation. Even if extended precision arithmetic is available and is used, every effort should be made to order the computations to minimize the error.

In the previous examples, we have seen how measurement errors or errors in our number representations can be propagated in our computations. In Sec. 3.4 studies of errors incurred by approximate modeling will be discussed. Such errors are also propagated in our computations.

Precautions must be taken so that the results of such studies are not masked by computational errors. In order to illustrate this danger, the following example is given to show how significant figures can be lost in a simple computation. The same problem will be used later in an error study.

The period (time for one revolution) of a near-Earth satellite is given by the formula

$$P = 84.48877a^{1.5} \tag{3.27}$$

where P is the period in minutes and a is the semi-major axis of the satellite orbit expressed in earth-radii. Let us assume that we need to calculate the period of a satellite in a circular orbit at an altitude of 100 statute miles above the Earth's surface. For a circular orbit, the semi-major axis is just the radius of the orbit, which is given by

$$a = 1. + (h/r) \tag{3.28}$$

where h is the satellite altitude and r is the earth radius, both given in the same units.

For this hypothetical problem, we can say that h is known to whatever precision we wish to specify. However, we must use some approximation for the earth radius such as $r = 3963.259$ statute miles. Since we know r only to seven figures, we are justified in carrying only seven figures in the ratio.

$$h/r = 0.02523176 \tag{3.29}$$

(Note that the leading zero is not counted as a significant figure. It is only required to position the decimal point.) In Eq. (3.28), the 1. is an exact quantity, therefore the number of places that can be justified in a is controlled by the number of places in h/r. By the rule of addition, we can justify carrying eight decimal places in the fractional part of the sum.

$$a = 1.02523176 \tag{3.30}$$

Hence, we know a to nine significant figures. However, some computers cannot carry this many significant figures in a single precision computation. Some types of computers can only carry numbers that are 24 binary bits in length in a single precision floating-point computation. Twenty-four binary bits is less than eight decimal digits. Therefore, in such a machine, the value calculated above and the subsequently calculated values can be carried out only to about seven decimal digits. Table 3.1 tabulates the step-by-step results if we carry only seven figures and if we carry all that can be justified at each step.

TABLE 3.1 COMPARISON OF COMPUTATIONS USING SEVEN FIGURES WITH COMPUTATIONS USING ALL FIGURES THAT ARE SIGNIFICANT

Quantity	*Constant Seven Figures*	*All that Are Significant*
a	1.025232	1.02523176
a^2	1.051101	1.05110016
a^3	1.077622	1.07762127
$\sqrt{a^3}$	1.038086	1.03808539
P	87.70661	87.70656

Note that the two results differ by 5 in the seventh figure although the assumption was made that all quantities used in the computation had errors of less than 0.5 in the seventh figure. The answer 87.70656 is the more nearly correct one and will be verified below. Therefore, to avoid introducing errors in computation, it is necessary to carry all figures that are significant. Sometimes it is more convenient to carry more figures than are significant. In the example, if a constant nine figures or more were carried at each step, the rounded result would have been the same. Although carrying extra figures is one way of preserving accuracy, it is sometimes better to change the way the problem is expressed mathematically so that accurate results can be computed using fewer figures. Some simple examples of this method were given earlier. In this problem, the need for extra digits arises when h/r is added to 1. If this addition can be avoided or postponed, the extra places may not be needed.

The three halves power of a can be found by use of a Taylor's series expansion (see Chap. 5). Let $h/r = x$; then, since $a = 1 + x$, we can write

$$(1 + x)^{1.5} = 1 + \frac{3}{2}x + \frac{3}{8}x^2 - \frac{1}{16}x^3 + \frac{3}{128}x^4 - \frac{3}{256}x^5 + \cdots \tag{3.31}$$

This series can be evaluated term by term. Since only multiplication and division are involved in each term, we are only justified in carrying seven figures in each. The results, excluding the 1., are

$$\begin{aligned}
\frac{3}{2}x &= 0.3784764 * 10^{-1} = 0.3784764 * 10^{-1} \\
\frac{3}{8}x^2 &= 0.2387406 * 10^{-3} = 0.0023874 * 10^{-1} \\
-\frac{1}{16}x^3 &= -0.1003974 * 10^{-5} = -0.0000100 * 10^{-1} \\
\frac{3}{128}x^4 &= 0.9499516 * 10^{-8} = 0.0000001 * 10^{-1} \\
-\frac{3}{256}x^5 &= -0.2396895 * 10^{-10} = -0.0000000 * 10^{-1}
\end{aligned} \tag{3.32}$$

If these terms are added, the number of figures that can be justified in the sum is determined by the number of figures in the first of these terms. The sum of the first four terms is $0.3808539 * 10^{-1}$ and the fifth and subsequent terms will not contribute anything to the significant figures that can be justified. This result still must be added to 1. and the result multiplied by a constant. However, the multiplication can be performed first on each operand and the results added.

$$84.48877 * 0.3808539 * 10^{-1} = 3.217788$$
$$84.48877 * 1.000000 = 84.48877$$

These two terms must be added and since the second term has fewer places in its fractional part, the first term should be rounded to the same number of places in its fractional part. The final result is

$$P = 87.70656 \tag{3.33}$$

which verifies the result given earlier. In this series of computations, no more than seven figures were ever required.

3.4 ERROR STUDIES

Section 3.2 considered two classes of errors, primarily computing but also measurement errors. Such errors, although not mistakes, are undesirable. We can also incur errors through "incorrect" mathematical models. In some cases this is undesirable but in other cases, it is desirable. Why would an "incorrect" model ever yield a desirable result? A simple model with a relatively large error may be preferable to a more complex model with a smaller error.

We have seen that it is inconvenient to analyze computational errors rigorously. Measurement and modeling errors are not as intractable. There are essentially two ways that such errors can be analyzed. Before discussing these methods, it is pertinent to define measurement and modeling errors for the purpose of this discussion.

The term *measurement errors* is intended to include not only actual errors in the measurement of physical quantities, but also the deviations from specified values due to manufacturing variations. The term *modeling errors* is intended to include errors arising from the mathematical formulation for a problem or a device and also errors arising from the inability to exactly mechanize a given mathematical formulation.

3.4.1 Direct Error Studies

The direct error study is the most straightforward and universal type of error study. It consists of determining a result by two different methods and then taking the difference between the two answers to determine the error. This method can be applied to all three classes of errors. In fact, we already have used it in determining a computing error.

The same example can be used to illustrate the effect of a measurement error. By assuming the Earth radius to be 3963.2585 instead of 3963.259, P can be calculated and compared with the P obtained previously to determine the error in P due to a 0.0005-mile error in r. However, the true result can be masked by computational errors. The direct error study becomes less precise as we try to determine the effects of small errors.

If we assume that for small errors the error in the result is proportional to the error in the input, we can determine the constant of proportionality for a large error and apply it to find the effects of a smaller error. That is, suppose that we have a process, whose result R can be expressed as a function of several independent variables.

$$R = f(x, y, \ldots) \tag{3.34}$$

We can usually say that the effect of a small error in one of the independent variables can be expressed as

$$\Delta R = k\Delta x \tag{3.35}$$

By finding ΔR for a given Δx, k can be calculated by

$$k = \Delta R/\Delta x \tag{3.36}$$

This value can be substituted in Eq. (3.35), which then can be used to calculate the values of ΔR for other Δx's.

If, in the previous example, Eqs. (3.28) and (3.27) were solved with $r = 3962.259$ ($\Delta r = 1.$), the result would be $P = 87.70737$. This means that the error in P is -0.00081 for the given error in r. Using (3.35), we may write

$$\Delta P = k\Delta r \tag{3.37}$$

Then, substituting the values above for ΔP and Δr, we obtain $k = -0.00081$ minutes/mile. Then, (3.37) becomes

$$\Delta P \doteq -0.00081\Delta r \tag{3.38}$$

Evaluating this for an error of $\Delta r = 0.0005$ miles, we find $\Delta P \doteq -0.00000041$ minutes. In order to obtain this result directly, it would have been necessary to extend the precision of our arithmetic to at least ten significant figures.

In a similar manner, we can perform what is called a *tolerance analysis*. Instead of an error in a measurement, we assume a variation in the value of a parameter. For example, we may wish to determine how the voltage gain of an amplifier varies due to variations in the value of a particular component, say a resistor. Since no two resistors can be manufactured with exactly the same resistance, no two amplifiers can be expected to have identical voltage gain. However, some variation is acceptable, and the designer must determine the specifications for the resistor. This can be done by solving the circuit equations with a nominal value of the resistor and then solving them again with the resistor value deviated by some amount. The comparison of results will aid in setting the specification for the resistor.

The direct error study can also be used to determine the effects of modeling errors. Suppose a device is designed to perform a specific function such as missile guidance. Several methods of missile guidance can be employed. Each method under consideration can be simulated numerically and the results compared with the desired results. Such a simulation would involve selecting a representative sample of test cases and then comparing the results case by case. Although this can be time consuming, it is much more efficient than building hardware and testing it.

3.4.2 Differential Error Studies

Differential error study lends itself only to the analysis of measurement (or similar) type errors. In Sec. 3.4.1 we computed a numerical approximation of the relation between the error in a parameter and the corresponding error in the result. We assumed that for small errors we could approximate the error by Eq. (3.35), where k was to be calculated by Eq. (3.36) for particular values of ΔR and Δx. Another method of calculating k is available. Note that by Eq. (3.36), k is approximately the derivative of R with respect to x. Therefore, we can write

$$\Delta R = \frac{dR}{dx}\Delta x \tag{3.39}$$

Then, if a mathematical relationship exists between R and x, the derivative (or partial derivative) can be calculated for certain values of the parameters.

In the previous example, P was related to r by Eqs. (3.28) and (3.27). Differentiating these, we obtain

$$\frac{dP}{dr} = -126.7332\frac{ha^{1/2}}{r^2} \tag{3.40}$$

Evaluating for a particular value of h and the nominal value of r gives

$$\left.\frac{dP}{dr}\right|_{h=100} = -0.00082 \tag{3.41}$$

The result is the same as that obtained by the direct method.

3.4.3 Multivariate Analysis

In Secs. 3.4.1 and 3.4.2 the subjects of direct and differential error studies have been presented. Although functions of more than one independent variable were shown in those sections, the analysis assumed that an error could occur in only one of the independent variables. Real problems are seldom this simple. However, direct and differential error studies can be performed on functions of more than one independent variable with errors occurring simultaneously in two or more of these variables. We refer to such studies as multivariate analysis or multivariate error studies.

Direct multivariate error studies are performed in a manner almost identical to the direct error studies discussed in Sec. 3.4.1. The function

$$Y = F(X_1, X_2, \cdots X_n) \tag{3.42}$$

can be evaluated for a given set of exact independent variables to yield an exact valuc of Y. Then each independent variable can be deviated by some amount ΔX and then substituted in (3.42)

$$Y + \Delta Y = F(X_1 + \Delta X_1, X_2 + \Delta X_2, \cdots, X_n + \Delta X_n) \tag{3.43}$$

to obtain the corresponding inexact value of Y, namely $Y + \Delta Y$. The result from (3.42) can be subtracted from (3.43) to obtain the error in Y for the given errors in the independent variables.

Differential error studies can also be performed on functions of more than one independent variable. In order to find the error in the dependent variable as a function of the errors in the independent variables, it is necessary to evaluate the total differential by the formula

$$\Delta Y = \frac{\partial F}{\partial X_1}\Delta X_1 + \frac{\partial F}{\partial X_2}\Delta X_2 + \cdots + \frac{\partial F}{\partial X_n}\Delta X_n \tag{3.44}$$

The partial derivatives required by Eq. (3.44) may be found by taking partial derivatives of the given function, Eq. (3.42), evaluating these for the exact values of parameters (independent variables), and substituting these values into (3.44) to simplify the differential formula to the form

$$\Delta Y = K_1\Delta X_1 + K_2\Delta X_2 + \cdots + K_n\Delta X_n \tag{3.45}$$

This equation can be readily evaluated for any combination of ΔX's. As suggested in Sec. 3.4.1, approximate values of the derivatives (in this case partial derivatives) can be found numerically. This is done by first evaluating Eq. (3.42) with the exact values of the independent variables, then evaluating (3.43) with only one nonzero ΔX, then taking the difference of the results to obtain a ΔY, and finally finding the ratio $\Delta Y/\Delta X$, which is an approximate derivative K, to use in (3.45). This process can be repeated to find a value for each K.

In an actual problem, we cannot state *a priori* in what combinations the ΔX's will occur. Therefore, it is necessary to have a technique(s) that will give us a realistic picture of how errors in the independent variables affect the final results. Two such techniques, statistical analysis and worst case analysis, are discussed in the following two subsections.

3.4.3.1 *Statistical Analysis*

Given a problem in which the independent variables X each may have an error ΔX, it is possible for the scientist or engineer to determine either by experiment or by specification the frequency distribution of such errors for each independent variable. Typically, the frequency distribution will be similar to that shown in Fig. 3.1.

Statistical analysis is the technique of finding a frequency distribution such as shown in Fig. 3.1 for the dependent variable(s) of a problem. In order to do this, the problem solver proceeds as follows. By a random process related to the error frequency distributions of each independent variable, a set of ΔX's are selected and the corresponding ΔY is calculated either by the direct or differential method. This process is repeated until enough ΔY's have been determined to give a meaningful frequency distribution of the ΔY's. This process gives a fairly realistic picture of how errors occurring randomly in the independent variables will affect the dependent variable(s). Some of the details of this process will be presented in the following paragraph.

Subroutines are available for most computers that will generate random numbers. Usually the frequency distribution is such that any number in some range, say 0 to 1, is equally likely. It is necessary to devise some method of using a random number from the subroutine to obtain a random error value. One technique is as follows. For the error frequency distribution curve for each independent variable, such as shown in Fig. 3.1, normalize the ordinates so that the maximum value is equal to 0.5. Then reflect the curve to the right of the vertical axis about the line $y = 0.5$. This will yield a curve such as shown in Fig. 3.2.

This curve can either be approximated by a formula or by a table of values. In either case, a random number can be determined by the computer subroutine. Then using this as the y value, the corresponding X can be determined from the appropriate formula or table. Even if the error curve is asymmetrical and/or the maximum frequency does not occur at $X = 0$, a curve such as that of Fig. 3.2 can be created such that X is a single

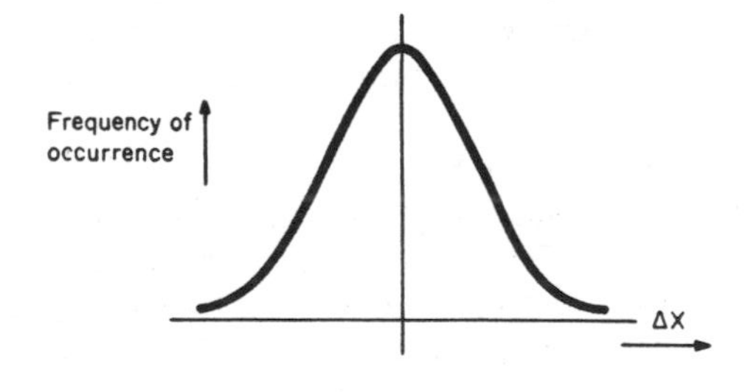

Fig. 3.1. Frequency distribution of errors occurring in independent variable X.

valued function of y which has a range of 0 to 1. Fig. 3.3 gives a generalized flowchart for performing a statistical error analysis.

This procedure simulates the combining of errors in the actual physical process being studied. If the sample size M is large, the results should closely approximate experimental results of the actual physical process. The computation of other quantities of interest, such as the root mean square value of the ΔY's, can easily be incorporated in the flowchart of Fig. 3.3.

It is possible to obtain other (but less informative) measures of the error distribution by simpler techniques. For example, the general term $(\partial F/\partial X_j)\,\Delta X_j$ of Eq. (3.44) approximates the error in the function caused by an error in one parameter. One simple technique is to add the absolute values of these terms using the maximum possible value of ΔX in each term.

$$|\Delta Y|_{max} = \left|\frac{\partial F}{\partial X_1}\Delta X_1\right| + \left|\frac{\partial F}{\partial X_2}\Delta X_2\right| + \cdots + \left|\frac{\partial F}{\partial X_n}\Delta X_n\right| \tag{3.46}$$

Thus, (3.46) gives a form of a worst case analysis which is the topic of Sec. 3.4.3.2. Another technique is to find the square root of the sum of the squares of the terms

$$\Delta Y = \sqrt{\left(\frac{\partial F}{\partial X_1}\Delta X_1\right)^2 + \left(\frac{\partial F}{\partial X_2}\Delta X_2\right)^2 + \cdots + \left(\frac{\partial F}{\partial X_n}\Delta X_n\right)^2} \tag{3.47}$$

Equation (3.47) can be evaluated either with the maximum possible values of the ΔX's or the most probable values of the ΔX's. The significance of the result is not always apparent because too many factors are involved such as the values of the ΔX's used and the distributions of the errors in each parameter. However, the result does provide a measure of the range of errors in Y and can be very useful when comparing the effects of using different nominal values for the independent variables.

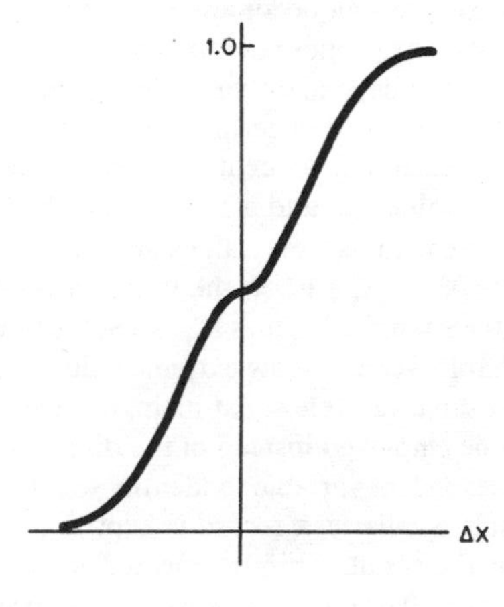

Fig. 3.2. Modified error frequency distribution curve.

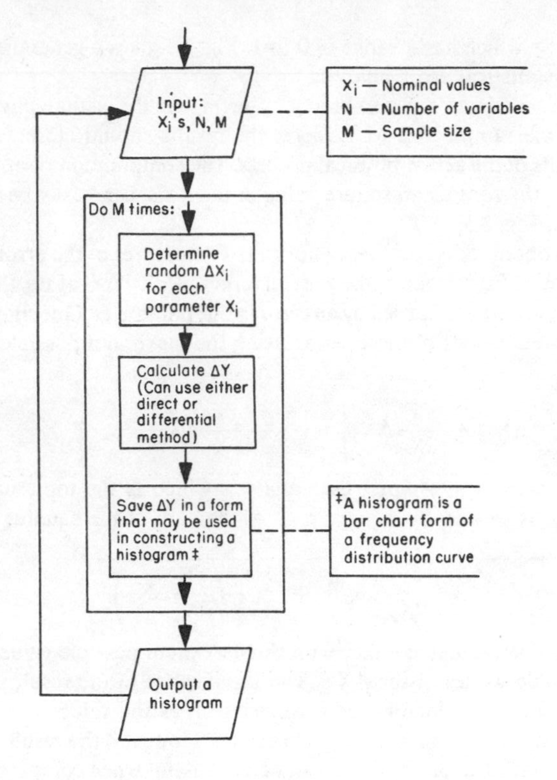

Fig. 3.3. Flowchart of a statistical error analysis.

3.4.3.2 *Worst Case Analysis*

Worst case analysis is a second technique for analyzing how errors in the independent variables affect the final results. The purpose of this type of analysis is to find the highest and lowest possible values of the dependent variable(s) subject to the limitation that the independent variables vary only within certain given limits. These values are found in the following manner. For each independent variable X, there is assigned a lowest possible value X_L, a nominal value X_N, and a highest possible value X_H. (Although it is not necessary, it is convenient to assume values such that the difference $X_H - X_N$ is the same as the difference $X_N - X_L$.) First, the function is evaluated with all independent variables equal to their nominal values X_N. Then the function is evaluated with one of the independent variables set at its low extreme value X_L, and then the function is again evaluated with that same variable set at its high extreme value X_H. (Note that the differential method can be employed instead of the direct method being described.) A record is kept for that independent variable to identify which of X_L, X_N, or X_H caused the highest value of the result. Similarly, a record is kept as to which of X_L, X_N, or X_H caused the lowest value of the result. This is repeated for each of the independent variables. Finally, the set of X's that individually caused the result to go high are used simultaneously to obtain a result which is assumed to be the highest possible answer.

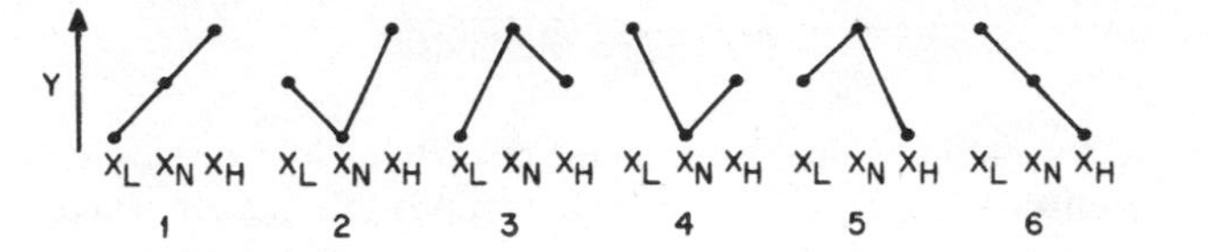

Fig. 3.4. Possible result distributions as a function of parameter values.

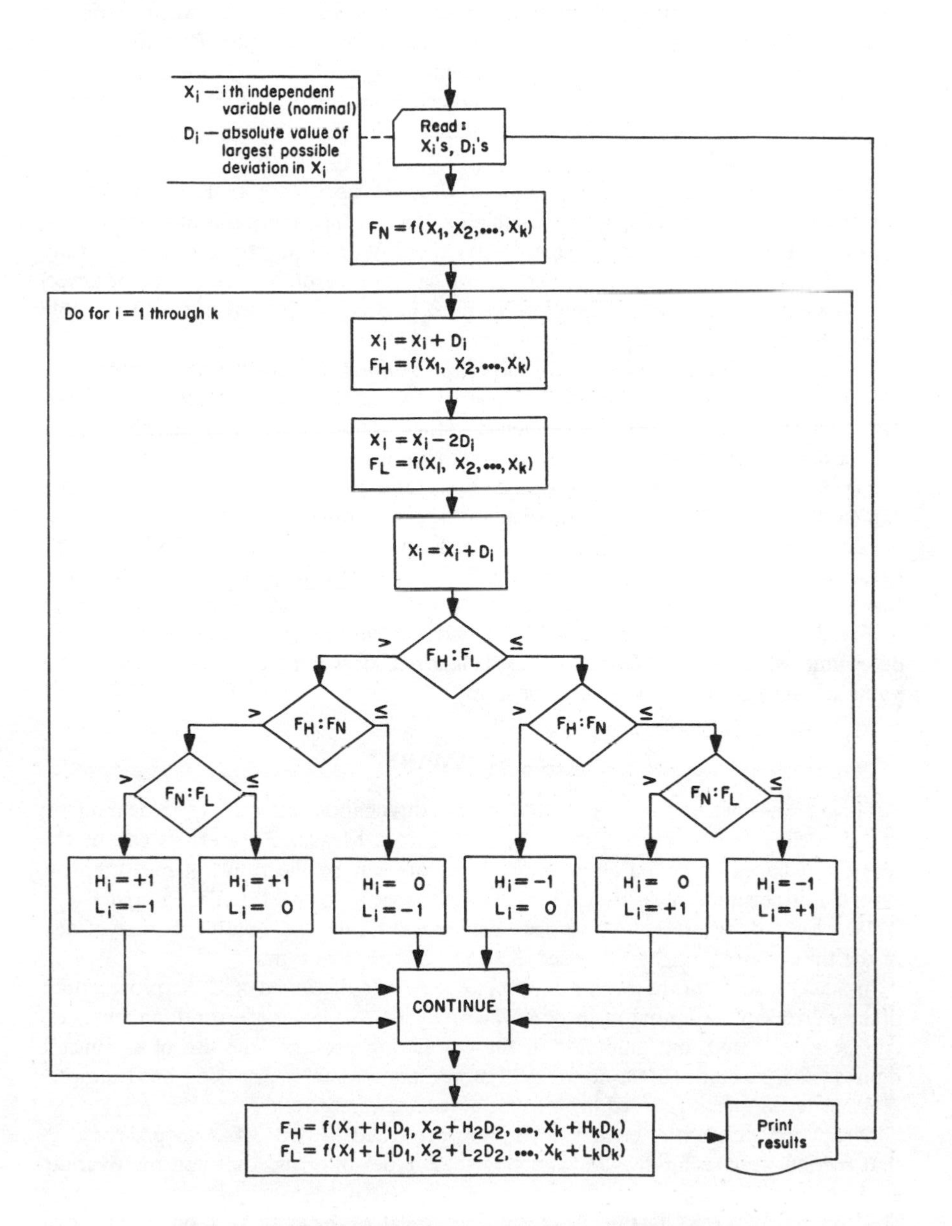

Fig. 3.5. Generalized flowchart for worst case analysis.

Also, the set of X's that individually caused the result to go low are used simultaneously to obtain a result that is assumed to be the lowest possible answer. These answers are pessimistic and do not give as complete a picture as the statistical analysis. However, the process is shorter and does give limits that can be useful in setting allowable tolerances on the independent variables. It is assumed that the function will behave in much the same manner when errors are made simultaneously in all independent variables as when errors are made one at a time in the independent variables. This implies that there is no cross-coupling between independent variables. Although this is not true in general, such effects are usually small if the errors are small. If the differential error method is used, it is further assumed that the error in the result is a linear function of the error in the independent variable.

The preceding description is adequate to describe worst case analysis if it is to be performed by a human using a desk calculator. If worst case analysis is to be performed by a computer, additional steps must be described. When using the direct error study method, it is possible that both X_L and X_H will cause the result Y to be higher (or lower) than the nominal value of Y. Therefore we must be prepared for any of the six possible cases illustrated in Fig. 3.4.

The Y values illustrated in Fig. 3.4 have been selected to show the six possible combinations of value positions relative to one another as a function of parameter value. No special significance should be attached to their actual values. Also, lines have been drawn between the points only to tie each set together.

Fig. 3.5 is a flowchart of a worst case analysis for a single function of k variables that can each deviate from their nominal values the same amount in both the positive and negative directions. The decision tree is for the purpose of determining which of the cases illustrated in Fig. 3.4 is applicable for each of the independent variables. The results of the decision tree correspond in left to right order with the cases illustrated in Fig. 3.4. The results of the decision tree are two multipliers that are to be used to determine whether a deviation is added, ignored, or subtracted from the nominal value to cause the result to go high or low.

3.5 SUMMARY

Three *classes* of errors have been discussed in this chapter: computing, measurement, and modeling errors. These classes are not distinct. Measurement errors can be the result of computing errors and computing errors can be the result of measurement errors. Furthermore, measurement errors are a special form of modeling errors.

Two *types* of errors have also been defined: absolute and relative. Errors of any of the three classes can be expressed as either of these two types.

In Sec. 3.3, some of the ways that errors occur were discussed, and the propagation of these errors in the computing process was analyzed. All classes of errors (and mistakes) can be propagated and amplified in the computing process. The use of significant figures to keep a record of the accuracy of computation was also discussed, and examples of other techniques to record or improve accuracy were given.

Direct and differential error studies were presented in Sec. 3.4. Also presented in that section were techniques for applying these types of error studies in multivariate analysis.

Although the topics of this chapter are interrelated, they can be grouped into two major categories: errors and error studies. In applying error analysis to computing errors, we are usually concerned with the order of magnitude of the errors propagated and how such errors can be avoided or reduced rather than with their actual values. To some extent this is also true of measurements errors. In some measurement error studies and in modeling error studies, the values of the errors are of prime importance. Therefore, we see that in one category we are more concerned with the theory of error analysis, while in the other category we are more concerned with the practice of error analysis.

EXERCISES

1. Calculate the sine of 2. radians by its Taylor's series expansion. Calculate each term to seven significant figures. Add the terms and compare the result with the nine significant figure value 0.909297427.

2. Calculate cos A − cos(0.1) to seven significant figures for A = 0., 0.05, 0.095, and 0.0995. First, use a table or a FORTRAN subroutine to find the cosines. Then calculate by using the identity

$$\cos A - \cos(0.1) = 2\sin\left(\frac{0.1 + A}{2}\right)\sin\left(\frac{0.1 - A}{2}\right)$$

3. Compute the period of an Earth satellite in a circular orbit for altitudes of 100 to 300 miles at 10-mile intervals. Derive a differential relation between period and altitude for a nominal altitude of 200 miles. Evaluate this relation for Δh from −100 miles to +100 miles at 10-mile intervals. Add each of these results to the nominal period for a 200-mile high orbit, and compare the approximate results with the exact results.

4. A drive-in movie screen has a vertical dimension of 20 feet, and its bottom edge is 10 feet above the ground. The angle A, subtended by the screen from a distance of x feet, is given by $\tan A = 20x/(x^2 + 300)$. Assume that the nominal viewing distance is 300 feet. Calculate the nominal value of A (in degrees). Also calculate, by both the direct and differential methods, the change in angle A from the nominal for the viewer positioned 1, 2, or 3 rows on either side of the nominal row. (Assume that rows are spaced 40 feet apart.)

5. The amount A at compound interest is given by the formula

$$A = P(1 + i)^n$$

where P is the principal, n is the number of interest-bearing periods, and i is the interest rate per period. Prepare a flowchart and write a FORTRAN program to calculate, by both the direct and differential methods, the change in amount ΔA for an incremental change in interest Δi. Write the program to accept as input any values of P, i, n, and Δi. Execute the program for the following sets of values of the parameters:

P	i	n	Δi
10000	0.005	240	0.001
5000	0.010	300	0.002
5000	0.015	48	0.005

6. The periodic payment p to pay off a loan in equal payments is given by

$$p = \frac{Pi}{1 - (1 + i)^{-n}}$$

where P is the amount of the loan, n is the number of payments, and i is the interest rate per payment period. Prepare a flowchart and write a FORTRAN program to calculate, by both the direct and differential methods, the change in payment Δp for an incremental change in interest Δi. Write the program to accept as input any values of P, i, n, and Δi. Execute the program using the data in Exercise 5.

7. The specific weight S of an object can be calculated by

$$S = \frac{A}{A - W}$$

where A is the weight of the object measured in air, and W is the weight of the object when submerged in water. Assuming that errors may occur in both measured quantities, derive a total differential formula to calculate the error ΔS that may occur in the calculated value of S. Prepare a flowchart and write a FORTRAN program that will calculate the error in S given A and W and errors in each. Execute the program for the following sets of inputs:

A	W	ΔA	ΔW
5	4	0.05	0.04
5	4	0.05	−0.04
10	9	0.10	0.09
10	9	0.10	−0.09
6	3	0.12	0.06
6	3	0.12	−0.06
7	1	0.03	0.01
7	1	0.03	−0.01

8. The focal length f of a thin lens is given by

$$\frac{1}{f} = (n - 1)\left(\frac{1}{r_1} + \frac{1}{r_2}\right)$$

where n is the index of refraction of the glass, and r_1 and r_2 are the radii of curvature of the two surfaces of the lens. Assume that errors of $\Delta n = +0.005, 0.0, -0.005$, $\Delta r_1 = +0.01, 0.0, -0.01$, and $\Delta r_2 = +0.01, 0.0, -0.01$ may occur in any combination. Calculate, by the direct method, the error in f for all 27 possible combinations of errors for the two cases: $n = 1.615$, $r_1 = r_2 = 20$, and $n = 1.360$, $r_1 = -5$, $r_2 = 10$.

9. Derive a total differential error formula for the equation given in Exercise 8. Using this formula, repeat Exercise 8 performing a differential rather than a direct error study.

10. Derive a total differential formula for the equation

$$Y = Ax^3\cos\theta - Bye^{-x}\sin\phi$$

assuming that errors may occur in the parameters x, θ, y, and ϕ.

11. Calculate the largest expected error using Eq. (3.46) for the equation in Exercise 10. Use the constants $A = 1$ and $B = 3$, and the nominal parameter values $x = 1$, $\theta = \pi/3$, $y = 2$, and $\phi = \pi/6$. Assume that each parameter may have an error of 0.1 percent.

12. Prepare a flowchart and write a FORTRAN program to perform an error frequency analysis of the equation in Exercise 7. Assume that errors in A and W both occur according to the frequency distribution given by

[*36R*]	*Error*
0	−0.05
1	−0.04
3	−0.03
6	−0.02
10	−0.01
15	0.00
21	0.01
26	0.02
30	0.03
33	0.04
35	0.05

where R is a random number in the range $0 \leq R < 1$, and the square brackets indicate the integer part of the product shown. Execute the program for $A = 5$ and $W = 4$ with a sample size of 100. Count and print out the number of errors in each of the following ranges: -45 through -22, -21 through -8, -7 through 7, 8 through 21, 22 through 45.

13. Using the technique described by Fig. 3.5, perform a worst case analysis on the equation and data in Exercise 8. Calculate the errors by the direct method. What are the least and greatest values for each case? Are these the least and greatest of all the values calculated for Exercise 8?

14. Using the technique described by Fig. 3.5, perform a worst case analysis on the equation in Exercise 10 using the data of Exercise 11. Calculate the errors by the direct method. What are the least and greatest values of Y?

15. The indicated power P for a one-cylinder heat engine can be approximately expressed in terms of its physical dimensions and shaft speed by

$$P = 2.9 \cdot 10^{-5}\left(1. + \frac{0.7854LD^2}{V_0}\right)LD^2R$$

where L is the length of the stroke in inches, D is the diameter of the cylinder in inches, V_0 is the minimum volume of the compressed gas in cubic inches, and R is the shaft speed in revolutions per minute. Prepare a flowchart and write a FORTRAN program to perform a worst case analysis using the direct method of calculating errors. Assume that any combination of parameter values may be input. Also assume that the high and low limits on the error of each parameter are $+1$ and -1 percent, respectively. Execute the above program for the following sets of parameter values:

L	D	V_0	R
3.0	3.0	3.0	4000
4.0	2.5	2.5	4500
2.5	3.5	2.8	5000

16. The total resistance of the circuit shown is given by $R = R_1R_2/(R_1 + R_2) + R_3$. For each of the following cases, find the nominal value of R and its tolerances (high and low limits) by worst case analysis using the direct method for calculating errors. The error in each resistor may be either plus or minus the amount shown.

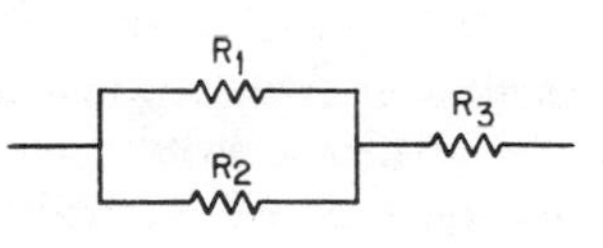

R_1	ΔR_1	R_2	ΔR_2	R_3	ΔR_3
100	10%	900	20%	110	10%
100	10%	400	20%	120	10%
100	10%	100	20%	150	10%
100	20%	400	10%	120	10%
100	20%	900	10%	110	10%
100	10%	900	10%	110	20%
100	10%	400	10%	120	20%
100	10%	100	10%	150	20%

4

Interpolation

4.1 INTRODUCTION

To *interpolate* means to estimate an intermediate term in a sequence. This chapter discusses methods for making such estimates. First, a few terms and concepts will be introduced to clarify the above definition.

Engineering data often is available only in tabular form. That is, for a discrete set of an independent variable such as

$$x_1, x_2, \ldots, x_n$$

there exists a corresponding set of function values

$$f(x_1), f(x_2), \ldots, f(x_n)$$

which are sometimes written more simply

$$f_1, f_2, \ldots, f_n$$

The independent variable (x in this case) is called the *argument* of the function.

Even when mathematical relationships exist (and are known) for engineering data or mathematical functions, it is sometimes more convenient to express them in tabular form.

If we require a value of a function corresponding to an argument value which is not a member of the tabulated arguments, then some method must be devised for estimating a function value from the given argument and the tabulated arguments and function values. If the given argument is inside the range of the tabulated arguments, the estimating method is called *interpolation*. If the given argument is outside the range of the tabulated arguments, the estimating method is called *extrapolation*. We will find that the methods used for interpolation can also be used for extrapolation, but the accuracy is usually less.

In this chapter, only two types of interpolation will be covered: graphical and polynomial. Nonpolynomial methods of interpolation exist, but will not be covered in this text. The polynomial methods that will be covered are Gregory–Newton forward, Gregory–Newton backward, Gauss forward, Gaus backward, Stirling, Everett, and Lagrange. Each of these methods has a different polynomial formula to be used and each (except the Lagrange method) requires tables that are prepared in a special manner for that method. In this chapter, the formulas and the preparation of the tables for each method are presented. To use these methods in a computer program, the techniques for selecting the table entries to be used must also be described. These techniques are discussed and example flowcharts are given in Secs. 4.3 and 4.8.

4.2 GRAPHICAL INTERPOLATION

One familiar method of interpolation is the graphical method. In this method, the tabulated values are carefully plotted on graph paper; then a smooth curve is passed through the plotted points. To obtain a function value, the argument value is located along the appropriate axis, the corresponding point on the curve is found, and the function value is read from the other axis.

The weaknesses of this method are

1. limited accuracy of plotting and reading the graph (usually poorer than 0.1 percent),
2. inaccuracy of the shape of the curve between plotted points.

If the data were originally obtained from a graph, the graphical method would probably be adequate. However, the graphical method is not practical to use in a computer solution.

4.3 LINEAR INTERPOLATION

In linear interpolation, we assume that the tabulated function can be represented by a series of straight lines joining consecutive tabulated points. The method is relatively simple to employ. For a given argument, the appropriate line segment is located and the function value is calculated.

4.3.1 Table Lookup

Although the computations required for linear interpolation are quite simple, the logic required in a computer program to determine which line segment to use is not trivial. This is done by a procedure called *table lookup*, which locates a pair of tabulated arguments that bracket the given argument. An example of one method of table lookup is given below. This same method with only minor modifications can also be used for other interpolation methods.

First we must assume that our computer contains two arrays of numbers: the x array (argument table) and the f array (function table). The individual elements of these arrays exist in a one-to-one correspondence; i.e., the ith entry in the x array x_i is the argument for the ith entry in the f array f_i. It is also necessary that the argument array be in monotonically increasing (or decreasing) order. The problem then is

> Given an argument a, locate a pair of argument values x_k, x_{k+1} such that $x_k \leqq a < x_{k+1}$. If $a < x_1$ or if $a > x_n$ (where x_n is the last table entry), then special provision for these circumstances must be included.

Figure 4.1 is a flowchart for this problem.

In Fig. 4.1, if $a < x_1$, the interval x_1, x_2 will be used to extrapolate a function value; if $a > x_n$, the interval x_{n-1}, x_n will be used to extrapolate a value. In some problems, it may be preferable to print an error message if the argument falls outside the tabulated argument range.

A step-by-step analysis of Fig. 4.1 follows. First an index k is set to 2. The argument a is compared with x_2. If a is less than x_2 (which also includes $a < x_1$), then we exit from the test loop. If $a \geqq x_2$, then k is increased to 3. A test is made to see if k is less than n;

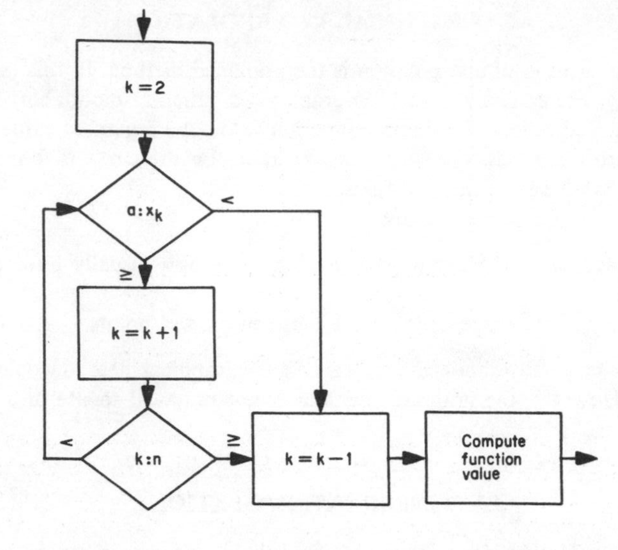

Fig. 4.1. Flowchart for table lookup.

if it is, the comparison of *a* and x_k is repeated either until we find a value such that $a < x_k$ or until the table is exhausted. In either event, k is reduced by 1. (This is only necessary to be consistent with the notation of Eq. (4.1) and other equations that follow.) Then finally a function value is calculated.

4.3.2 Function Computation

The reader is probably familiar with linear interpolation. Most science and engineering students are introduced to interpolation when they have to use logarithmic and trigonometric tables in a high school trigonometry course. Usually interpolation is introduced as a tool with little attention given to the theory. The student is instructed to

1. find two table arguments that bracket the given argument (i.e., $x_k \leqq a < x_{k+1}$),
2. find the distance of the given argument from the first table argument expressed as a fraction of the distance between the two table arguments, i.e.,

$$\frac{a - x_k}{x_{k+1} - x_k}$$

3. multiply this fraction by the difference between the two corresponding table function values and add to the first of these table function values.

Although this method is seldom presented as a mathematical formula, the formula is

$$y = f_k + \left(\frac{a - x_k}{x_{k+1} - x_k}\right)(f_{k+1} - f_k) \tag{4.1}$$

where y is the function value corresponding to the argument *a*.

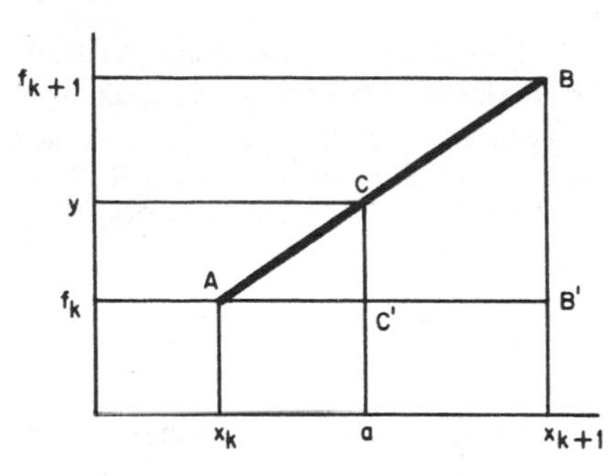

Fig. 4.2. Geometry of linear interpolation.

Intuitively, this method seems reasonable. If the given argument is say one fourth of the way between two table arguments, the corresponding function value is probably about one fourth of the way between the corresponding two table function values.

Let us now derive the formula for linear interpolation and compare it with Eq. (4.1). Consider Fig. 4.2. In this figure, a straight line joins the two table points (x_k, f_k) and (x_{k+1}, f_{k+1}). It is desired to find the function value y that corresponds to the argument *a*. By similar triangles,

$$\frac{CC'}{C'A} = \frac{BB'}{B'A}$$

which can be written

$$\frac{y - f_k}{a - x_k} = \frac{f_{k+1} - f_k}{x_{k+1} - x_k} \tag{4.2}$$

Solving for y,

$$y = f_k + \left(\frac{a - x_k}{x_{k+1} - x_k}\right)(f_{k+1} - f_k)$$

This equation is identical to (4.1). Furthermore, we could have written (4.2) directly from the analytic geometry formula for the equation of a line passing through two points. To complete the flowchart of Fig. 4.1, it is only necessary to insert Eq. (4.1) in the final box.

4.3.3 Critical Tables

The only limitation the previous discussion has put on the set of numbers $x_1, x_2, \ldots, x_n$, is that they be in a monotonic sequence. Most interpolation schemes require that the table arguments be given at equal intervals. Since the formula for linear interpolation requires information about only one interval, the intervals need not be equal. This is a tremendous advantage, since it allows us to construct what is known as a critical table. Figure 4.3 illustrates a function with varying degrees of curvature. This function is linear from A to B; therefore, it would only be necessary to tabulate the function at the end points. From B to C the curvature is large, so this segment would have to be subdivided into many small segments for linear interpolation to be accurate. The segment from C to D is of moderate curvature, so the interval size required would not have to be as small as between B and C. If equal size intervals were required, then all intervals would have to be as small as required between B and C. Thus, a critical table can significantly reduce the amount of storage required for storing a table.

4.3.4 Computing the Table Argument

If a table does have the arguments at equal intervals, it is not necessary to store the argument array, and the table arguments to be used in the computation can be computed.

If the intervals are equal, we can write

$$x_k = x_1 + (k - 1)h \tag{4.3}$$

where h is the constant argument interval. To find a value of k such that $x_k \leqq a < x_{k+1}$, it is only necessary to calculate

$$k = \left[\frac{a - x_1}{h}\right] + 1 \tag{4.4}$$

where the square brackets indicate "largest integer contained in." We then can use this value of k in Eq. (4.3) to find x_k and x_{k+1} and then use these and k to evaluate (4.1). Or by substitution, (4.1) can be rewritten as

$$y = f_k + \left(\frac{a - x_1}{h} - k + 1\right)(f_{k+1} - f_k) \tag{4.5}$$

or

$$y = f_{k+1} + \left(\frac{a - x_1}{h} - k\right)(f_{k+1} - f_k) \tag{4.6}$$

It may still be necessary to include some tests to determine, and to take appropriate action, if the argument *a* falls outside the range of the table. The flowcharting of this table lookup technique is left as an exercise.

4.4 HIGHER-ORDER INTERPOLATION

In the previous section, linear interpolation was discussed. Linear interpolation can always yield sufficiently accurate results if the interval size of the table is small enough. However, in many cases, either only a few table values are known or the table requires too much computer memory. Therefore, it is often necessary to employ a more sophisticated interpolation method that can deliver greater accuracy with fewer table entries.

One such method is polynomial interpolation. Linear interpolation is a special case of polynomial interpolation where the order is one. Although other interpolation methods can be derived, none of them will be covered in this text.

First, let us derive a second-order polynomial interpolation formula. A general form for such a polynomial is

$$y = A_0 + A_1 a + A_2 a^2 \tag{4.7}$$

where *a* is the argument and y is the corresponding function value. The coefficients

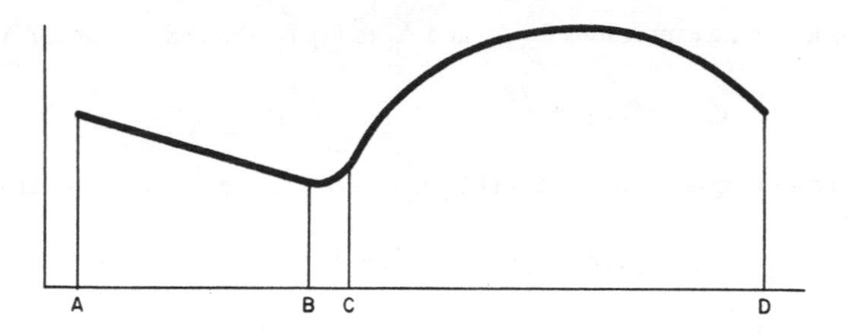

Fig. 4.3. Graph of function to be tabulated in a critical table.

A_0, A_1, and A_2 can be determined from any three sets of table entries. For convenience, let us pick (x_{k-1}, f_{k-1}), (x_k, f_k), and (x_{k+1}, f_{k+1}). Then we can write

$$\begin{aligned} f_{k-1} &= A_0 + A_1x_{k-1} + A_2x_{k-1}^2 \\ f_k &= A_0 + A_1x_k + A_2x_k^2 \\ f_{k+1} &= A_0 + A_1x_{k+1} + A_2x_{k+1}^2 \end{aligned} \tag{4.8}$$

Although these equations can be solved for A_0, A_1, and A_2, we will find the expressions for these quantities to be very unwieldy, and therefore, it is most practical to consider using only equally spaced arguments when using higher-order interpolation. For equally spaced table arguments, we can write:

$$\begin{aligned} x_{k-1} &= x_k - h \\ x_{k+1} &= x_k + h \end{aligned} \tag{4.9}$$

Then Eqs. (4.8) can be rewritten as

$$f_{k-1} = A_0 + A_1x_k - A_1h + A_2x_k^2 - 2A_2hx_k + A_2h^2 \tag{4.10a}$$

$$f_k = A_0 + A_1x_k + A_2x_k^2 \tag{4.10b}$$

$$f_{k+1} = A_0 + A_1x_k + A_1h + A_2x_k^2 + 2A_2hx_k + A_2h^2 \tag{4.10c}$$

Eliminating A_0 by subtracting these equations in pairs yields

$$f_k - f_{k-1} = A_1h + 2A_2hx_k - A_2h^2 \tag{4.11a}$$

$$f_{k+1} - f_k = A_1h + 2A_2hx_k + A_2h^2 \tag{4.11b}$$

Eliminating A_1h by subtraction,

$$f_{k+1} - 2f_k + f_{k-1} = 2A_2h^2 \tag{4.12}$$

Therefore,

$$A_2 = \frac{f_{k+1} - 2f_k + f_{k-1}}{2h^2} \tag{4.13}$$

Substituting this expression for A_2 in Eq. (4.11b) and solving for A_1,

$$A_1 = \frac{f_{k+1} - f_{k-1}}{2h} - \frac{x_k}{h^2}(f_{k+1} - 2f_k + f_{k-1}) \tag{4.14}$$

Using the foregoing expressions for A_2 and A_1 in Eq. (4.10b) and solving for A_0,

$$A_0 = f_k - \frac{x_k}{2h}(f_{k+1} - f_{k-1}) + \frac{x_k^2}{2h^2}(f_{k+1} - 2f_k + f_{k-1}) \tag{4.15}$$

Finally, substituting (4.13), (4.14), and (4.15) into (4.7) gives the rather complicated form

$$\begin{aligned} y = f_k &- \frac{x_k}{2h}(f_{k+1} - f_{k-1}) + \frac{x_k^2}{2h^2}(f_{k+1} - 2f_k + f_{k-1}) \\ &+ \left[\frac{f_{k+1} - f_{k-1}}{2h} - \frac{x_k}{h^2}(f_{k+1} - 2f_k + f_{k-1})\right] a + \left(\frac{f_{k+1} - 2f_k + f_{k-1}}{2h^2}\right) a^2 \end{aligned} \tag{4.16}$$

In actual practice, numeric values can be found for each of A_0, A_1, and A_2 before evaluating Eq. (4.7) with a particular value of *a*. However, Eq. (4.16) is given here in order to show how it may be rewritten in a simpler form. Regrouping terms so those of like order are together,

$$\begin{aligned} y = f_k &+ \left(\frac{f_{k+1} - f_{k-1}}{2h}\right) a - \left(\frac{f_{k+1} - f_{k-1}}{2h}\right) x_k + \left(\frac{f_{k+1} - 2f_k + f_{k-1}}{2h^2}\right) a^2 \\ &- \left(\frac{f_{k+1} - 2f_k + f_{k+1}}{h^2}\right) a x_k + \left(\frac{f_{k+1} - 2f_k + f_{k-1}}{2h^2}\right) x_k^2 \end{aligned} \tag{4.17}$$

or

$$y = f_k + \left(\frac{f_{k+1} - f_{k-1}}{2h}\right)(a - x_k) + \left(\frac{f_{k+1} - 2f_k + f_{k-1}}{2h^2}\right)(a^2 - 2ax_k + x_k^2) \tag{4.18}$$

which simplifies to

$$y = f_k + \left(\frac{f_{k+1} - f_{k-1}}{2}\right)\left(\frac{a - x_k}{h}\right) + \left(\frac{f_{k+1} - 2f_k + f_{k-1}}{2}\right)\left(\frac{a - x_k}{h}\right)^2 \tag{4.19}$$

Equation (4.19) can be applied to find a function value corresponding to *a*, using any value of k in the range $2 \leqq k \leqq n - 1$. However, the result will be most accurate if a value of k is found such that *a* is near the center of the table points used by the interpolation formula. In this case, k should be found such that $|x_k - a| \leqq 0.5h$. The techniques of Sec. 4.3.1 are still applicable, but must be modified for this particular interpolation method. Since, in this section, we are only considering equally spaced arguments, k can be computed by techniques similar to those of Sec. 4.3.4. Care must be taken that the k to be used does not result in a call for a function value not in the table.

Equation (4.19) suggests that if (4.7) had been originally written in powers of $(a - x_k)/h$ instead of in powers of *a*, the solution for the coefficients would have been greatly simplified. There are a number of ways that polynomials can be written. Some of these serve as the basis for the interpolating methods given in later sections.

4.5 DIFFERENCE TABLES

Before discussing specific methods, it is pertinent to introduce the concept and nomenclature of difference tables, since they will be needed in most of the methods to be discussed. Table 4.1 shows the literals for the table arguments and table function values. In addition, it also shows the first and second differences of the function values. The first difference is found by subtracting one table function entry from the following table function entry. Symbolically we express this as

$$\Delta f_i = f_{i+1} - f_i \tag{4.20}$$

Equation (4.20) defines, in general, the first *forward difference* of a function. Likewise, the second differences are found by taking first differences of the first differences. Symbolically this is expressed as

$$\Delta^2 f_i = \Delta f_{i+1} - \Delta f_i = f_{i+2} - 2f_{i+1} + f_i \tag{4.21}$$

Higher-order differences can be similarly defined. Therefore, in the symbolism of Eqs. (4.20) and (4.21), Table 4.1 becomes Table 4.2. Note that differences with the same subscript appear on a diagonal.

TABLE 4.1 DIFFERENCE TABLE (LITERAL)

x_1	f_1		
		f_2-f_1	
x_2	f_2		$(f_3-f_2) - (f_2-f_1) = f_3 - 2f_2 + f_1$
		f_3-f_2	
x_3	f_3		$(f_4-f_3) - (f_3-f_2) = f_4 - 2f_3 + f_2$
		f_4-f_3	
x_4	f_4		$(f_5-f_4) - (f_4-f_3) = f_5 - 2f_4 + f_3$
		f_5-f_4	
x_5	f_5		

Using the given symbolism and substituting,

$$u = \frac{a - x_k}{h} \tag{4.22}$$

Equation (4.19) can be further simplified to

$$y = f_k + \left(\frac{\Delta f_k + \Delta f_{k-1}}{2}\right)u + \frac{\Delta^2 f_{k-1}}{2}u^2 \tag{4.23}$$

We will find that most interpolating polynomials can be expressed in a form similar to Eq. (4.23).

4.6 GREGORY–NEWTON FORWARD INTERPOLATION

In Sec. 4.4 we derived a second-order interpolation formula. The derivation was lengthy and it is obvious that the derivation of a higher-order formula would be even more lengthy. The final form does suggest a short cut. However, this short cut is still

TABLE 4.2 FORWARD DIFFERENCES

x_1	f_1		
		Δf_1	
x_2	f_2		$\Delta^2 f_1$
		Δf_2	
x_3	f_3		$\Delta^2 f_2$
		Δf_3	
x_4	f_4		$\Delta^2 f_3$
		Δf_4	
x_5	f_5		

not the simplest approach. In the Gregory–Newton method, we start with the following expression for our polynomial formula:

$$y = A_0 + A_1(a - x_k) + A_2(a - x_k)(a - x_{k+1}) + A_3(a - x_k)(a - x_{k+1})(a - x_{k+2}) + \cdots + A_N(a - x_k)(a - x_{k+1})(\cdots)(a - x_{k+N-1}) \quad (4.24)$$

In this form, for a polynomial of degree N, we still have terms in all powers of *a* up through N. Furthermore, by algebraic manipulation, (4.24) can be rewritten in standard polynomial form. Now if we substitute N + 1 sets of table values in (4.24), we can write the following set of N + 1 equations:

$$\begin{aligned} f_k &= A_0 \\ f_{k+1} &= A_0 + A_1(x_{k+1} - x_k) \\ f_{k+2} &= A_0 + A_1(x_{k+2} - x_k) + A_2(x_{k+2} - x_k)(x_{k+2} - x_{k+1}) \\ &\vdots \\ f_{k+N} &= A_0 + A_1(x_{k+N} - x_k) + A_2(x_{k+N} - x_k)(x_{k+N} - x_{k+1}) + \cdots \\ &\quad + A_N(x_{k+N} - x_k)(\cdots)(x_{k+N} - x_{k+N-1}) \end{aligned} \quad (4.25)$$

If again we assume equal intervals, in x, then,

$$x_p - x_q = (p - q)\,h \quad (4.26)$$

where h is the interval and p and q are any integers. Using (4.26) in (4.25),

$$\begin{aligned} f_k &= A_0 \\ f_{k+1} &= A_0 + A_1(h) \\ f_{k+1} &= A_0 + A_1(2h) + A_2(2h)(h) \\ &\vdots \\ f_{k+N} &= A_0 + A_1(Nh) + A_2(Nh)((N - 1)\,h) + \cdots + A_N(Nh)((N - 1)h)(\cdots)(h) \end{aligned} \quad (4.27)$$

These equations are easily solved for $A_0, A_1, A_2, \ldots, A_N$, and in fact become

$$\begin{aligned} A_0 &= f_k \\ A_1 &= (f_{k+1} - f_k)/h \\ A_2 &= (f_{k+2} - 2f_{k+1} + f_k)/(2h^2) \\ &\vdots \\ A_N &= \frac{(f_{k+N} - Nf_{k+N-1} + \cdots + f_k)}{[Nh(N-1)h(\cdots)h]} \end{aligned} \tag{4.28}$$

A detailed study of Eqs. (4.28) will show that the numerators are differences of various orders and the denominators are of the form $i!h^i$. Equations (4.28) may be written as

$$\begin{aligned} A_0 &= f_k \\ A_1 &= \Delta f_k/(1!h) \\ A_2 &= \Delta^2 f_k/(2!h^2) \\ &\vdots \\ A_N &= \Delta^N f_k/(N!h^N) \end{aligned} \tag{4.29}$$

Substituting (4.29) into (4.24) yields

$$\begin{aligned} y = f_k &+ \frac{\Delta f_k}{1!h}(a - x_k) + \frac{\Delta^2 f_k}{2!h^2}(a - x_k)(a - x_{k+1}) + \cdots \\ &+ \frac{\Delta^N f_k}{N!h^N}(a - x_k)(a - x_{k+1})(\cdots)(a - x_{k+N-1}) \end{aligned} \tag{4.30}$$

Using Eqs. (4.22) and (4.26), we can derive the more general relation

$$\frac{a - x_{k+j}}{h} = u - j \tag{4.31}$$

which may be used in (4.30) to give

$$y = f_k + \frac{\Delta f_k}{1!}u + \frac{\Delta^2 f_k}{2!}u(u-1) + \cdots + \frac{\Delta^N f_k}{N!}u(u-1)(\cdots)(u-N+1) \tag{4.32}$$

Equation (4.32) is the Gregory–Newton forward interpolation formula. Note that if only the first two terms are retained, it is identical to linear interpolation. To use the Gregory–Newton formula, we employ a procedure similar to the one described earlier. That is, determine a k such that $x_k \leqq a \leqq x_{k+1}$ (this implies $0 \leqq u < 1$, but we will see in Sec. 4.8 a better method for selecting k); if $k > M - N$, where M is the number of entries in the table and N is the order of the interpolation formula, then set $k = M - N$; then evaluate Eq. (4.32). In the event that all the necessary differences are not available, then either a different method must be used or a k may be selected that violates the first condition ($x_k \leqq a < x_{k+1}$), but satisfies the second (all $\Delta^i f_k$ exist).

4.6.1 Examples of Gregory–Newton Forward Interpolation

A certain drive-in movie screen subtends an angle θ when viewed from a distance x. Table 4.3 gives the corresponding x's and θ's. Also included in the table are the first, second, and third differences of θ. The problem is to find values of θ for various values of x using the Gregory–Newton method.

For $x = 280$, use $k = 3$. Then $x_3 = 260$, $u = 0.5$, $\theta_3 = 4.379347$, $\Delta\theta_3 = -0.577907$, $\Delta^2\theta_3 = 0.134234$, and $\Delta^3\theta_3 = -0.041774$. Then by Eq. (4.32),

$$\begin{aligned}\theta &= 4.379347 + (-0.577907)(0.5) + \frac{0.134234}{2}(0.5)(-0.5) \\ &\quad + \frac{(-0.041774)}{6}(0.5)(-0.5)(-1.5) \\ &= 4.379347 - 0.2889535 - 0.0167793 - 0.0026109\end{aligned}$$

The first-order approximation (sum of the first two terms) is $\theta = 4.090394$
The second-order approximation (sum of the first three terms) is $\theta = 4.073614$
The third-order approximation (sum of the first four terms) is $\theta = 4.071003$
The exact answer is $\theta = 4.070097$

For $x = 350$, we would select $k = 5$. But in so doing, we would not have a value for $\Delta^3\theta_5$. Therefore, select $k = 4$. Then $u = 1.25$, $\theta_4 = 3.801440$, $\Delta\theta_4 = -0.443673$, $\Delta^2\theta_4 = 0.092460$, $\Delta^3\theta_4 = -0.026110$. Then by Eq. (4.32),

$$\begin{aligned}\theta &= 3.801440 + (-0.443673)(1.25) + \frac{0.092460}{2}(1.25)(0.25) \\ &\quad + \frac{(-0.026110)}{6}(1.25)(0.25)(-0.75)\end{aligned}$$

$$\begin{aligned}\theta &= 3.801440 - 0.5545913 + 0.0144469 + 0.0010199 \\ &= 3.246849 \text{ (first order)} \\ &= 3.261296 \text{ (second order)} \\ &= 3.262316 \text{ (third order)}\end{aligned}$$

The exact answer is 3.262520. The answer to this problem has a smaller error than the answer to the previous problem. This is true partly because the argument for the second problem comes closer to a table entry than the argument for the first problem. Section 4.7 develops a method that permits using $k = 5$ in the second problem of this section.

4.7 GREGORY–NEWTON BACKWARD INTERPOLATION

In Sec. 4.6 an interpolation formula was derived based on the interpolating polynomial of Eq. (4.24). This form was chosen so that when the independent variable is exactly equal to certain table arguments, some terms are zero. In that formula, the table arguments for which this occurs have indices greater than or equal to the base index k. In a like manner, we can work backward from the base index. Thus,

$$\begin{aligned}y &= A_0 + A_1(a - x_k) + A_2(a - x_k)(a - x_{k-1}) + \cdots \\ &\quad + A_N(a - x_k)(a - x_{k-1})(\cdots)(a - x_{k-N+1})\end{aligned} \tag{4.33}$$

TABLE 4.3 DIFFERENCE TABLE FOR AN EXAMPLE

i	$x(ft)$	$\theta(degrees)$	$\Delta\theta$	$\Delta^2\theta$	$\Delta^3\theta$
1	180	6.282489			
			−1.119883		
2	220	5.162606		0.336624	
			−0.783259		−0.131272
3	260	4.379347		0.205352	
			−0.577907		−0.071118
4	300	3.801440		0.134234	
			−0.443673		−0.041774
5	340	3.357767		0.092460	
			−0.351213		−0.026110
6	380	3.006554		0.066350	
			−0.284863		
7	420	2.721691			

From this equation, we can write N + 1 equations by using N + 1 data points to find the N + 1 coefficients.

$$
\begin{aligned}
f_k &= A_0 \\
f_{k-1} &= A_0 + A_1(x_{k-1} - x_k) \\
f_{k-2} &= A_0 + A_1(x_{k-2} - x_k) + A_2(x_{k-2} - x_k)(x_{k-2} - x_{k-1}) \\
&\vdots \\
f_{k-N} &= A_0 + A_1(x_{k-N} - x_k) + A_2(x_{k-N} - x_k)(x_{k-N} - x_{k-1}) + \cdots \\
&\quad + A_N(x_{k-N} - x_k)(x_{k-N} - x_{k-1})(\cdots)(x_{k-N} - x_{k-N+1})
\end{aligned}
\tag{4.34}
$$

If we make the same assumption about equal intervals and use Eq. (4.26) in Eq. (4.34), we obtain

$$
\begin{aligned}
f_k &= A_0 \\
f_{k-1} &= A_0 + A_1(-h) \\
f_{k-2} &= A_0 + A_1(-2h) + A_2(-2h)(-h) \\
&\vdots \\
f_{k-N} &= A_0 + A_1(-Nh) + A_2(-Nh)(-N+1)h + \cdots \\
&\quad + A_N(-Nh)((-N+1)h)(\cdots)(-h)
\end{aligned}
\tag{4.35}
$$

These can be solved to obtain

$$\begin{aligned} A_0 &= f_k \\ A_1 &= (f_k - f_{k-1})/h \\ A_2 &= (f_k - 2f_{k-1} + f_{k-2})/(2h^2) \\ &\vdots \\ A_N &= \frac{(f_k - Nf_{k-1} + \cdots + f_{k-N})}{[Nh(N-1)h(\cdots)h]} \end{aligned} \quad (4.36)$$

Equations (4.36) look similar to (4.28), but the successive differences do not have the same subscript. Writing in a form similar to (4.29),

$$\begin{aligned} A_0 &= f_k \\ A_1 &= \Delta f_{k-1}/(1!h) \\ A_2 &= \Delta^2 f_{k-2}/(2!h^2) \\ &\vdots \\ A_N &= \Delta^N f_{k-N}/(N!h^N) \end{aligned} \quad (4.37)$$

This form can be used in Eq. (4.33) to obtain the final interpolation formula. However, to avoid confusion, a new difference definition is introduced. The following symbolism is used to define *backward* differences:

$$\begin{aligned} \nabla f_i &= f_i - f_{i-1} \\ \nabla^2 f_i &= \nabla f_i - \nabla f_{i-1} \end{aligned} \quad (4.38)$$

and in general,

$$\nabla^j f_i = \nabla^{j-1} f_i - \nabla^{j-1} f_{i-1}$$

We can also show that

$$\nabla^j f_i = \Delta^j f_{i-j} \quad (4.39)$$

Then Eqs. (4.37) can be written as

$$\begin{aligned} A_0 &= f_k \\ A_1 &= \nabla f_k/(1!h) \\ A_2 &= \nabla^2 f_k/(2!h^2) \\ &\vdots \\ A_N &= \nabla^N f_k/(N!h^N) \end{aligned} \quad (4.40)$$

Substituting Eqs. (4.40) into (4.33) and using the relation of (4.31), we get the following Gregory–Newton backward interpolation formula:

$$y = f_k + \frac{\nabla f_k u}{1!} + \frac{\nabla^2 f_k}{2!} u(u+1) + \cdots + \frac{\nabla^N f_k}{N!} u(u+1)(\cdots)(u+N-1) \quad (4.41)$$

Backward interpolation is more convenient to use near the *end* of a table than forward interpolation. Consider the second example of Sec. 4.6.1. For the argument given, we had no third difference for $k = 5$. In order to obtain an answer, we selected $k = 4$. If, instead, we select $k = 7$ and use backward interpolation, then $u = (350 - 420)/40 = -1.75$, and use of third-order interpolation leads to the answer 3.262316 degrees. This is exactly the same answer as obtained with forward interpolation. At first this seems surprising, but a little analysis shows that the answers should be identical. In the forward interpolation example, we used information from table points 4, 5, 6, and 7 to form a third-degree polynomial. In the backward interpolation example, we used information from table points 7, 6, 5, and 4 to form a third-degree polynomial. Since these are the same points and the polynomial is of the same degree, we can show that the resulting polynomials are identical.

If we use $k = 6$ and backward interpolation, we will obtain a slightly different result because a different set of data points will be used. The result will be 3.262928, which has a greater error than the previous result. In the first case, the table arguments used were 300, 340, 380, and 420. In the latter case, the table arguments used were 260, 300, 340, and 380. It can be seen that the argument of the example 350, is more nearly in the center of the table arguments for the first case than for the second.

4.8 SELECTING THE TABLE ENTRIES

When using interpolation methods of second or higher order, the selection of the parameter k is not quite as simple as when using linear interpolation. In the case of linear interpolation, we have indicated that k should be selected so that $x_k \leq a < x_{k+1}$, if possible. (If not possible, it means that we must extrapolate.)

In the higher-order interpolation methods, we have seen that, even though the methods can be stated in terms of differences, actually three or more table entries are involved. The question arises as to which set of table entries should be used for a given argument. We have also seen that for a given argument we can select more than one set of table entries and still have the argument bracketed. However, the results obtained using different sets of table entries are not equally accurate.

In the methods under discussion, the approximating polynomial of order N passes exactly through $N + 1$ table points. Therefore, it would be wise to select a set of table points such that one of them has an argument as near as possible to the given argument. Still assuming equally spaced table arguments, we can evaluate the expression $(a - x_1)/h + 1$ and round it to the nearest integer j. Using this, we can find a tentative value of u by

$$u^* = \frac{a - x_1}{h} + 1 - j \tag{4.42}$$

The quantity u^* may turn out to be positive or negative, but its magnitude will be less than or equal to one half. This procedure will locate the table argument x_j that is nearest to the given argument.

We will state here, without proof, that the interpolating polynomial deviates least from the function being approximated near the center of the data points being used. Therefore, in the case of an even-order interpolating polynomial (which uses an odd number of table points), x_j should be the center point of the table arguments used.

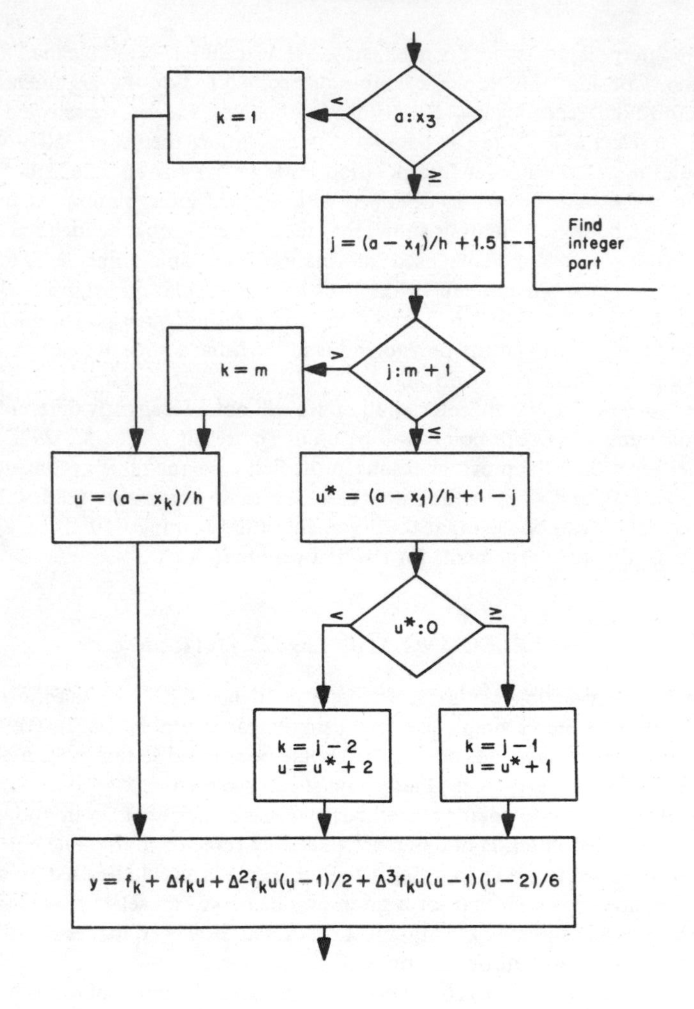

Fig. 4.4. Flowchart for third-order Gregory-Newton forward interpolation.

For an odd-order interpolating polynomial (which uses an even number of table points), the table arguments to be used should be selected such that the given argument lies between the center pair of the table arguments.

To determine k and u for *forward* interpolation: For an even-order interpolation of order N, set $k = j - N/2$ and $u = u^* + N/2$. For an odd-order interpolation of order N, if $u^* > 0$, set $k = j - (N - 1)/2$ and $u = u^* + (N - 1)/2$; if $u^* < 0$, set $k = j - (N + 1)/2$ and $u = u^* + (N + 1)/2$.

To determine k and u for *backward* interpolation: For an even-order interpolation of order N, set $k = j + N/2$ and $u = u^* - N/2$. For an odd-order interpolation of order N, if $u^* > 0$, set $k = j + (N + 1)/2$ and $u = u^* - (N + 1)/2$; if $u^* < 0$, set $k = j + (N - 1)/2$ and $u = u^* - (N - 1)/2$.

In addition to these rules, we must observe the limits of the table and adjust k and u accordingly. These rules seem unwieldy, but in an actual computer program, the type of interpolation and its order would be predetermined, and therefore the actual programming would be relatively simple. To illustrate, consider Fig. 4.4, which is a flowchart of a third-order Gregory–Newton forward interpolation. In this example, $N = 3$. Therefore, four table values of the function will be represented in the differences. For maximum accuracy, we would like to choose the table entries such that the given argument *a* falls between the second and third table arguments of the corresponding set of four table arguments. Expressed mathematically, this means that, for this example, we would like to find a k such that $x_{k+1} \leqq a < x_{k+2}$, if at all possible. This means that $k = 1$ for any *a* up to but not including x_3. This explains the first test. This test also allows for extrapolation below the range of the table. Next, j is calculated as described earlier in this section. What is illustrated is a convenient way of rounding positive numbers by computer. Assuming there are m table arguments with m corresponding function values, first differences, second differences, and third differences, the last set of entries actually contains information from function values f_m, f_{m+1}, f_{m+2}, and f_{m+3}. Therefore, the last set can be used to interpolate or extrapolate values for which $a > x_{m+1}$. For any $j \leqq m + 1$, a, k, and u are computed, as shown, for use in the Gregory–Newton formula.

4.9 OTHER INTERPOLATION METHODS

We have seen how polynomial interpolation formulas can be developed using table values selected in different manners. These by no means exhaust the possibilities. In this section, we will confine our investigation to methods classified as *central difference* methods. The central difference methods are appealing because they appear to be easier to use. This is certainly true in the case of hand computation. Whether this is still true in the case of machine computation depends on too many factors to allow an objective analysis. The primary advantage is that when we construct a difference table, the differences to be used are found, essentially, along a horizontal line extending to the right of the base table argument.

TABLE 4.4 FORWARD CENTRAL DIFFERENCES

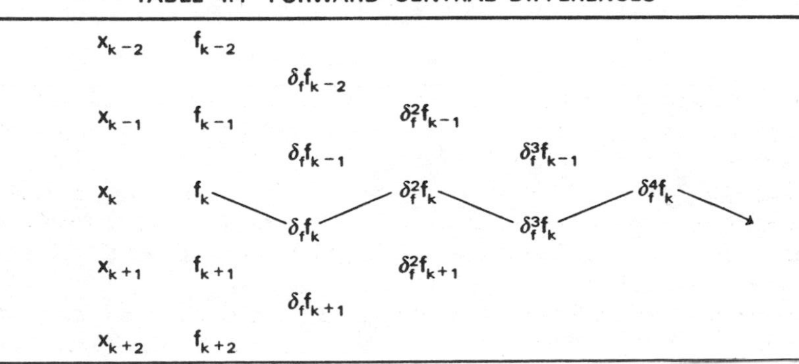

x_{k-2}	f_{k-2}				
		$\delta_f f_{k-2}$			
x_{k-1}	f_{k-1}		$\delta_f^2 f_{k-1}$		
		$\delta_f f_{k-1}$		$\delta_f^3 f_{k-1}$	
x_k	f_k		$\delta_f^2 f_k$		$\delta_f^4 f_k$
		$\delta_f f_k$		$\delta_f^3 f_k$	
x_{k+1}	f_{k+1}		$\delta_f^2 f_{k+1}$		
		$\delta_f f_{k+1}$			
x_{k+2}	f_{k+2}				

4.9.1 Gauss Forward Central Difference Method

Consider Table 4.4. The differences in this table are found by the expressions

$$\delta_f f_i = f_{i+1} - f_i$$
$$\delta_f^2 f_i = \delta_f f_i - \delta_f f_{i-1}$$

and in general,

$$\begin{aligned} \delta_f^j f_i &= \delta_f^{j-1} f_{i+1} - \delta_f^{j-1} f_i \qquad j \text{ odd} \\ \delta_f^j f_i &= \delta_f^{j-1} f_i - \delta_f^{j-1} f_{i-1} \qquad j \text{ even} \end{aligned} \tag{4.43}$$

These are called *forward central differences.*

Using the differences indicated by the arrowed path in Table 4.4, we can derive the following interpolation formula:

$$\begin{aligned} y = f_k &+ \frac{\delta_f f_k}{1!} u + \frac{\delta_f^2 f_k}{2!} u(u-1) + \frac{\delta_f^3 f_k}{3!} u(u-1)(u+1) \\ &+ \frac{\delta_f^4 f_k}{4!} u(u-1)(u+1)(u-2) + \cdots \end{aligned} \tag{4.44}$$

This is called the *Gauss forward central difference method.*

4.9.2 Gauss Backward Central Difference Method

In a similar manner, we can construct Table 4.5. The differences in Table 4.5 are found by the expressions

$$\delta_b f_i = f_i - f_{i-1}$$
$$\delta_b^2 f_i = \delta_b f_{i+1} - \delta_b f_i$$

and in general,

$$\begin{aligned} \delta_b^j f_i &= \delta_b^{j-1} f_i - \delta_b^{j-1} f_{i-1} \qquad j \text{ odd} \\ \delta_b^j f_i &= \delta_b^{j-1} f_{i+1} - \delta_b^{j-1} f_i \qquad j \text{ even} \end{aligned} \tag{4.45}$$

These are called *backward central differences.*

TABLE 4.5 BACKWARD CENTRAL DIFFERENCES

x_{k-2}	f_{k-2}				
		$\delta_b f_{k-1}$			
x_{k-1}	f_{k-1}		$\delta_b^2 f_{k-1}$		
		$\delta_b f_k$		$\delta_b^3 f_k$	
x_k	f_k		$\delta_b^2 f_k$		$\delta_b^4 f_k$
		$\delta_b f_{k+1}$		$\delta_b^3 f_{k+1}$	
x_{k+1}	f_{k+1}		$\delta_b^2 f_{k+1}$		
		$\delta_b f_{k+2}$			
x_{k+2}	f_{k+2}				

Using the differences indicated by the arrowed path in Table 4.5 we can derive the following interpolation formula:

$$y = f_k + \frac{\delta_b f_k}{1!} u + \frac{\delta_b^2 f_k}{2!} u(u+1) + \frac{\delta_b^3 f_k}{3!} u(u+1)(u-1) + \frac{\delta_b^4 f_k}{4!} u(u+1)(u-1)(u+2) + \cdots \tag{4.46}$$

This is called the *Gauss backward central difference interpolation method.*

4.9.3 Stirling Central Difference Method

In the preceding two methods, the odd-order central differences had to be selected either from the line below or the line above that of the base table argument. Table 4.6 gives the central differences in Stirling's method. The differences are found by the following expressions

$$\delta f_i = \frac{(f_{i+1} - f_i) + (f_i - f_{i-1})}{2} = \frac{f_{i+1} - f_{i-1}}{2}$$

$$\delta^2 f_i = (f_{i+1} - f_i) - (f_i - f_{i-1}) = f_{i+1} - 2f_i + f_{i-1}$$

or in general for $j \geqq 3$,

$$\begin{aligned} \delta^j f_i &= \frac{\delta^{j-1} f_{i+1} - \delta^{j-1} f_{i-1}}{2} && j \text{ odd} \\ \delta^j f_i &= \delta^{j-2} f_{i+1} - 2\delta^{j-2} f_i + \delta^{j-2} f_{i-1} && j \text{ even} \end{aligned} \tag{4.47}$$

These generalizations are based on the definitions

$$\delta^j f_i = \frac{(\delta_f^j f_i + \delta_b^j f_i)}{2} \qquad j \text{ odd}$$

$$\delta^j f_i = \delta_f^j f_i = \delta_b^j f_i = \delta_f^{j-1} f_i - \delta_b^{j-1} f_i \qquad j \text{ even}$$

The interpolation formula is given by

$$y = f_k + \frac{\delta f_k}{1!} u + \frac{\delta^2 f_k}{2!} u^2 + \frac{\delta^3 f_k}{3!} u(u^2 - 1) + \frac{\delta^4 f_k}{4!} u^2(u^2 - 1) + \frac{\delta^5 f_k}{5!} u(u^2 - 1)(u^2 - 2^2) + \cdots \tag{4.48}$$

This is called the *Stirling central difference interpolation formula.* This formula always involves an odd number of ordinate values no matter how many terms are used. If the formula is truncated after an even number of terms, the number of ordinates involved will be the same as the number involved when one more term of the formula is included. If an even number of terms is used, the resulting interpolation formula will not pass exactly through the end points of the set of ordinates involved in the formula. Therefore, if Stirling's formula is to be used, it is recommended that Eq. (4.48) be truncated only after an odd number of terms.

Note that the second-order interpolation formula, Eq. (4.19), derived in Sec. 4.4, is exactly the first three terms of Stirling's formula. This is because, in that derivation,

TABLE 4.6 CENTRAL DIFFERENCES

x_{k-2}	f_{k-2}				
x_{k-1}	f_{k-1}	δf_{k-1}	$\delta^2 f_{k-1}$		
x_k	f_k	δf_k	$\delta^2 f_k$	$\delta^3 f_k$	$\delta^4 f_k$ →
x_{k+1}	f_{k+1}	δf_{k+1}	$\delta^2 f_{k+1}$		
x_{k+2}	f_{k+2}				

three points were used and the subscripting was selected so that there was symmetry about the center table entry.

In the three central difference methods discussed, the quantity u is still calculated by Eq. (4.22) and the base index k must be selected by some meaningful process. In the case of the central difference methods, the selection of k is simpler. The easiest rule to follow is the one given first, namely, find k such that $x_k \leqq a < x_{k+1}$. A method that will probably yield better accuracy, in some cases, is that which was described in Sec. 4.8, where k was found such that $|u| \leqq 0.5$.

4.9.4 Everett Central Difference Method

The Everett interpolation method has the advantage in that only even-order differences are used. However, two consecutive values of each of these differences must be used in the formula. The difference table is formed just as in any of the preceding three methods (even-order differences are the same regardless of which method is used), and then the following ***Everett central difference formula*** is applied:

$$\begin{aligned} y = \; & E_{00} f_k + E_{10}\delta^2 f_k + E_{20}\delta^4 f_k + \cdots + E_{(N/2)0}\delta^N f_k \\ & + E_{01} f_{k+1} + E_{11}\delta^2 f_{k+1} + E_{21}\delta^4 f_{k+1} + \cdots + E_{(N/2)1}\delta^N f_{k+1} \end{aligned} \tag{4.49}$$

The Everett coefficients E depend on the argument *a* in the following manner: By Eq. (4.22), $u = (a - x_k)/h$; then t is defined by

$$t = 1 - u \tag{4.50}$$

Then

$$\begin{aligned} E_{00} &= t \\ E_{10} &= \frac{1}{3!} t(t^2 - 1) = \binom{t+1}{3} \\ E_{20} &= \frac{1}{5!} t(t^2 - 1)(t^2 - 2^2) = \binom{t+2}{5} \\ &\vdots \\ E_{j0} &= \binom{t+j}{2j+1} \end{aligned} \tag{4.51}$$

and

$$\begin{aligned} E_{01} &= u \\ E_{11} &= \frac{1}{3!}u(u^2 - 1) = \binom{u+1}{3} \\ E_{21} &= \frac{1}{5!}u(u^2 - 1)(u^2 - 2^2) = \binom{u+2}{5} \\ &\vdots \\ E_{j1} &= \binom{u+j}{2j+1} \end{aligned} \tag{4.52}$$

where

$$\binom{p}{q} = \frac{p(p-1)(p-2)(\cdots)(p-q+1)}{q!} \tag{4.53}$$

As in the other methods, k should be selected so that $x_k \leqq a < x_{k+1}$. And in all methods we must be sure that the value of k selected requires only differences that are available.

4.10 DIVIDED DIFFERENCES

In all of the second- or higher-order interpolation methods discussed thus far, we have assumed equal argument intervals to simplify the algebra. This is actually not necessary, only convenient. For any one of the methods, a formula can be derived even if the argument intervals are not equal. However, such a formula is dependent on the actual table argument values. By sufficient algebraic manipulation, the formula can be put into a form very similar to those arrived at by assuming equal intervals.

For example, if we define forward *divided* differences as follows:

$$\begin{aligned} \Delta_d f_k &= \frac{f_{k+1} - f_k}{x_{k+1} - x_k} \\ \Delta_d^2 f_k &= \frac{\Delta_d f_{k+1} - \Delta_d f_k}{x_{k+2} - x_k} \end{aligned} \tag{4.54}$$

and in general,

$$\Delta_d^j f_k = \frac{\Delta_d^{j-1} f_{k+1} - \Delta_d^{j-1} f_k}{x_{k+j} - x_k}$$

we can then write the Gregory–Newton forward interpolation formula as follows:

$$\begin{aligned} y = f_k &+ \Delta_d f_k(a - x_k) + \Delta_d^2 f_k(a - x_k)(a - x_{k+1}) + \cdots \\ &+ \Delta_d^N f_k(a - x_k)(a - x_{k+1})(\cdots)(a - x_{k+N-1}) \end{aligned} \tag{4.55}$$

Formulas for the other methods can also be derived.

4.11 LAGRANGE INTERPOLATION

In the previous methods, a polynomial formulation was devised so that the simultaneous equations to be solved were in triangular form. In the Lagrange method, the polynomial is formed so that the equations to be solved are in diagonal form. This means that the polynomial coefficients are just table function values. The formula is

$$y = f_k \frac{(a - x_{k+1})(a - x_{k+2})(\cdots)(a - x_{k+N})}{(x_k - x_{k+1})(x_k - x_{k+2})(\cdots)(x_k - x_{k+N})} + f_{k+1} \frac{(a - x_k)(a - x_{k+2})(\cdots)(a - x_{k+N})}{(x_{k+1} - x_k)(x_{k+1} - x_{k+2})(\cdots)(x_{k+1} - x_{k+N})} + \cdots + f_{k+N} \frac{(a - x_k)(a - x_{k+1})(\cdots)(a - x_{k+N-1})}{(x_{k+N} - x_k)(x_{k+N} - x_{k+1})(\cdots)(x_{k+N} - x_{k+N-1})} \tag{4.56}$$

4.12 REVIEW OF METHODS

In all of the methods summarized below, a base index k must be located and then the modified argument u calculated by

$$u = \frac{a - x_k}{h} \tag{4.22}$$

4.12.1 Gregory–Newton Forward

$$\Delta f_i = f_{i+1} - f_i \tag{4.20}$$

$$\Delta^2 f_i = \Delta f_{i+1} - \Delta f_i \tag{4.21}$$

$$\Delta^j f_i = \Delta^{j-1} f_{i+1} - \Delta^{j-1} f_i$$

$$y = f_k + \frac{\Delta f_k}{1!} u + \frac{\Delta^2 f_k}{2!} u(u-1) + \cdots + \frac{\Delta^N f_k}{N!} u(u-1)(\cdots)(u-N+1) \tag{4.32}$$

4.12.2 Gregory–Newton Backward

$$\nabla f_i = f_i - f_{i-1}$$

$$\nabla^2 f_i = \nabla f_i - \nabla f_{i-1} \tag{4.38}$$

$$\nabla^j f_i = \nabla^{j-1} f_i - \nabla^{j-1} f_{i-1}$$

$$y = f_k + \frac{\nabla f_k}{1!} u + \frac{\nabla^2 f_k}{2!} u(u+1) + \cdots + \frac{\nabla^N f_k}{N!} u(u+1)(\cdots)(u+N-1) \tag{4.41}$$

4.12.3 Gauss Forward

$$\begin{aligned} \delta_f f_i &= f_{i+1} - f_i \\ \delta_f^2 f_i &= \delta_f f_i - \delta_f f_{i-1} \\ \delta_f^j f_i &= \delta_f^{j-1} f_{i+1} - \delta_f^{j-1} f_i \qquad j \text{ odd} \\ \delta_f^j f_i &= \delta_f^{j-1} f_i - \delta_f^{j-1} f_{i-1} \qquad j \text{ even} \end{aligned} \tag{4.43}$$

$$y = f_k + \frac{\delta_f f_k}{1!} u + \frac{\delta_f^2 f_k}{2!} u(u-1) + \frac{\delta_f^3 f_k}{3!} u(u-1)(u+1) + \frac{\delta_f^4 f_k}{4!} u(u-1)(u+1)(u-2) + \cdots \tag{4.44}$$

4.12.4 Gauss Backward

$$
\begin{aligned}
\delta_b f_i &= f_i - f_{i-1} \\
\delta_b^2 f_i &= \delta_b f_{i+1} - \delta_b f_i \\
\delta_b^j f_i &= \delta_b^{j-1} f_i - \delta_b^{j-1} f_{i-1} \qquad j \text{ odd} \\
\delta_b^j f_i &= \delta_b^{j-1} f_{i+1} - \delta_b^{j-1} f_i \qquad j \text{ even}
\end{aligned}
\tag{4.45}
$$

$$
\begin{aligned}
y = f_k &+ \frac{\delta_b f_k}{1!} u + \frac{\delta_b^2 f_k}{2!} u(u+1) + \frac{\delta_b^3 f_k}{3!} u(u+1)(u-1) \\
&+ \frac{\delta_b^4 f_k}{4!} u(u+1)(u-1)(u+2) + \cdots
\end{aligned}
\tag{4.46}
$$

4.12.5 Stirling

$$\delta f_i = \frac{\delta_f f_i + \delta_b f_i}{2} = \frac{f_{i+1} - f_{i-1}}{2}$$

$$\delta^2 f_i = \delta_f f_i - \delta_b f_i = f_{i+1} - 2f_i + f_{i-1}$$

For $j \geqq 3$,

$$
\begin{aligned}
\delta^j f_i &= \frac{\delta_f^j f_i + \delta_b^j f_i}{2} = \frac{\delta^{j-1} f_{i+1} - \delta^{j-1} f_{i-1}}{2} \qquad j \text{ odd} \\
\delta^j f_i &= \delta_f^j f_i = \delta_b^j f_i = \delta_f^{j-1} f_i - \delta_b^{j-1} f_i \qquad j \text{ even}
\end{aligned}
\tag{4.47}
$$

$$
\begin{aligned}
y = f_k &+ \frac{\delta f_k}{1!} u + \frac{\delta^2 f_k}{2!} u^2 + \frac{\delta^3 f_k}{3!} u(u^2 - 1) + \frac{\delta^4 f_k}{4!} u^2(u^2 - 1) \\
&+ \frac{\delta^5 f_k}{5!} u(u^2 - 1)(u^2 - 2^2) + \cdots
\end{aligned}
\tag{4.48}
$$

4.12.6 Everett

$$t = 1 - u \tag{4.50}$$

$$E_{j0} = \binom{t+j}{2j+1} \tag{4.51}$$

$$E_{j1} = \binom{u+j}{2j+1} \tag{4.52}$$

$$
\begin{aligned}
y = \; & E_{00} f_k + E_{10} \delta^2 f_k + E_{20} \delta^4 f_k + \cdots + E_{(N/2)0} \delta^N f_k \\
& + E_{01} f_{k+1} + E_{11} \delta^2 f_{k+1} + E_{21} \delta^4 f_{k+1} + \cdots + E_{(N/2)1} \delta^N f_{k+1}
\end{aligned}
\tag{4.49}
$$

4.13 SUMMARY

In this chapter, polynomial interpolation formulas have been presented. No judgment has been made as to which of these methods is best. In fact, no error analysis has been given. This has been deferred to Chap. 5, since the subject matter of that chapter is needed for such an analysis.

The judgment as to which method to use is left to the reader. Each method has advantages to recommend it over other methods. In Sec. 4.7, an example was given that yielded a result by the Gregory–Newton backward interpolation method that was identical to a result obtained in Sec. 4.6 by the Gregory–Newton forward interpolation method. This was explained by the fact that the two examples used exactly the same table entries and the same degree polynomials. In general, two different polynomial interpolation methods can be shown to be exactly equivalent if they use the same table entries and polynomials of the same degree.

The selection of the method to use is usually based on the determination of which method has the most convenient form for the given problem. Advantages of each of the four different methods are listed below:

1. The Gregory–Newton forward interpolation method is convenient to use when the argument is near the beginning of the table.
2. The Gregory–Newton backward interpolation method is convenient to use when the argument is near the end of the table.
3. The modified argument needed for the Stirling interpolation method is easy to calculate by the use of Eq. (4.42).
4. Everett linear interpolation is convenient to use with a desk calculator since the coefficients E_{00} and E_{01} can be formed mentally and no differences need be calculated.

This list illustrates what is meant by "convenient form." Other advantages exist for the methods above and for other methods. By comparing methods, the reader must determine for himself which is the best to use on a particular problem.

EXERCISES

1. From a table of sines tabulated at intervals of 0.001 radians, select a set of argument values such that the error in the sine by linear interpolation at the midpoint of each interval is less than 0.0005. Select 0.000 as the first table argument and continue with increasing arguments until $\pi/2$ is exceeded.

2. Prepare a flowchart for determining the index k for linear interpolation by the method of Sec. 4.3.4. Include provisions for the argument falling outside the range of the table arguments.

3. Derive a second-order interpolation formula that passes exactly through the points (x_{k-1}, f_{k-1}), (x_k, f_k), (x_{k+2}, f_{k+2}). Assume $x_{k+j} = x_k + jh$.

4. Derive a fifth-order Gregory–Newton forward interpolation formula that uses all the values from the following table of inverse tangents.

x	$tan^{-1}x$
0.0	0.00000000
0.2	0.19739556
0.4	0.38050638
0.6	0.54041950
0.8	0.67474094
1.0	0.78539816

5. Evaluate the formula derived in Exercice 4 for the values 0.1, 0.3, 0.5, 0.7, and 0.9. Find the exact values of the inverse tangent for these arguments, and calculate the error in the computed values.

6. Simplify the formula derived in Exercise 4 to obtain a fifth-order polynomial in standard form with the argument A as the independent variable.

7. Evaluate the polynomial derived in Exercise 6 for the arguments given in Exercise 5. Also calculate the error for each argument.

8. Derive a fifth-order Gauss forward interpolation formula that uses the table given in Exercise 4.

9. Evaluate the formula derived in Exercise 8 for the arguments given in Exercise 5. Also calculate the error for each argument.

10. Prepare a flowchart and write a FORTRAN program to perform the following. Using Table 4.3 and third-order Gregory–Newton forward interpolation, calculate a value of θ for each x from 180 feet through 420 feet at intervals of 5 feet. Calculate the exact value for each x by the formula

$$\theta = 57.29578 \tan^{-1}\left(\frac{20x}{x^2 + 300}\right)$$

Also calculate the error (exact minus approximate) for each of these points.

11. Using the table calculated in Exercise 10, linearly interpolate between pairs of function values which have arguments that are exact multiples of 10 feet to obtain approximate function values corresponding to the intervening points. Also calculate the error for each of these points. Are these values as accurate as those obtained for the same points in Exercise 10?

12. Prepare a table of sines that are tabulated at intervals of 0.02 radians from 0. through 1.60 radians. Also prepare tables of the first, second, and third forward differences of this table. Prepare a flowchart and write a FORTRAN program using these tables and using Gregory–Newton forward interpolation to find first–, second–, and third-order approximations of the sine function for the arguments −0.01, 0.005, 0.46, 0.55, 0.78539816, 1.01, 1.2345, 1.57079632, 1.65, and 1.67. The program should allow for the fact that each table of differences is of different length. Therefore, the handling of arguments near the ends of the tables should be different for each order interpolation. Also compute the error for each input.

13. Repeat Exercise 12 using Stirling's central difference method instead of the Gregory–Newton method. In this exercise, use only that portion of each table for which there is a third difference for each entry for all orders of interpolation. This means that the argument range for each table will be 0.04 through 1.56 inclusive.

14. Write a computer program (flowchart and code) that uses the sine table and second difference table prepared in Exercise 12 and uses Everett's interpolation method to find an approximation to the sine function through second differences. Execute using the data in Exercise 12.

15. Write a third-order Lagrange interpolation formula, and simplify it by using the assumption that argument values are spaced at equal intervals.

16. Draw a flowchart and code a program in FORTRAN using the formula derived in Exercise 15 to interpolate a sine value from a table of sines. Execute the program using the table of sines and the arguments from Exercise 12.

5

Taylor's Series

5.1 INTRODUCTION

In Chap. 4, the use of polynomials as interpolating formulas was presented. In that chapter it was shown that if a polynomial is written in a special form, Eq. (4.24), its coefficients can be found easily, Eqs. (4.25). Taylor's series, a special polynomial form, is introduced in Sec. 5.2 which also contains examples of the use of this polynomial form for function approximation and interpolation (an important use of Taylor's series). To show the parallelism between the form of the interpolating polynomials and Taylor's series, Eq. (4.24) is repeated below, but with $k = 0$.

$$\begin{aligned} P(x) = a_0 + a_1(x - x_0) + a_2(x - x_0)(x - x_1) + \cdots \\ + a_n(x - x_0)(x - x_1)(\cdots)(x - x_{n-1}) \end{aligned} \tag{5.1}$$

In this form, if the tabular values $x_0, x_1, \ldots, x_n$ and their corresponding function values $P(x_0), P(x_1), \ldots, P(x_n)$ are substituted into the equation one pair at a time, $n + 1$ equations, which are in triangular form, will result. That is,

$$\begin{aligned} P(x_0) &= a_0 \\ P(x_1) &= a_0 + a_1(x_1 - x_0) \\ &\vdots \\ P(x_n) &= a_0 + a_1(x_n - x_0) + \cdots + a_n(x_n - x_0)(x_n - x_1)(\cdots)(x_n - x_{n-1}) \end{aligned} \tag{5.2}$$

The solution of Eqs. (5.2) for the a_i's is very convenient. The first equation is solved for a_0, which is substituted in the second equation, which is then solved for a_1, etc.

Section 5.3 contains an analysis of the truncation error in the use of Taylor's series. Section 5.4 presents the expansion of a function of two independent variables in a Taylor's series. Another important use of Taylor's series is presented in Sec. 5.5 where it is used to determine the truncation error (also called inherent error) of other computing methods. Also given in Sec. 5.5 is an example of the derivation of the truncation error of an interpolation method given in Chap. 4. Taylor's series is used extensively in Chaps. 6, 10, 11, and 12 for the derivation of truncation error expressions. Note that truncation error is one of the forms of computing errors discussed in Sec. 3.3.1.

5.2 TAYLOR'S SERIES

In this section, we will derive the Taylor's series for a function of a single independent variable and give examples of two of its applications. In Chap. 4 we derived several polynomial forms for approximating functions of one variable. The coefficients of these polynomials are found by using known values of the function and the corresponding values of the independent variable. A set of tabular values for a given function may not be available, but we may know or be able to obtain the values of the function and several of its derivatives for one value of the independent variable. For such a function, it is convenient to write a polynomial approximation in the following form:

$$P(x) = a_0 + a_1(x - x_0) + a_2(x - x_0)^2 + \cdots + a_n(x - x_0)^n \tag{5.3}$$

Differentiating (5.3) n times gives

$$\begin{aligned} P'(x) &= a_1 + 2a_2(x - x_0) + 3a_3(x - x_0)^2 + \cdots + na_n(x - x_0)^{n-1} \\ P''(x) &= 2a_2 + 2(3)a_3(x - x_0) + 3(4)a_4(x - x_0)^2 + \cdots + n(n-1)a_n(x - x_0)^{n-2} \\ P'''(x) &= 2(3)a_3 + 2(3)(4)a_4(x - x_0) + \cdots + n(n-1)(n-2)a_n(x - x_0)^{n-3} \\ &\vdots \\ P^{(n)}(x) &= n(n-1)(n-2)(\cdots)(2)(1)a_n \end{aligned} \tag{5.4}$$

When x_0 is substituted into Eqs. (5.3) and (5.4), they simplify to

$$\begin{aligned} P(x_0) &= a_0 \\ P'(x_0) &= a_1 \\ P''(x_0) &= 2a_2 \\ P'''(x_0) &= 3!a_3 \\ &\vdots \\ P^{(n)}(x_0) &= n!a_n \end{aligned} \tag{5.5}$$

In general, $a_j = P^j(x_0)/j!$. Therefore, given a function f(x), for which we know the value of the function and n derivatives of the function at a single point, we can write Eq. (5.3) in the form

$$T(x) = f(x_0) + (x - x_0)f'(x_0) + (x - x_0)^2 f''(x_0)/2! + \cdots + (x - x_0)^n f^n(x_0)/n! \tag{5.6}$$

This is called *Taylor's series.* For the special case $x_0 = 0$, (5.6) reduces to

$$T(x) = f(0) + xf'(0) + x^2 f''(0)/2! + \cdots + x^n f^{(n)}(0)/n! \tag{5.7}$$

which is called *Maclaurin's series.*

There are two methods that may be used to evaluate a Taylor's series. These are illustrated in Fig. 5.1. Both methods can be adapted for evaluating other series also. Assume that values are known or can be computed for the function and its first $N - 1$ derivatives evaluated at some point x_0. Let $C_1 = f(x_0)$ and $C_i = f^{(i-1)}(x_0)$ for $i \geq 2$.

Both of these methods have at least one advantage and one disadvantage. The method of Fig. 5.1(a) can be performed open ended. That is, instead of stopping the process when i exceeds N, the process may be terminated when the term to be added C_iZ becomes sufficiently small. The disadvantage of this method is that there may be

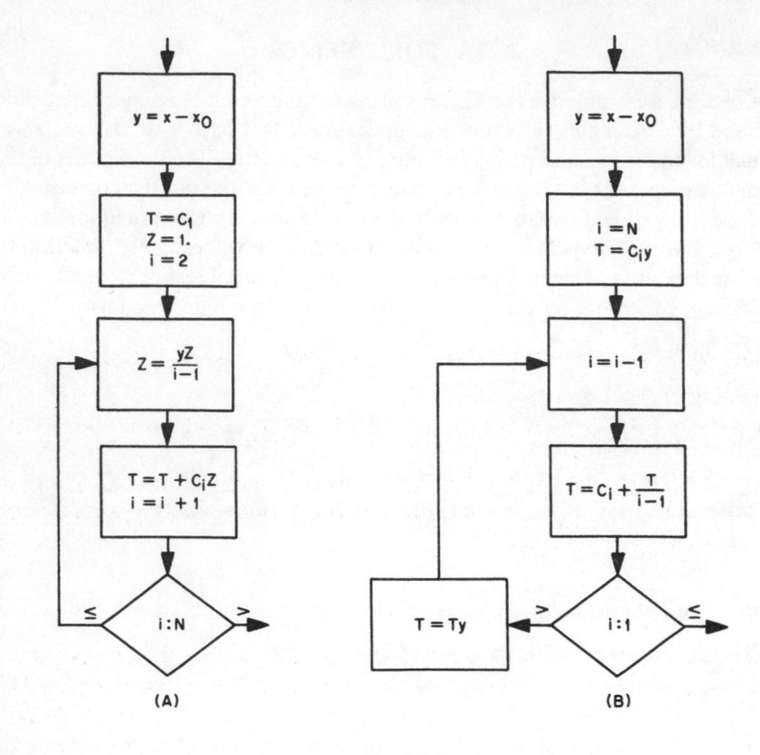

Fig. 5.1. Two methods of evaluating a Taylor's series.

loss of accuracy due to adding of terms of different orders of magnitude. This was discussed in Sec. 3.3. The method of Fig. 5.1(b) has the advantage of producing a more accurate result than the first method, and since it contains fewer arithmetic operations, it can be executed more rapidly (for the same number of terms) than the first method. The disadvantage of the second method is that the number of terms to be evaluated must be known in advance of the evaluation.

5.2.1 Example of Use of Taylor's Series to Approximate a Function

Suppose that we wish to approximate the cube root with a series. Therefore, the function and its first three derivatives are given by

$$
\begin{aligned}
f(x) &= x^{1/3} \\
f'(x) &= \frac{1}{3}x^{-2/3} \\
f''(x) &= -\frac{2}{3}\left(\frac{1}{3}\right)x^{-5/3} \\
f'''(x) &= -\frac{5}{3} - \left(\frac{2}{3}\right)\left(\frac{1}{3}\right)x^{-8/3}
\end{aligned}
\tag{5.8}
$$

If $x_0 = 1$, we have

$$f(x_0) = 1$$
$$f'(x_0) = 1/3$$
$$f''(x_0) = -2/9$$
$$f'''(x_0) = 10/27$$

and the Taylor's series, neglecting terms involving fourth and higher derivatives, is given by

$$T(x) = 1 + (x-1)/3 - (x-1)^2/9 + 5(x-1)^3/81 \tag{5.9}$$

The cube root of 0.5 is 0.7937 and Eq. (5.9) gives 0.7978. Therefore, the actual error is -0.0041.

This equation is suitable for use when x is in the neighborhood of 1. How large the neighborhood is depends on the accuracy desired. The accuracy of Taylor's series approximations will be discussed in Sec. 5.2.2. In general, the accuracy can be improved either by carrying more terms in the approximation or by expanding additional polynomials about other values of x_0.

5.2.2 Example of Use of Taylor's Series as an Interpolation Method

Taylor's series can be used as the basis for an interpolation method. It is only necessary to expand the given function by Eq. (5.6) to the desired number of terms and to tabulate for a set of x_0's the corresponding function values $f(x_0)$ and the necessary derivative values $f'(x_0)$, $f''(x_0)$, etc. Then the approximate function and the tabulated values can be used in a manner analogous to the interpolating methods given in Chap. 4.

As a specific example, consider the function

$$f(x) = \sin x$$

whose derivatives are

$$f'(x) = \cos x$$
$$f''(x) = -\sin x$$
$$f'''(x) = -\cos x$$
$$\vdots \qquad \vdots$$

Therefore, the Taylor's series expansion is

$$f(x) = \sin x_0 + (x - x_0)\cos x_0 - (x - x_0)^2 \sin x_0/2! - (x - x_0)^3 \cos x_0/3! + (x - x_0)^4 \sin x_0/4! + \cdots \tag{5.10}$$

Although this approximation can be simplified further, it would lead us away from the purpose of this example.

Many tables of trigonometric functions contain values of the sine and cosine on the same line as the argument value. With such a table and a desk calculator, it is convenient to apply Eq. (5.10) to interpolate for a sine (or cosine) value not in the table. Suppose that the table has values at intervals of 0.0001 in the argument. This means that $x - x_0$ will always be less than 0.0001 and since $\sin x_0$ and $\cos x_0$ can never exceed 1.0, no term after the second of Eq. (5.10) can be as great as 0.000000005. Therefore, only two terms of (5.10) are needed with such a table to interpolate to eight-place accuracy.

5.3 TRUNCATION ERROR OF TAYLOR'S SERIES APPROXIMATIONS

There are several ways of deriving the mathematical form of the error incurred in truncating a Taylor's series. However, none of these derivations is direct. In each case some artifice must be employed to attain the desired result. Although indirect derivations are not as desirable as direct ones, they are, nonetheless, valid.

The exact Taylor's series representation of a function can be written as

$$f(x) = \sum_{i=0}^{n} (x - x_0)^i f^{(i)}(x_0)/i! + R_n \tag{5.11}$$

where R_n is the remainder term that is necessary to exactly balance the equation. In order to derive a mathematical expression for R_n, we must assume a form for it and prove that that form is the required expression. Therefore, assume

$$R_n = \int_{x_0}^{x} \frac{(x - t)^n}{n!} f^{n+1}(t)\,dt \tag{5.12}$$

This can be integrated by parts by the formula

$$\int u\,dv = uv - \int v\,du \tag{5.13}$$

Let $u = (x - t)^n/n!$ and $dv = f^{(n+1)}(t)dt$; then $du = (x - t)^{n-1}/(n - 1)!$ and $v = f^{(n)}(t)$. Therefore, substituting and evaluating over the limits given,

$$R_n = -\frac{(x - x_0)^n}{n!} f^{(n)}(x_0) + \int_{x_0}^{x} \frac{(x - t)^{n-1}}{(n - 1)!} f^{(n)}(t)\,dt \tag{5.14}$$

or

$$R_n = -(x - x_0)^n f^{(n)}(x_0)/n! + R_{n-1} \tag{5.15}$$

Solving for R_{n-1},

$$R_{n-1} = (x - x_0)^n f^{(n)}(x_0)/n! + R_n \tag{5.16}$$

Expressions for R_{n-2}, R_{n-3}, etc. can be derived in a similar manner. However, from Eq. (5.12), we have

$$R_0 = \int_{x_0}^{x} f'(t)\,dt = f(x) - f(x_0) \tag{5.17}$$

Therefore,

$$f(x) = f(x_0) + R_0 \tag{5.18}$$

Then, applying (5.16) for $n = 1$ and substituting in (5.18),

$$f(x) = f(x_0) + (x - x_0)f'(x_0) + R_1 \tag{5.19}$$

Again using (5.16) but with $n = 2$ and substituting in (5.19),

$$f(x) = f(x_0) + (x - x_0)f'(x_0) + (x - x_0)^2 f''(x_0)/2 + R_2 \tag{5.20}$$

It can be seen that Eqs. (5.18), (5.19), and (5.20) are of the form of (5.11), and since each application of (5.16) adds a term of the form $(x - x_0)^j f^j(x_0)/j!$, then in general, all equations so derived will be of the form of (5.11). Therefore, (5.12) is a valid expression for the remainder term.

This form for the remainder term is not very useful because it is difficult to evaluate. However, another form, called the *Lagrange remainder*, is more useful. This form is derived by applying the mean-value theorem for integrals. That theorem states that

$$\int_a^b f(x)\,g(x)\,dx = f(\xi)\int_a^b g(x)\,dx \tag{5.21}$$

where $a \leqq \xi \leqq b$, $f(x)$ and $g(x)$ are continuous functions on the interval $a \leqq x \leqq b$, and $g(x)$ does not change sign on the interval. Equation (5.12) is of the form of (5.21) and satisfies the necessary conditions if $f^{(n+1)}(t)$ is continuous on the interval. Therefore,

$$R_n = f^{(n+1)}(\xi)\int_{x_0}^{x}\frac{(x-t)^n}{n!}dt = -\,f^{(n+1)}(\xi)\frac{(x-x_0)^{n+1}}{(n+1)!} \tag{5.22}$$

where $x_0 \leqq \xi \leqq x$.

For the example from Sec. 5.2.1,

$$R_3 = -\,f^{iv}(\xi)\frac{(x-1)^4}{24} \tag{5.23}$$

Since we do not know the value of ξ that will make (5.23) exact, we must use the worst case as the upper bound for the error. Since

$$f^{iv}(x) = -\frac{80}{81}x^{-11/3} \tag{5.24}$$

the largest value for $|f^{iv}(x)|$ will occur for the smallest value of x in the interval. Therefore, if Eq. (5.10) is to be used to approximate the cube root of x on the interval $1-\delta \leqq x \leqq 1+\delta$, then the greatest absolute value of the fourth derivative occurs for $x = 1-\delta$. Then,

$$|R_3| \leqq \frac{10}{243}(1-\delta)^{-11/3}\delta^4 \tag{5.25}$$

If $\delta = 0.5$, then,

$$|R_3| \leqq \frac{10}{243}(0.5)^{1\;3} \doteq 0.0327 \tag{5.26}$$

The actual error is -0.0041 which satisfies the condition (5.26).

5.4 FUNCTIONS OF TWO VARIABLES

The Taylor's series discussed in previous sections was for functions of a single independent variable. A Taylor's series can also be derived for functions of more than one variable. That derivation will be omitted here, but the results will be presented. The series can be written as

$$\begin{aligned} f(x,y) = f(x_0,y_0) &+ \left[(x-x_0)\frac{\partial f(x_0,y_0)}{\partial x} + (y-y_0)\frac{\partial f(x_0,y_0}{\partial y}\right] \\ &+ \left[(x-x_0)^2\frac{\partial^2 f(x_0,y_0)}{\partial x^2} + 2(x-x_0)(y-y_0)\frac{\partial^2 f(x_0,y_0)}{\partial x\partial y}\right. \\ &\left. + (y-y_0)^2\frac{\partial^2 f(x_0,y_0)}{\partial y^2}\right]\Big/2! + \cdots + R_n \end{aligned} \tag{5.27}$$

Noting that the bracketed terms are similar to binomial expansions, we can adopt the notation

$$\left[(x - x_0)\frac{\partial}{\partial x} + (y - y_0)\frac{\partial}{\partial y}\right]^n f(x_0, y_0) = \sum_{j=0}^{n} \binom{n}{j} (x - x_0)^j (y - y_0)^{n-j} \frac{\partial^n f(x_0, y_0)}{\partial x^j \partial y^{n-j}} \tag{5.28}$$

where

$$\binom{n}{j} = \frac{n!}{j!(n - j)!}$$

This permits writing Eq. (5.27) in the more compact form

$$f(x, y) = \sum_{j=0}^{n} \frac{1}{j!}\left[(x - x_0)\frac{\partial}{\partial x} + (y - y_0)\frac{\partial}{\partial y}\right]^j f(x_0, y_0) + R_n \tag{5.29}$$

The remainder term can be expressed as

$$R_n = \frac{\left[(x - x_0)\frac{\partial}{\partial x} + (y - y_0)\frac{\partial}{\partial y}\right]^{n+1} f(x_0 + \theta(x - x_0), y + \theta(y_0 - y_0))}{(n + 1)!} \tag{5.30}$$

where $0 \leq \theta \leq 1$.

5.4.1 Example of a Taylor's Series Expansion of a Function of Two Variables

Suppose we wish to expand the function

$$f(x, y) = (\sqrt{x} + y)^{1/2} \tag{5.31}$$

about the point $x_0 = 1$, $y_0 = 1$. If we take this function (5.31) only through second derivatives, we have

$$\begin{aligned}
\frac{\partial f}{\partial x} &= \frac{1}{4}(\sqrt{x} + y)^{-1/2} x^{-1/2} \\
\frac{\partial f}{\partial y} &= \frac{1}{2}(\sqrt{x} + y)^{-1/2} \\
\frac{\partial^2 f}{\partial x^2} &= -\frac{1}{16}(\sqrt{x} + y)^{-1/2} x^{-1}\left[\frac{1}{\sqrt{x} + y} + \frac{2}{\sqrt{x}}\right] \\
\frac{\partial^2 f}{\partial x \partial y} &= -\frac{1}{8}(\sqrt{x} + y)^{-3/2} x^{-1/2} \\
\frac{\partial^2 f}{\partial y^2} &= -\frac{1}{4}(\sqrt{x} + y)^{-3/2}
\end{aligned} \tag{5.32}$$

Evaluating these at x_0, y_0,

$$\begin{aligned}
\left(\frac{\partial f}{\partial x}\right)_0 &= \sqrt{2}/8 \\
\left(\frac{\partial f}{\partial y}\right)_0 &= \sqrt{2}/4 \\
\left(\frac{\partial^2 f}{\partial x^2}\right)_0 &= -5\sqrt{2}/64 \\
\left(\frac{\partial^2 f}{\partial x \partial y}\right)_0 &= -\sqrt{2}/32 \\
\left(\frac{\partial^2 f}{\partial y^2}\right)_0 &= -\sqrt{2}/16
\end{aligned} \tag{5.33}$$

The Taylor's series approximation becomes

$$\begin{aligned}
T(x, y) = \sqrt{2} &+ [(x-1)\sqrt{2}/8 + (y-1)\sqrt{2}/4] \\
&+ [(x-1)^2(-5\sqrt{2})/64 + 2(x-1)(y-1)(-\sqrt{2})/32 \\
&+ (y-1)^2(-\sqrt{2})/16]/2
\end{aligned} \tag{5.34}$$

or

$$\begin{aligned}
T(x, y) = \sqrt{2}[64 + 8(x-1) + 16(y-1) - (5/2)(x-1)^2 - 2(x-1)(y-1) \\
- 2(y-1)^2]/64
\end{aligned} \tag{5.35}$$

Evaluating at $x = 0.5$, $y = 1.5$, we obtain the exact result

$$f(0.5, 1.5) = 1.4856 \tag{5.36}$$

and the approximate result

$$T(0.5, 1.5) = 1.4888 \tag{5.37}$$

These results are fairly close and illustrate that the Taylor's expansion can be used to approximate functions of more than one variable. However, this is not the only application. We will also see that it is useful in the derivation and error analysis of other numerical methods.

5.5 USE OF TAYLOR'S SERIES IN ERROR ANALYSIS

Taylor's series is useful in approximating functions to obtain numerical values. This application was illustrated in Secs. 5.2.1 and 5.4.1. However, Taylor's series is also useful in the error analysis of numerical methods. In subsequent chapters, Taylor's series will be used to determine the inherent error of the methods discussed.

In Chap. 4, no such error analysis was included, since Taylor's series had not yet been introduced. However, an error expression for each of the numerical interpolation methods discussed can be derived for any order interpolation polynomial. In the following subsection, an example will be given for one order of one method. The technique employed is also applicable for other orders and other methods.

5.5.1 Error Analysis of Third-Order Gregory–Newton Forward Interpolation

The formula for third-order Gregory–Newton forward interpolation is

$$Y_{GN}^{(3)} = f_k + \Delta f_k\, u + \Delta^2 f_k\, u(u-1)/2 + \Delta^3 f_k\, u(u-1)(u-2)/6 \tag{5.38}$$

where $u = (a - x_k)/h$. The differences can be expressed in terms of ordinates by

$$\begin{aligned} \Delta f_k &= f_{k+1} - f_k \\ \Delta^2 f_k &= f_{k+2} - 2f_{k+1} + f_k \\ \Delta^3 f_k &= f_{k+3} - 3f_{k+2} + 3f_{k+1} - f_k \end{aligned} \tag{5.39}$$

Then f_{k+1}, f_{k+2}, and f_{k+3} can be represented exactly by expanding each in a Taylor's series about the point x_k, f_k.

$$\begin{aligned} f_{k+1} &= f_k + hf'_k + h^2 f''_k/2 + h^3 f'''_k/6 + h^4 f^{iv}(\xi_1)/24 \\ f_{k+2} &= f_k + 2hf'_k + 4h^2 f''_k/2 + 8h^3 f'''_k/6 + 16h^4 f^{iv}(\xi_2)/24 \\ f_{k+3} &= f_k + 3hf'_k + 9h^2 f''_k/2 + 27h^3 f'''_k/6 + 81h^4 f^{iv}(\xi_3)/24 \end{aligned} \tag{5.40}$$

Substituting (5.40) into (5.39) gives

$$\begin{aligned} \Delta f_k &= hf'_k + h^2 f''_k/2 + h^3 f'''_k/6 + h^4 f^{iv}(\xi_1)/24 \\ \Delta^2 f_k &= h^2 f''_k + h^3 f'''_k + h^4[8f^{iv}(\xi_2) - (f^{iv}(\xi_1)]/12 \\ \Delta^3 f_k &= h^3 f'''_k + h^4[27f^{iv}(\xi_3) - 16f^{iv}(\xi_2) + f^{iv}(\xi_1)]/8 \end{aligned} \tag{5.41}$$

Equation (5.38) can be expressed in terms of f_k and its derivatives by use of (5.41). Thus,

$$\begin{aligned} Y_{GN}^{(3)} = f_k &+ uhf'_k + u^2h^2 f''_k/2 + u^3h^3 f'''_k/6 \\ &+ \frac{uh^4}{48}\Big[u^2[27f^{iv}(\xi_3) - 16f^{iv}(\xi_2) + f^{iv}(\xi_1)] \\ &- u[81f^{iv}(\xi_3) - 64f^{iv}(\xi_2) + 5f^{iv}(\xi_1)] \\ &+ 54f^{iv}(\xi_3) - 48f^{iv}(\xi_2) + 6f^{iv}(\xi_1)\Big] \end{aligned} \tag{5.42}$$

The function itself can be expressed exactly in terms of a Taylor's series

$$y = f_k + (a - x_k)f'_k + (a - x_k)^2 f''_k/2 + (a - x_k)^3 f'''_k/6 + (a - x_k)^4 f^{iv}(\xi_4)/24 \tag{5.43}$$

But since $a - x_k = uh$, (5.43) can be written as

$$y = f_k + uhf'_k + u^2h^2 f''_k/2 + u^3h^3 f'''_k/6 + u^4h^4 f^{iv}(\xi_4)/24 \tag{5.44}$$

We note from comparing (5.44) with (5.42) that the first four terms of each are exactly alike. Since the error in the third-order Gregory–Newton formula is

$$E_{GN}^{(3)} = y - Y_{GN}^{(3)} \tag{5.45}$$

then

$$\begin{aligned} E_{GN}^{(3)} = \frac{uh^4}{48}\Big[2u^3 f^{iv}(\xi_4) &- u^2[27f^{iv}(\xi_3) - 16f^{iv}(\xi_2) + f^{iv}(\xi_1)] \\ &+ u[81f^{iv}(\xi_3) - 64f^{iv}(\xi_2) + 5f^{iv}(\xi_1)] \\ &- 54f^{iv}(\xi_3) + 48f^{iv}(\xi_2) - 6f^{iv}(\xi_1)\Big] \end{aligned} \tag{5.46}$$

We cannot find the values of the $f^{iv}(\xi_j)$'s, but we do know that $x_k \leqq \xi_1 \leqq x_{k+1}$, $x_k \leqq \xi_2 \leqq x_{k+2}$, $x_k \leqq \xi_3 \leqq x_{k+3}$, and $x_k \leqq \xi_4 \leqq a$. Note that ξ_1, ξ_2, and ξ_3 depend only on the interval size and are independent of the argument. However, ξ_4 does depend on the argument. If we assume that $f^{iv}(\xi)$ is constant for all u in the range $0 \leqq u \leqq 3$, Eq. (5.46) can be written as

$$E_{GN}^{(3)} = \frac{h^4 f^{iv}(\xi_4)}{24}(u^4 + Bu^3 + Cu^2 + Du) \tag{5.47}$$

Moreover, since we know that the interpolation polynomial passes exactly through the points x_k, f_k; x_{k+1}, f_{k+1}; x_{k+2}, f_{k+2}; x_{k+3}, f_{k+3}; the error is zero for $u = 0, 1, 2, 3$. This information can be used to find B, C, and D of Eq. (5.47).

$$E_{GN}^{(3)} = \frac{h^4 f^{iv}(\xi_4)}{24}(u^4 - 6u^3 + 11u^2 - 6u) \tag{5.48}$$

Since the polynomial

$$P(u) = u^4 - 6u^3 + 11u^2 - 6u \tag{5.49}$$

has four real zeros, then it must have three extrema. These are

$$\begin{aligned} P\left(\frac{3-\sqrt{5}}{2}\right) &= -1. \\ P\left(\frac{3}{2}\right) &= 0.5625 \\ P\left(\frac{3+\sqrt{5}}{2}\right) &= -1. \end{aligned} \tag{5.50}$$

Therefore, subject to the assumptions made earlier, we can write the approximate condition

$$|E_{GN}^{(3)}| \leqq \frac{h^4 f^{iv}(\xi_4)}{24} \tag{5.51}$$

If the fourth derivative does not change markedly on the interval $0 \leqq u \leqq 3$, then the error function, Eq. (5.48), will have nearly the same shape as the polynomial (5.49). This is illustrated in Fig. 5.2. From the curve we see that the deviation from zero is least in the center of the curve (i.e., $1 \leqq u \leqq 2$). This supports, but does not prove, the contention made in Chap. 4 that the base index k should be selected such that the argument is as near as possible to the center of the points used in the interpolation.

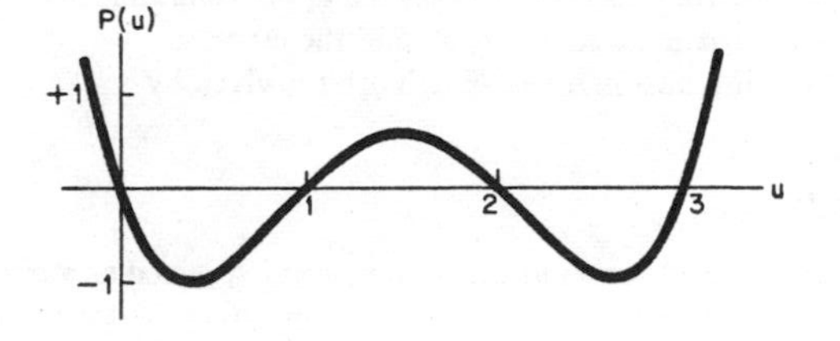

Fig. 5.2. Graph of $P(u) = u^4 - 6u^3 = 11u^2 - 6u$.

5.6 SUMMARY

In this chapter, the derivation and error analysis of Taylor's series has been presented. In addition, two of its applications have been discussed. The first application involved the use of Taylor's series to approximate functions. The reader is probably familiar with the Maclaurin's series [Eq. (5.7)] expansions of the elementary functions such as sine, cosine, exponential, etc. Taylor's series, which is more general, can also be used to approximate elementary and other functions. In fact, Taylor's series can form the basis for a valuable interpolation technique (Sec. 5.2.2) which was not covered in Chap. 4. In addition to using Taylor's series to find an approximate value for a function, the series approximation can be differentiated or integrated term by term in order to approximate the derivative or integral of the given function.

The second application of Taylor's series involved its use for the error analysis of other numerical methods. This technique of error analysis will be used in subsequent chapters of this text. Another application of Taylor's series which has *not* been covered in this chapter is its use in the solution of equations. This application will be discussed briefly in Sec. 6.9.

EXERCISES

1. Expand the function

$$f(t) = e^{-t^2/2}$$

in a Taylor's series about the point $t_0 = 0$. Derive enough terms to deduce the general form.

2. What is the remainder term of the series derived in Exercise 1 if only the first three nonzero terms are retained? What is the maximum value of the remainder over the interval $0 \leq t \leq 1$?

3. Expand the function $y = e^x \sin x$ in a Taylor's series about the point $x = 0$. Find the first six nonzero terms.

4. In terms of x, write the remainder term R_7 for the series in Exercise 3. What is the maximum value of this remainder over the interval $-\pi/2 \leq x \leq \pi/2$?

5. Expand the function $y = \tan^{-1} x$ in a Taylor's series about the point $x = 1$. Retain only the terms through the fifth power. Evaluate the series for $x = 1.1$. What is the actual error?

6. Table 4.3 was computed by the formula $\tan \theta = 20x/(x^2 + 300)$. Expand this function in a Taylor's series about $x = 300$. Carry the expansion through the third-order term. Compute θ, by your series, for $x = 280$ and $x = 350$. Compare with the interpolated values given in Sec. 4.6.1.

7. Derive the Taylor's series expansion of the function $f(x) = (1. + x)^{1/2}$ expanded about the point $x = 0$. Derive enough terms to assure four decimal places of accuracy when $x = 0.5$.

8. Prepare a flowchart and write and execute a computer program that will use the approximation in Exercise 7 to find the square roots of all numbers from 0.5 to 1.5 at intervals of 0.1. Also find the exact square root and the error in the approximation. Print each argument, its exact square root, its approximate square root, and the error.

9. The period of a point mass in a near-Earth orbit is given by

$$P = K\left(\frac{1 + x}{1 - e}\right)^{1.5}$$

where K is a constant, x is the lowest altitude in the orbit expressed as a fraction of the Earth's radius, and e is the eccentricity of the orbit. Expand P by means of Taylor's series in two variables, x and e, about the point $x = 0, e = 0$. Find all terms through the third power of the independent variables.

10. The total resistance for two resistors in parallel is given by

$$F(R_1,R_2) = \frac{R_1R_2}{R_1 + R_2}$$

Expand this function in two variables about the point $R_1 = R_0$, $R_2 = R_0$. Find only the first three terms of the series.

11. Prepare a flowchart and write a computer program that will a) evaluate the approximate function derived in Exercise 10, b) evaluate the exact function, and c) calculate the error between the two. The program should print out the input parameters, the approximate and exact function values, and the error. Execute this program for each of R_0, R_1, and R_2, taking on each of the values 95, 100, and 105 in all 27 possible combinations.

12. Derive the upper bound for the error term for a second-order Gregory–Newton forward interpolation formula.

13. Derive the upper bound for the error term for a second-order Stirling's interpolation formula.

14. Derive the upper bound for the error term for a third-order Stirling's interpolation formula.

15. Use the Gregory–Newton third-order forward interpolation formula to find expressions for interpolated values for $u = 4/3$ and $u = 5/3$ in terms of f_k, f_{k+1}, f_{k+2}, and f_{k+3}. Using these two expressions, find, by linear interpolation, an expression for an interpolated value at $u = 3/2$. Determine the inherent error in the above derived expression.

6

Roots of Equations

6.1 INTRODUCTION

The determination of the roots of equations is a common problem encountered in engineering and science. This chapter identifies two types of equations, polynomial and transcendental, and in Sec. 6.2 defines two methods of numerical solution for their roots. Section 6.3 gives properties of polynomial equations that aid in finding the roots of such equations. Direct methods for finding the roots of some polynomial equations are presented in Sec. 6.4. Sections 6.6 through 6.10 give various indirect methods for finding roots of equations. The methods presented are binary search, false position, secant, iteration, Newton's, and Bairstow's. Also included are analyses of the convergence criteria for the indirect methods and their rates of convergence.

The *roots* of an equation are those values of some parameter, say x, for which the equation is satisfied. Stating it in another way, the equation

$$f(x) = 0 \tag{6.1}$$

is satisfied only for certain values of x. These values of x are called the roots. Examples of equations of the form of (6.1) are

$$3x^3 + 14x^2 + 13x - 6 = 0 \tag{6.2}$$

$$1.0/x - 1.1 - \sin x = 0 \tag{6.3}$$

Equation (6.2) is an example of a *polynomial* equation, whose roots are 1/3, −2, −3. Equation (6.3) is an example of a *transcendental* equation. This equation has three real roots whose approximate values are 0.60056222, 4.17815720, 5.14987300. An equation such as (6.3) can be written as a function

$$f(x) = 1.0/x - 1.1 - \sin x \tag{6.4}$$

The *zeros* of this function will be the *roots* of (6.3). This function can be plotted as in Fig. 6.1. The zeros are the values of x where the function crosses the x-axis.

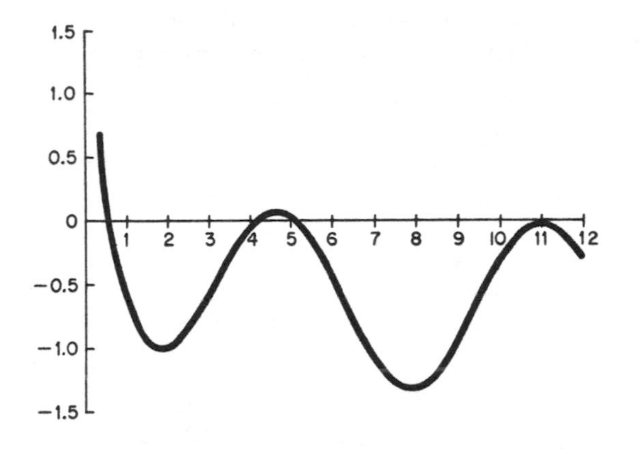

Fig. 6.1. Graph of $f(x) = 1./x - 1.1 - \sin x$

6.2 TYPES OF METHODS

There are two types of numerical methods that can be employed for finding the roots of equations: direct and indirect.

A *direct* (*closed*) numerical method is a process for finding a solution(s) to a problem by executing a sequence of arithmetical and logical operations in a nonrepetitive fashion. Although a direct method may involve alternative procedures that are selected on the basis of values obtained during computation, each step that is executed is usually executed only once and the answer(s) obtained is essentially exact.

An *indirect* (*open*) numerical method is a process for obtaining a solution(s) to a problem by executing a sequence of arithmetical and logical operations, a portion of which may be repeated. The answer(s) obtained is approximate and is improved with each repetition of the operations. The process is terminated when two successive approximations agree with each other to within some allowable error.

In this chapter and in Chaps. 7, 8, and 12, both direct and indirect numerical methods will be presented.

6.3 PROPERTIES OF POLYNOMIAL EQUATIONS

A frequently occurring equation is the polynomial. This type of equation has the general form

$$A_n x^n + A_{n-1} x^{n-1} + \cdots + A_2 x^2 + A_1 x + A_0 = 0 \tag{6.5}$$

where $A_n \neq 0$. Considerable analysis of the properties of polynomials has been performed. Therefore, certain tools are available to aid us in determining the roots of polynomial equations. First, we know that an nth-degree polynomial has n roots. These roots can be either real or complex. In most physical problems, the coefficients A_i of the polynomial are all real, in which case all complex roots appear in *complex conjugate* pairs of the form

$$r = a + bi \qquad \text{and} \qquad r^* = a - bi$$

Also, all odd-order polynomials have at least one real root.

Equation (6.5) can be factored and expressed in terms of the roots r_i as follows:

$$A_n(x - r_1)(x - r_2)(\cdots)(x - r_{n-1})(x - r_n) = 0 \tag{6.6}$$

Obviously, A_n can be factored out of either (6.5) or (6.6), and without any loss of generality, any polynomial equation can be written in the form

$$x^n + a_{n-1}x^{n-1} + \cdots + a_2x^2 + a_1x + a_0 = 0 \tag{6.7}$$

where $a_i = A_i/A_n$. If all factors of (6.6), except A_n, are multiplied and terms of like powers collected, relationships between the roots and the coefficients of (6.7) can be derived:

$$\begin{aligned}
a_{n-1} &= -(r_1 + r_2 + \cdots + r_n)\\
a_{n-2} &= r_1r_2 + r_1r_3 + \cdots + r_1r_n + r_2r_3 + \cdots + r_2r_n + \cdots + r_{n-1}r_n\\
&\;\;\vdots\\
a_1 &= (-1)^{n-1}(r_2r_3r_4\cdots r_n + r_1r_3r_4\cdots r_n + r_1r_2r_4\cdots r_n + \cdots + r_1r_2r_3\cdots r_{n-1})\\
a_0 &= (-1)^n(r_1r_2r_3\cdots r_n)
\end{aligned} \tag{6.8}$$

Equations (6.8) are more easily expressed in words:

$$\begin{aligned}
a_{n-1} &= -(\text{sum of all roots})\\
a_{n-2} &= \text{sum of all products of roots taken two at a time}\\
a_{n-3} &= -(\text{sum of all products of roots taken three at a time})\\
&\;\;\vdots\\
a_1 &= (-1)^{n-1}(\text{sum of all products of roots taken } n-1 \text{ at a time})\\
a_0 &= (-1)^n(\text{product of all the roots})
\end{aligned}$$

These are called Newton's relations (sometimes called Birge–Vieta relations). They can be helpful in determining some facts about the roots before solving for them. For example, consider the third-degree equation (6.2). By the foregoing rules, $a_0 = -2 = -(r_1r_2r_3)$. Therefore, $r_1r_2r_3 = 2$. There are only three conditions under which the above condition is satisfied.

1. three positive real roots,
2. one positive real root and two negative real roots,
3. one positive real root and two complex conjugate roots.

(*Note:* The product of complex conjugate roots is always positive.) From these, one fact is common to all, namely, there is at least one positive real root. Other useful relations can also be derived.

It is difficult to derive general relationships that are useful for all polynomials, but this is not usually necessary. It is frequently possible to derive specific relationships for individual problems. For example, an actual problem may require finding a root of a third-degree polynomial equation for various values of the coefficients. From physical considerations, we may know that the equation will only have one real root and we are only interested in cases where it is negative. From the previous example, it is easy to see then that if $a_0 < 0$, the polynomial would have no negative real root and could be rejected. Only in the cases where $a_0 > 0$ would we need to continue the solution and then only for the one real root.

Additional information about the signs of the roots can also be obtained by applying Descartes' rule of signs. The derivation of this rule is given in many algebra books and will not be given here. The statement of the rule is:

The number of positive real roots of a polynomial is equal to, or less than by an even integer, the number of sign changes in the coefficients.

The number of negative real roots of a polynomial is equal to, or less than by an even integer, the number of sign changes in the coefficients, if x *is replaced by* $-x$.

To illustrate this rule, consider Eq. (6.2) whose coefficients are $+3$, $+14$, $+13$, -6. In this series there is only one sign change. Therefore, there is, at most, only one positive real root. Since we previously showed that there is at least one positive real root, we can now say there is one and only one positive real root. If x is replaced by $-x$, the coefficients become -3, $+14$, -13, -6. In this set there are two sign changes. Therefore, there are either two negative roots or no negative roots. These conclusions are in agreement with the preceding conclusions 2 and 3, which were arrived at by consideration of only the coefficient a_0 and the order of the equation.

Another special relationship that may be useful in some cases is a method for determining an upper bound for the roots. We can write

$$r_1^2 + r_2^2 + \cdots + r_n^2 = (r_1 + r_2 + \cdots + r_n)^2 - 2(r_1r_2 + r_1r_3 + \cdots + r_1r_n + r_2r_3 + \cdots + r_2r_n + \cdots + r_{n-1}r_n) \tag{6.9}$$

which by the relations (6.8) can be rewritten as

$$r_1^2 + r_2^2 + \cdots + r_n^2 = (a_{n-1})^2 - 2(a_{n-2}) \tag{6.10}$$

This relation is exact and holds for all cases. However, it is not always useful in general. If we know that all roots are real, then we can say

$$r_{max}^2 \leqq r_1^2 + r_2^2 + \cdots + r_n^2 \tag{6.11}$$

because we know that the square of the root of greatest magnitude is certainly less than or equal to the sum of the squares of all roots. Then, by Eq. (6.10),

$$|r_{max}| \leqq \sqrt{(a_{n-1})^2 - 2(a_{n-2})} \tag{6.12}$$

Applying this to (6.2),

$$|r_{max}| \leqq \sqrt{(14/3)^2 - 2(13/3)} \doteq 3.62$$

Therefore, we know that no root exceeds 3.62 in absolute magnitude.

Methods of solving special polynomials exist and are discussed in the following sections.

6.4 DIRECT SOLUTIONS OF POLYNOMIALS

In this section, direct methods will be given for finding the roots of some polynomial equations. Later sections will give indirect methods for finding the roots of other equations. Although direct methods are available for use on some equations, it is possible and sometimes preferable to use indirect methods on these equations also.

6.4.1 Quadratic Equations

The solution of quadratic (second-degree polynomial) equations is relatively simple but deserves some special attention as it can be an important phase in solving higher-order equations. The quadratic equation can always be reduced to the general form

$$x^2 + a_1x + a_0 = 0 \tag{6.13}$$

It has two roots given by

$$\begin{aligned} r_1 &= -a_1/2 + \sqrt{(a_1/2)^2 - a_0} \\ r_2 &= -a_1/2 - \sqrt{(a_1/2)^2 - a_0} \end{aligned} \tag{6.14}$$

If the discriminant $(a_1/2)^2 - a_0$ is positive, the roots are real; if the discriminant is zero, the roots are real and equal; if the discriminant is negative, the roots are complex.

In an actual problem, we may know in advance whether the roots are real or complex and which of the two roots to solve for.

6.4.2 Cubic Equations

A general method also exists for finding the roots of cubic (third-degree polynomial) equations. The cubic equation may always be expressed in the general form

$$x^3 + a_2x^2 + a_1x + a_0 = 0 \tag{6.15}$$

This equation can be reduced to the form

$$y^3 + b_1y + b_0 = 0 \tag{6.16}$$

by the substitution

$$x = y - a_2/3 \tag{6.17}$$

Then, the new coefficients b_1 and b_0 are given by

$$\begin{aligned} b_1 &= (3a_1 - a_2^2)/3 \\ b_0 &= (2a_2^3 - 9a_2a_1 + 27a_0)/27 \end{aligned} \tag{6.18}$$

A discriminant d^2 can be calculated by

$$d^2 = (b_1/3)^3 + (b_0/2)^2 \tag{6.19}$$

If

$d^2 > 0$ there will be one real root and two conjugate complex roots.

$d^2 = 0$ there will be three real roots, of which at least two will be equal.

$d^2 < 0$ there will be three real and unequal roots.

The following intermediate quantities are useful to calculate:

$$\begin{aligned} A &= (-b_0/2 + d)^{1/3} \\ B &= (-b_0/2 - d)^{1/3} \end{aligned} \tag{6.20}$$

Then the roots of the transformed equation (6.16) become

$$\begin{aligned} y_1 &= A + B \\ y_2 &= -(A + B)/2 + (A - B)\sqrt{-3/2} \\ y_3 &= -(A + B)/2 - (A - B)\sqrt{-3/2} \end{aligned} \tag{6.21}$$

Then the roots of the original equation (6.15) can be found by substituting the values of Eqs. (6.21) in (6.17).

In the case where $d^2 < 0$, d is imaginary, which leads to complex numbers in the evaluation of (6.20). In order to find the cube root of a complex number, it is necessary to express the number in polar form. Rather than do this, it is possible to algebraically modify the process for this special case. Instead of evaluating (6.20), find

$$\cos\phi = -(b_0/2)(-3/b_1)^{3/2} \tag{6.22}$$

Then the roots of the modified equation (6.16) are given by

$$\begin{aligned} y_1 &= 2(-b_1/3)^{1/2}\cos(\phi/3) \\ y_2 &= 2(-b_1/3)^{1/2}\cos(\phi/3 + 2\pi/3) \\ y_3 &= 2(-b_1/3)^{1/2}\cos(\phi/3 + 4\pi/3) \end{aligned} \tag{6.23}$$

Then the roots of the original equation (6.15) can be found by applying (6.17).

Applying this method to Eq. (6.2) gives

$$\begin{aligned} b_1 &= -2.9259259 \\ b_0 &= -1.2126200 \\ \cos\phi &= 0.62947955 \\ \phi &= 0.88991310 \\ \cos(\phi/3) &= 0.95632471 \\ \cos(\phi/3 + 2\pi/3) &= -0.73130714 \\ \cos(\phi/3 + 4\pi/3) &= 1.9751543 \\ 2(-b_1/3)^{1/2} &= 1.9751543 \end{aligned}$$

$$\begin{aligned} y_1 &= 1.8888889 \rightarrow x_1 = 0.33333333 \\ y_2 &= -1.4444444 \rightarrow x_2 = -3.0000000 \\ y_3 &= -0.44444444 \rightarrow x_3 = -2.0000000 \end{aligned}$$

The final roots x_i agree with those given earlier.

6.4.3 Quartic Equations

A method also exists for finding the roots of a quartic (biquadratic, or fourth-degree polynomial) equation. Since any polynomial equation can be expressed as a product of linear terms, as in Eq. (6.6), it follows that for polynomials of second or higher degree, these factors can be multiplied in pairs such that the polynomial is expressed as a product of linear and quadratic terms or quadratic terms only. Although it is possible for the quadratic terms to have complex coefficients, it is desirable to multiply together those factors containing complex conjugate roots so that the resulting quadratic factors have real coefficients.

Thus, it is possible to express the quartic equation

$$x^4 + a_3x^3 + a_2x^2 + a_1x + a_0 = 0 \tag{6.24}$$

in the form

$$(x^2 + b_1x + b_0)(x^2 + c_1x + c_0) = 0 \tag{6.25}$$

By multiplying out (6.25), we can show that

$$a_3 = b_1 + c_1 \tag{6.26}$$

$$a_2 = b_0 + b_1c_1 + c_0 \tag{6.27}$$

$$a_1 = b_0c_1 + b_1c_0 \tag{6.28}$$

$$a_0 = b_0c_0 \tag{6.29}$$

Solving these equations for b_1, b_0, c_1, and c_0 will allow finding the roots of (6.25), which will also be the roots of (6.24). Equations (6.26) through (6.29) can be solved, but the process is not simple. Solving (6.26) and (6.28) simultaneously yields

$$b_1 = \frac{a_3b_0 - a_1}{b_0 - c_0} \tag{6.30}$$

$$c_1 = \frac{a_1 - a_3c_0}{b_0 - c_0} \tag{6.31}$$

Substituting these in (6.27) gives

$$(b_0 + c_0 - a_2)(b_0 - c_0)^2 + a_1a_3(b_0 + c_0) - a_3^2b_0c_0 - a_1^2 = 0 \tag{6.32}$$

In order to find b_0 and c_0 separately, Eq. (6.29) must be solved simultaneously with (6.32). This is done most easily by introducing a new variable

$$u = b_0 + c_0 \tag{6.33}$$

Since

$$(b_0 - c_0)^2 = (b_0 + c_0)^2 - 4b_0c_0 \tag{6.34}$$

and using (6.29) and (6.33),

$$(b_0 - c_0)^2 = u^2 - 4a_0 \tag{6.35}$$

Then, using Eqs. (6.29), (6.33), and (6.35) in (6.32),

$$(u - a_2)(u^2 - 4a_0) + a_1a_3u - a_3^2a_0 - a_1^2 = 0 \tag{6.36}$$

or

$$u^3 - a_2u^2 + (a_1a_3 - 4a_0)u + 4a_0a_2 - a_3^2a_0 - a_1^2 = 0 \tag{6.37}$$

Equation (6.37) can be solved by the methods of Sec. 6.4.2. However, there are three roots for (6.37). In order to determine b_0 and c_0 individually, (6.29) and (6.33) must be solved simultaneously which leads to the quadratic

$$c_0^2 - uc_0 + a_0 = 0 \tag{6.38}$$

In order for Eq. (6.38) *not* to have complex roots (we do not want any of b_1, b_0, c_1, c_0 to be complex), we must select a root of (6.37) such that

$$u^2 \geqq 4a_0 \tag{6.39}$$

After selecting a root of (6.37) that satisfies condition (6.39), Eq. (6.38) can be used to find c_0; (6.29) to find b_0; and (6.30) and (6.31) to find b_1 and c_1, respectively. These then can be used to solve each of the quadratic factors of Eq. (6.25). If $u^2 = 4a_0$, then $b_0 = c_0$ and therefore Eqs. (6.30) and (6.31) cannot be used to find b_1 and c_1. However, a quadratic equation that can be solved for b_1 (or c_1) can be derived from Eqs. (6.26) and (6.27).

6.4.4 Higher-Order Polynomials

No general direct methods are available for solving polynomial equations of order greater than four. However, in some cases the roots of higher-order polynomial equations can be found using methods for lower-order polynomials. For example, although the equation

$$x^6 + a_2x^4 + a_1x^2 + a_0 = 0 \tag{6.40}$$

is a sixth-order polynomial, it is a third-degree polynomial in x^2.

Occasionally, a trick may be employed. For example, the equation

$$x^6 - x^5 + x^4 - x^3 + x^2 - x + 1 = 0 \tag{6.41}$$

if multiplied by $x + 1$ yields the equation

$$x^7 + 1 = 0 \tag{6.42}$$

To solve this, we need only find the seven one-seventh roots of -1. One of these will be -1 which we introduced. The remaining six will be the roots of (6.41).

The trouble with tricks is that finding an appropriate one is more of an art than science. Also, a trick that will work for one problem may not work for another. However, the scientist or engineer is usually only concerned with solving a particular problem at any one time. Therefore, any special techniques (tricks) that will aid him are perfectly acceptable.

If no special techniques are known, the indirect methods discussed in later sections can be employed.

6.5 TRANSCENDENTAL EQUATIONS

In Sec. 6.1, an example of a transcendental equation was given [Eq. (6.3)]. Transcendental equations can take numerous forms, and therefore no general direct methods exist for finding the roots of such equations. The roots of such equations and also higher-order polynomial equations are usually found by indirect methods.

6.6 SEARCHING

To find the roots of any equation, we must find the values of a parameter, say x, for which the given equation is satisfied. Both polynomial and transcendental equations can be put in the form

$$f(x) = 0 \tag{6.1}$$

Therefore, by repeatedly evaluating the function f(x) with different values of x, we can find one or more values for which (6.1) is satisfied. Selection of the trial values of x must be done in some methodical manner. One method is to select a starting value x_s, an ending value x_F, and an increment size Δx. Then evaluate the function for x_s and for each of $x_s + \Delta x, x_s + 2\Delta x, \ldots, x_F$. If one or more roots exist in the interval, the sign of the tabulated function will usually change in the vicinity of each root. As an example, suppose we use the function

$$f(x) = 3x^3 + 14x^2 + 13x - 6 \tag{6.43}$$

to aid in finding the roots of Eq. (6.2). Table 6.1 lists the values of this function for $x_s = -3.6$, $x_F = 3.6$, and $\Delta x = 0.5$. From the table, it appears that there is a root (the function goes through zero) between −3.1 and −2.6; there appears to be another root between −2.1 and −1.6; and there appears to be another root between −0.1 and 0.4. In Sec. 6.2, a method was suggested for determining an upper limit on the roots of a polynomial equation. Application of Eq. (6.12) to (6.2) gave us approximately ±3.6, hence, these values were selected as limits for the above search.

TABLE 6.1

x	*f(x)*	*x*	*f(x)*
−3.6	−11.328	0.4	1.632
−3.1	−1.133	0.9	19.227
−2.6	2.112	1.4	47.872
−2.1	0.657	1.9	89.817
−1.6	−3.248	2.4	147.312
−1.1	−7.353	2.9	222.607
0 6	9.408	3.4	317.952
−0.1	−7.163		

Equation (6.12) will yield a value which will be greater than or equal to, in absolute value, the largest root of a polynomial equation having all real roots. It *may* yield such a value for polynomial equations with complex roots. Therefore, even if there is no assurance that all roots are real, (6.12) may still be better than a guess for the limits of a search.

This method will fail to detect an even number of roots that occur within one interval Δx. Also, it will fail to detect a root of even multiplicity (the function is tangent to the x-axis).

6.6.1 Binary Search Method

The searching method described in the first paragraph of this section gives only very rough values for the location of roots. The smaller the search interval, the better the result. However, it is not practical to use a small search increment over the entire search range. A way around this is to vary the search increment so that it is smaller and smaller as we approach the root. Such a method is given below.

1. Evaluate the function at equal intervals in x until two successive function values f_{n-1} and f_n, have opposite signs.

2. Evaluate the function at the midpoint $x_m = (x_{n-1} + x_n)/2$ of the last interval. $f_m = f(x_m)$.
3. If the value of f_m agrees in sign with the value of f_{n-1}, replace the value of f_{n-1} with the value of f_m and replace the value of x_{n-1} with the value of x_m. If the value of f_m agrees in sign with the value of f_n, replace the value of f_n with the value of f_m and the value of x_n with the value of x_m.
4. If $|f_m|$ is less than some acceptable error, stop. Otherwise return to step 2 and repeat.

In step 4 we can test on the interval size $|x_n - x_{n-1}|$ to stop the process as it may be easier to set the accuracy desired for x more easily than that desired for f. Figure 6.2 is a fowchart of this process including some additional steps needed to guard against the case that $f = 0$ and to provide some intermediate printouts.

Figure 6.3 gives a FORTRAN program† and the results for finding a root of Eq. (6.2). This program can easily be changed to find the root of any equation by changing the arithmetic statement function in the first line of the program and by modifying the starting value, the increment, and the allowable error given in the second, third, and fourth lines, respectively. If only a different root of the same equation is desired, it is only necessary to alter the starting value given in the second line of the program.

This method, *binary search*, will assure finding a root to the desired accuracy in N halvings of the interval. The value of N can be determined from the inequality

$$2^{-N} \leqq \varepsilon/\Delta x \tag{6.44}$$

In the example given, $\varepsilon = 10^{-7}$ and $\Delta x = 0.5$, which leads to $N = 23$. Figure 6.3 gives 24 lines of results, indicating that the interval has been halved 23 times. This method will always require N halvings of the interval plus the initial search unless the initial search happens to hit exactly on a root.

6.6.2 False Position Method (Regula Falsi)

Another modification to the search method is called the *false position method*, *regula falsi*, or *method of interpolation*. This method's advantage is that the result *may* be obtained more quickly than by the first method.

In the false position method, a search is again performed to locate two values of the argument x for which the function values have opposite signs. Then by linear interpolation, we determine an approximate root. Fig. 6.4 shows a portion of a function for which a root is to be determined.

The search arguments x_1 and x_2 lead to the function values f_1 and f_2 with opposite signs. By linear interpolation, we can write

$$y = f_1 + \left(\frac{a - x_1}{x_2 - x_1}\right)(f_2 - f_1) \tag{6.45}$$

†Since it is not the purpose of this text to instruct the reader in the use of FORTRAN, most computing methods are illustrated by flowcharts only. In this chapter, some of the root finding methods are also illustrated with FORTRAN programs to show how easily such programs can be modified for use with different equations.

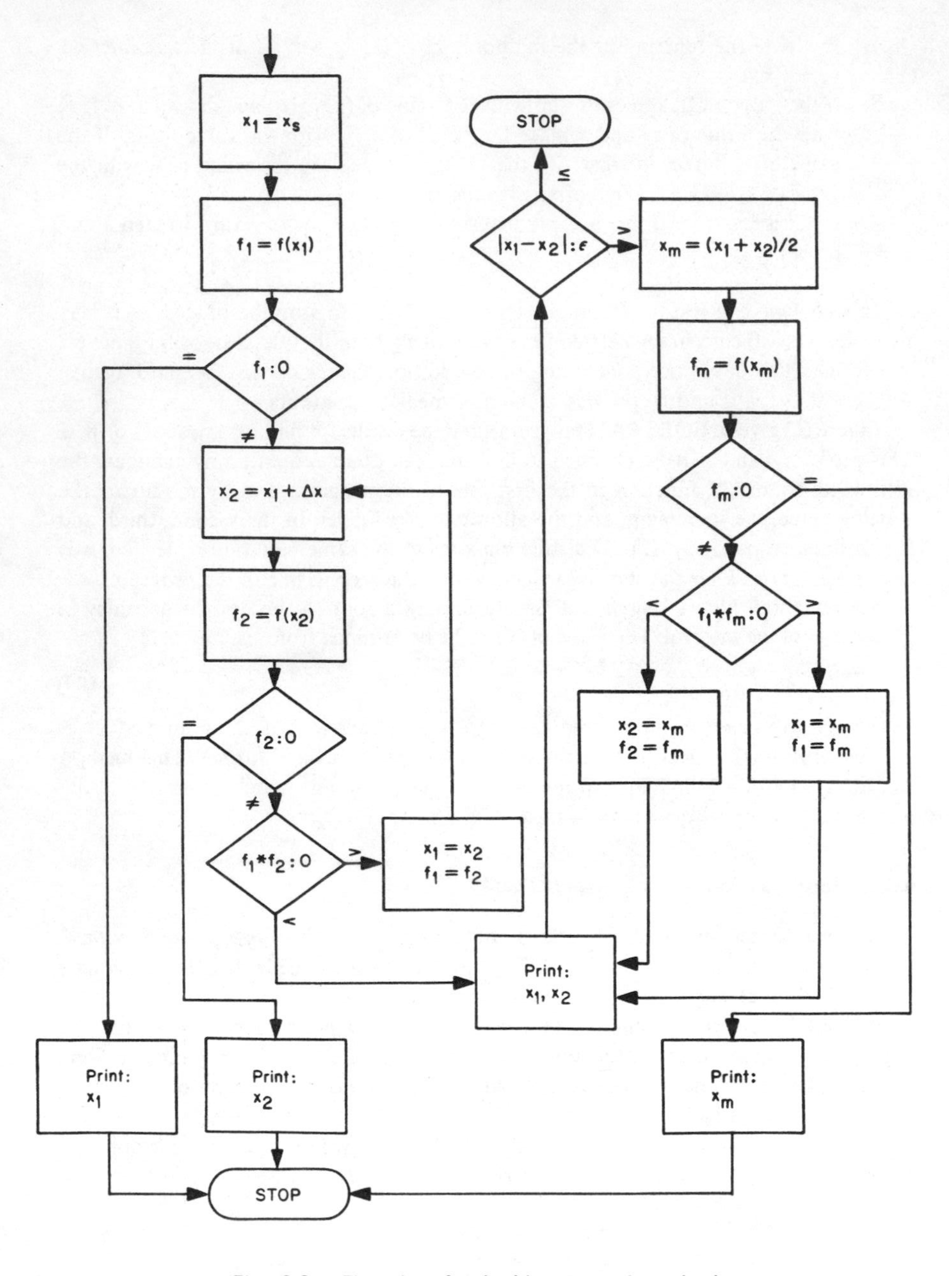

Fig. 6.2. Flowchart for the binary search method.

Program

```
      POLY(X)=((3.*X+14.)*X+13.)*X-6.
      XS=-3.6
      DX=.5
      EPS=.0000001
      X1=XS
      F1=POLY(X1)
      IF(F1) 1,2,1
    1 X2=X1+DX
      F2=POLY(X2)
      IF(F2) 3,4,3
    3 IF(F1*F2) 5,5,6
    6 X1=X2
      FI=F2
      GO TO 1
    5 PRINT 10, X1, X2
   10 FORMAT(2F12.7)
      IF(ABS(X1-X2)-EPS) 7,7,8
    8 XM=(X1+X2)/2.
      FM=POLY(XM)
      IF(FM) 9,11,9
    9 IF(F1*FM) 12,12,13
   12 X2=XM
      F2=FM
      GO TO 5
   13 X1=XM
      F1=FM
      GO TO 5
    2 PRINT 10, X1
      GO TO 7
    4 PRINT 10, X2
      GO TO 7
   11 PRINT 10, XM
    7 STOP
      END
```

Results

x_1	x_2
-3.1000000	-2.6000000
-3.1000000	-2.8500000
-3.1000000	-2.9750000
-3.0375000	-2.9750000
-3.0062500	-2.9750000
-3.0062500	-2.9906250
-3.0062500	-2.9984375
-3.0023437	-2.9984375
-3.0003906	-2.9984375
-3.0003906	-2.9994140
-3.0003906	-2.9999023
-3.0001465	-2.9999023
-3.0000244	-2.9999023
-3.0000244	-2.9999633
-3.0000244	-2.9999939
-3.0000091	-2.9999939
-3.0000015	-2.9999939
-3.0000015	-2.9999977
-3.0000015	-2.9999996
-3.0000005	-2.9999996
-3.0000001	-2.9999996
-3.0000001	-2.9999998
-3.0000001	-2.9999999
-3.0000001	-3.0000000

Fig. 6.3. Program and results for binary search method.

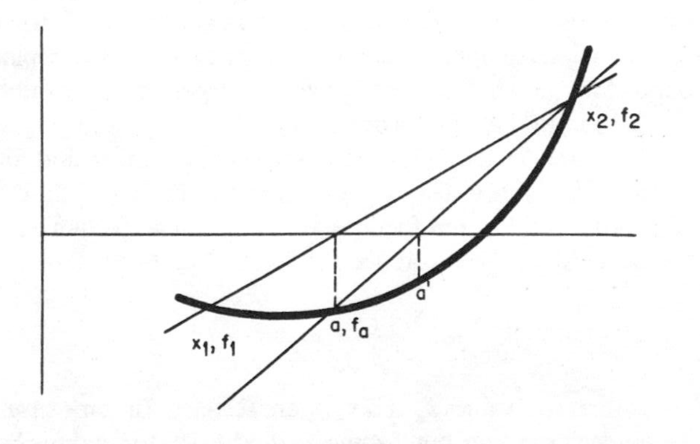

Fig. 6.4. Illustration of linear interpolation to find a root.

But we desire to find an a such that $y = 0$. Therefore,

$$a = x_1 - f_1\left(\frac{x_2 - x_1}{f_2 - f_1}\right) \tag{6.46}$$

When this value of a is used in evaluating the function, the value obtained f_a will not necessarily be zero. By a procedure similar to that of Sec. 6.6.1, the values of a and f_a can be substituted for one of the sets x_1, f_1 or x_2, f_2 depending on whether f_a agrees in sign with f_1 or f_2. Then with this new set of points a new a can be determined. A flowchart of this procedure is shown in Fig. 6.5.

Figure 6.6 gives a FORTRAN program and the results for finding a root of Eq. 6.2 by the false position method. This program also can be modified easily to obtain other roots of this or other equations. Note that in the example of this method, the result is obtained more quickly than in the previous example. Also note that x_1 remains at one value throughout the run. This does not happen in all cases.

The preceding example is incomplete in that it does not provide for the eventuality that $f_2 - f_1 = 0$. Such a situation can result in failure for the process to converge to an answer. Also, neither of the foregoing examples provide for terminating the search if no root is located after a reasonable number of tries.

6.7 SECANT METHOD

The false position method suggests the *secant method*, which is based on interpolating (or extrapolating): Given two points on the function x_1, f_1 and x_2, f_2, we can pass a straight line through these points and determine where it intersects the x-axis for an approximate value of a root. The actual root may lie between the two points as in Fig. 6.4 or it may be external to the two points as in Fig. 6.7.

Then, from Eq. (6.46) we can determine a and hence f_a. The point thus obtained can be substituted for one of the points x_1, f_1 or x_2, f_2. There are several schemes for making this substitution. One has already been given in Section 6.6.2, where that method presumes that f_1 and f_2 have opposite signs. Another method is to replace x_1, f_1 or x_2, f_2 with a, f_a depending on whether f_1 or f_2 has the larger absolute value. This process would then continue to find a new a from the current x_1, f_1 and x_2, f_2, etc.

The scheme discussed in the following is somewhat simpler and although it may falter at the beginning, it will usually produce a result. In this method, the equivalent of Eq. (6.46) is used, but the x and f values are cycled in a mechanical fashion that is independent of their values. The modified version of Eq. (6.46) is

$$x_{n+1} = x_{n-1} - f(x_{n-1})\frac{x_n - x_{n-1}}{f(x_n) - f(x_{n-1})} \tag{6.46a}$$

To start the procedure, two values x_n and x_{n-1} are selected. In some cases, this selection may have to be made at random, but if some method exists for making better estimates, then it should be used. The function is then evaluated for these arguments yielding $f(x_n)$ and $f(x_{n-1})$. Then, x_{n+1} can be found by Eq. (6.46a) and $f(x_{n+1})$ determined. Each time a new x is determined, a test is applied to determine if the approximation is sufficiently accurate to terminate the procedure. One test that can be used is to compare consecutive values of x to see if they agree to within some acceptable error. Another method is to test the value of the function to see if it is acceptably close to zero. If the

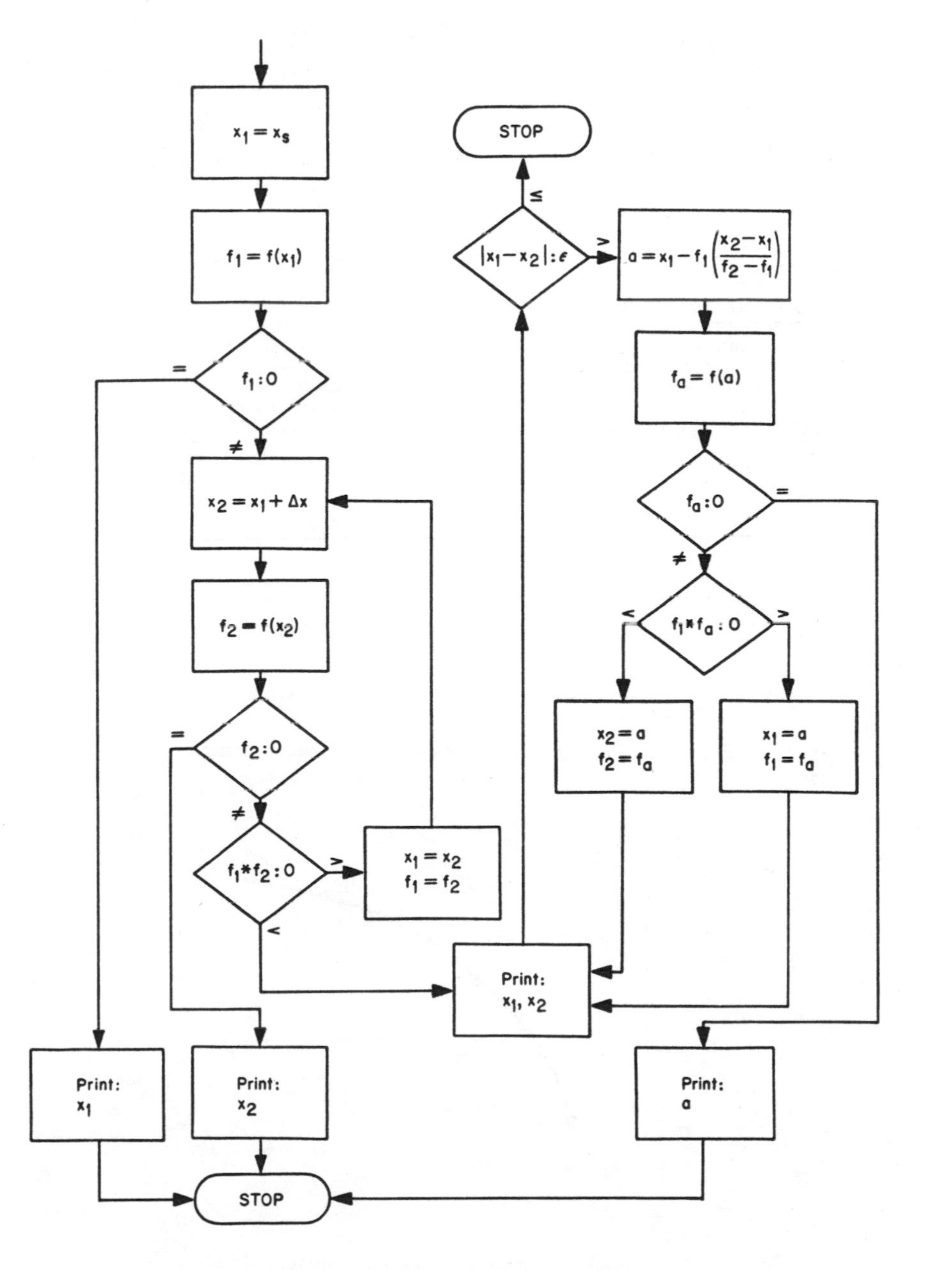

Fig. 6.5. Flowchart for the false position method.

Program

```
      POLY(X)=((3.*X+14.)*X+13.)*X-6.
      XS=-3.6
      DX=.5
      EPS=.0000001
      X1=XS
      F1=POLY(X1)
      IF(F1) 1,2,1
      1 X2=X1+DX
      F2=POLY(X2)
      IF(F2) 3,4,3
    3 IF(F1*F2) 5,5,6
    6 X1=X2
      F1=F2
      GO TO 1
    5 PRINT 10, X1,X2
   10 FORMAT(2F12.7)
      IF(ABS(X1-X2)-EPS) 7,7,8
    8 A=X1-F1*(X2-X1)/(F2-F1)
      FA=POLY(A)
      IF(FA) 9,11,9
    9 IF(F1*FA) 12,12,13
   12 X2=A
      F2=FA
      GO TO 5
   13 X1=A
      F1=FA
      GO TO 5
    2 PRINT 10, X1
      GO TO 7
    4 PRINT 10, X2
      GO TO 7
   11 PRINT 10, A
    7 STOP
      END
```

Results

x_1	x_2
-3.1000000	-2.6000000
-3.1000000	-2.9254237
-3.1000000	-2.9905824
-3.1000000	-2.9988845
-3.1000000	-2.9998689
-3.1000000	-2.9999846
-3.1000000	-2.9999982
-3.1000000	-2.9999998
-3.1000000	-3.0000000
-3.0000000	-3.0000000

Fig. 6.6. Program and results for false position method.

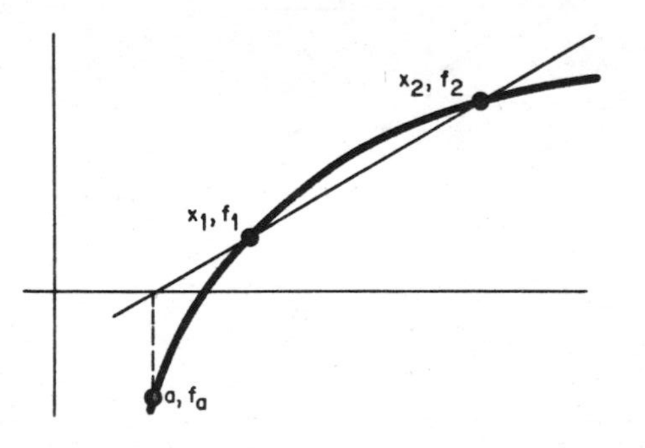

Fig. 6.7. Illustration of secant method.

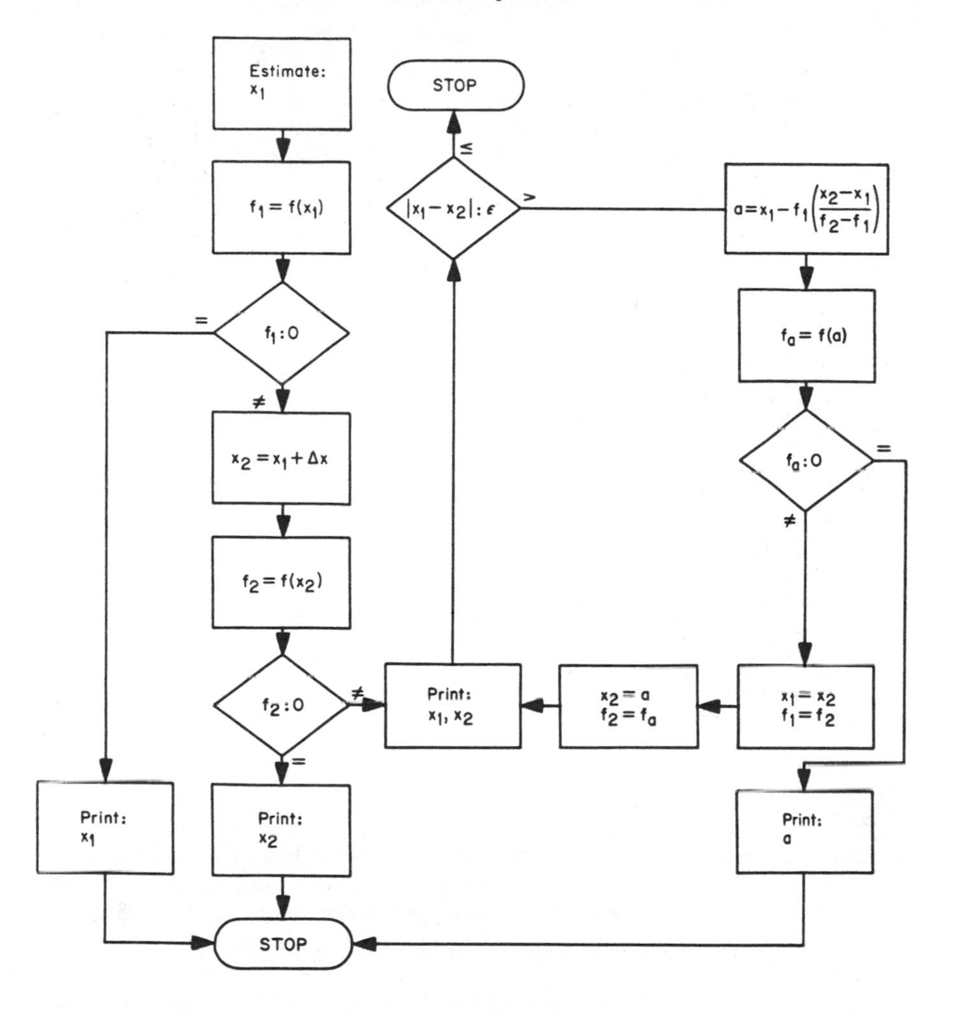

Fig. 6.8. Flowchart for secant method.

process must be continued to achieve the desired accuracy, then the next step is to cycle the x's and f's in the following manner. Replace x_{n-1} by x_n, $f(x_{n-1})$ by $f(x_n)$, x_n by x_{n+1}, and $f(x_n)$ by $f(x_{n+1})$. Then, (6.46a) is again evaluated, etc. Figure 6.8 gives a flowchart of this procedure. The notation given in the flowchart is that of Eq. (6.46) and not of (6.46a) so that it can be compared easily with the flowchart of Fig. 6.5.

In both the binary search method and false position method, we were assured of reaching a solution if we ever located two function values having opposite signs. Although the secant method uses the same equation as the false position method, it is not as easy to prove that successive approximations will converge to a solution. In fact, it is possible

Program

```
      POLY(X)=((3.*X+14.)*X+13.)*X-6.
      XS=-3.6
      DX=.5
      EPS=.0000001
      X1=XS
      F1=POLY(X1)
      IF(F1) 1,2,1
    1 X2=X1+DX
      F2=POLY(X2)
      IF(F2) 5,4,5
    5 PRINT 10, X1,X2
   10 FORMAT(2F12.7)
   90 IF(ABS(X1-X2)-EPS) 7,7,8
    8 A=X1-F1*(X2-X1)/(F2-F1)
      FA=POLY(A)
      IF(FA) 9,11,9
    9 X1=X2
      F1=F2
      X2=A
      F2=FA
      GO TO 5
    2 PRINT 10, X1
      GO TO 7
    4 PRINT 10, X2
      GO TO 7
   11 PRINT 10, A
    7 STOP
      END
```

Results

x_1	x_2
-3.6000000	-3.1000000
-3.1000000	-3.0444335
-3.0444335	-3.0050046
-3.0050046	-3.0002745
-3.0002745	-3.0000018
-3.0000018	-3.0000000
-3.0000000	-3.0000000

Fig. 6.9. Program and results for secant method.

that, for some functions, the secant method will not converge to a solution. To develop the convergence criterion for the secant method requires concepts that will be introduced in the next two sections. Therefore, the convergence criterion will be deferred to Sec. 6.9, where it will be developed and compared to the convergence criterion of a similar method.

Figure 6.9 gives a FORTRAN program and the results for finding a root of Eq. (6.2) by the secant method. This method can easily be modified to find other roots of (6.2) or to find roots of other equations. Both the binary search and false position methods will usually converge to a result once two function values of opposite signs have been found. We do not have this same assurance with the secant method. Since the process may not converge to a result, Fig. 6.8 should include some test to terminate the process and print an error message if too many repetitions of the loop are encountered. There are tests for determining if this process will converge to a root. A discussion of these tests will be given in Sec. 6.9.2.

6.8 ITERATION METHOD

The word *iteration*, as applied to computer techniques, means a repetitive process for performing some computation. Actually, in this sense, the root solving methods already discussed are iterative procedures. However, the method now to be described is called the *iteration method* because the formulation of the equation to be solved pro-

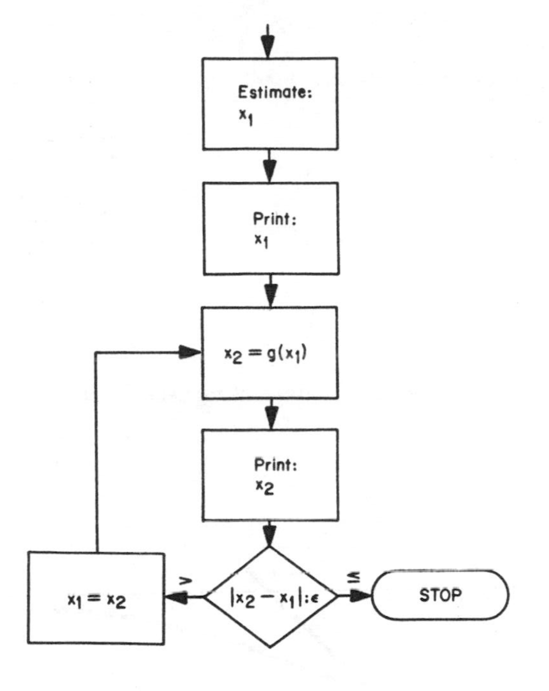

Fig. 6.10. Flowchart of iteration method.

vides the method also. It is sometimes called the *direct substitution method* or *method of successive approximations.*

Equation (6.1) can be manipulated so that it is expressed in the form

$$x = g(x) \tag{6.47}$$

Now this equation is only satisfied for certain values of x. However, we can use it as a basis for a root-solving scheme in which we write

$$x_{n+1} = g(x_n) \tag{6.48}$$

In this scheme, we estimate a root x_1, find $x_2 = g(x_1)$, and repeat using x_2, etc. The process is terminated when two consecutive x's agree to within some specified error.

Suppose, for example, Eq. (6.2) is rewritten in the form

$$x_{n+1} = 6/13 - (14/13)x_n^2 - (3/13)x_n^3 \tag{6.49}$$

Then the process can be flowcharted as in Fig. 6.10.

Figure 6.11 gives a FORTRAN program and the results of using (6.49) in the procedure flowcharted in Fig. 6.10.

The simplicity of this method is very attractive; however, this method is probably the one most likely to *fail* to converge to a result. This does *not* mean it should not be used, but rather is given as a warning so that precautions are taken to insure convergence or to stop the process and print an error message if it fails to converge.

Program

```
      POLY(X)=(6.-X*X*(14.+3.*X))/13.
      XS=-3.6
      EPS=.0000001
      X1=XS
      PRINT 10, X1
   10 FORMAT(F12.7)
    1 X2=POLY(X1)
      PRINT 10, X2
      IF(ABS(X1-X2)-EPS) 7,7,8
    3 X1=X2
      GO TO 1
    2 STOP
      END
```

Results

X
-3.6000000
-2.7286154
-2.8683356
-2.9528071
-2.9869064
-2.9968075
-2.9992531
-2.9998271
-2.9999600
-2.9999908
-2.9999979
-2.9999995
-2.9999999
-2.9999999

Fig. 6.11. Program and results for iteration method.

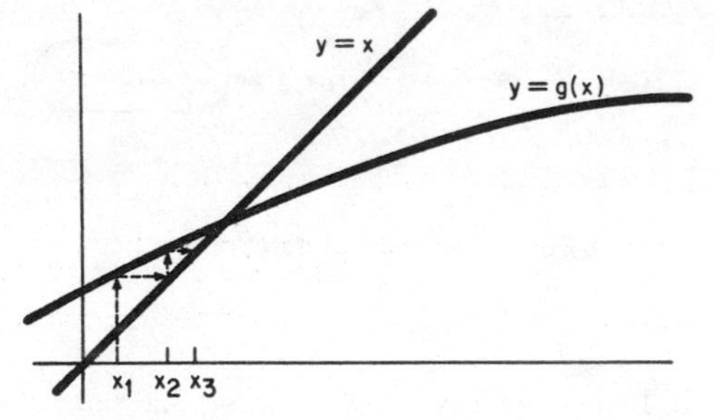

Fig. 6.12. Illustration of convergence of iteration method—positive slope.

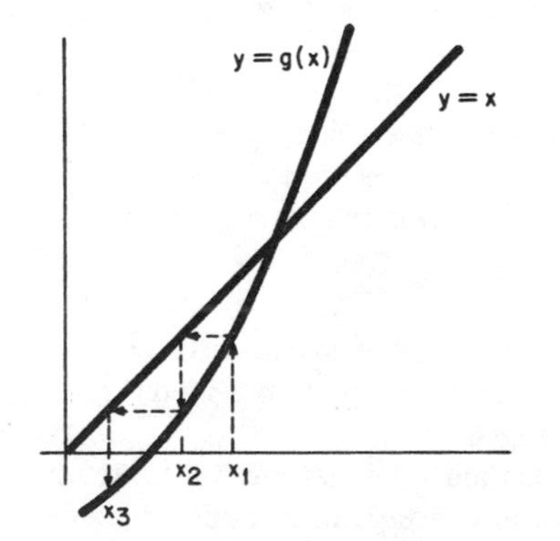

Fig. 6.13. Illustration of divergence of iteration method—positive slope.

6.8.1 Convergence Criterion for the Iteration Method

The conditions of convergence are most easily understood by showing the process graphically. In Fig. 6.12, a plot is shown of $y = x$ and $y = g(x)$. The value of the abscissa (or ordinate) of the point where the curves intersect is a root of an equation of the form (6.47). The dashed path of Fig. 6.12 illustrates how the process converges to a solution. In Fig. 6.12, we can see that a guess x_1 leads to a function value y that, when substituted in $y = x$, gives a new x, (x_2), which can be used to determine a new function value, etc. For the figure given, the process converges. We find that even if the first guess were greater than the solution, the process would still converge. Now consider Fig. 6.13. In this example, it can be seen that the process is diverging from the solution. The difference between the two examples is that in Fig. 6.12, the slope of $y = g(x)$ is less than that of $y = x$ and in Fig. 6.13, the slope of $y = g(x)$ is greater than the slope of $y = x$.

Consider next Fig. 6.14. In Fig. 6.14(A), the process is converging with the sign of the error alternating; in Fig. 6.14(B), the process is diverging. In the former case, the slope of $y = g(x)$ is greater than that of the normal to $y = x$; while in the other case, the slope of $y = g(x)$ is less than that of the normal to $y = x$. The generalization that can be inferred from these examples is that the slope of g(x) must be less, in absolute value, than 1.

$$|g'(x)| < 1 \tag{6.50}$$

This condition must be valid in the neighborhood of the solution. (A more rigorous proof of this will be given in Sec. 6.11). It is possible to obtain convergence even when the slope is greater than unity for the initial guesses and still iterate into a region where the slope is less than 1. This actually occurred in the example of Fig. 6.11. In that case,

$$g'(x) = (-28/13)x - (9/13)x^2 \tag{6.51}$$

Evaluating this for the first guess gives

$$g'(-3.6) = -1.218\ldots \tag{6.52}$$

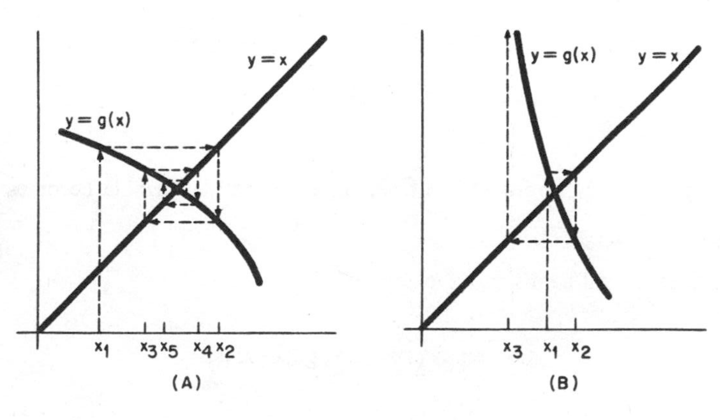

Fig. 6.14. Illustration of (A) convergence and (B) divergence of iteration method—negative slope.

But at the solution,

$$g'(-3) = 0.2307\ldots \tag{6.53}$$

If an equation of the form (6.47) can be differentiated, it is possible to determine if the condition (6.50) is satisfied. This can be done if we have a reasonably good approximation of the root to use in evaluating $g'(x)$.

Suppose that the condition (6.50) is not satisfied. In that case, it is possible to reformulate the given equation so that the condition is satisfied. The equations

$$\begin{aligned} x &= -0.3x^3 - 1.4x^2 - 0.3x + 0.6 \\ x &= (6 - 13x - 3x^3)/(14x) \\ x &= ((6 - 13x - 3x^3)/14)^{1/2} \\ x &= (6 - 13x - 14x^2)/(3x^2) \end{aligned} \tag{6.54}$$

all are different formulations of (6.2). The possible formulations are endless, but usually one can be found that will converge to a solution.

The closer $g'(x)$ is to zero, the faster the iteration will converge to the root. This suggests that we might be able to improve convergence by reformulating our equation at each step of the iteration so that $g'(x) = 0$. One way of doing this is as follows. Given a function

$$f(x) = 0 \tag{6.1}$$

add to both sides Kx:

$$Kx + f(x) = Kx \tag{6.55}$$

Dividing by K,

$$x + f(x)/K = x \tag{6.56}$$

puts the equation in the form of (6.47) with

$$g(x) = x + f(x)/K \tag{6.57}$$

Differentiating gives

$$g'(x) = 1 + f'(x)/K \tag{6.58}$$

Then, for a given guess x_n, we can find $g'(x_n) = 0$ which leads to

$$K = -f'(x_n) \tag{6.59}$$

Substituting this in Eq. (6.56) and rearranging, the iteration formula becomes

$$x_{n+1} = x_n - f(x_n)/f'(x_n) \tag{6.60}$$

This modification will be discussed in Sec. 6.9.

6.9 NEWTON'S ITERATION

In Sec. 6.7, the secant method was presented. In that method, a line was passed through two points on a function and extended (if necessary) to determine where it crossed the x-axis. This x value was assumed to be an approximation to a root of the function. If it was not sufficiently close to the actual root of the function, then that x value was used to determine a new point on the function, which when used with one of the previous points could determine a new secant line.

An interesting property of a secant line is that its slope is equal to that of a tangent to the curve at some point between the two points of intersection. This is, of course, assuming that the curve and its first derivative are continuous between the points of intersection. This suggests that instead of using two points to determine a secant line, we use instead one point and the slope of the tangent at that point. In other words, we extend the tangent to the curve at a point to intersect with the x-axis. *Newton's iteration* method may be described as the *tangent method.* Consider Fig. 6.15. By analytic geometry, the equation of a line in terms of one point $[x_n, f(x_n)]$ and its slope $f'(x_n)$ is given by

$$y = f(x_n) + (x - x_n)f'(x_n) \tag{6.61}$$

Solving this equation for $y = 0$ gives

$$x = x_n - f(x_n)/f'(x_n) \tag{6.62}$$

The x determined by this equation can be used in the equation to determine a new x. Writing this in iterative form yields

$$x_{n+1} = x_n - f(x_n)/f'(x_n) \tag{6.63}$$

which is identical to (6.60). This then is Newton's rule which we have arrived at by two different processes. Note that the flowchart in Fig. 2.6 is based on Newton's iteration, where $f(r) = N - r^2$, $f'(r) = 2r$, and hence $r_{n+1} = 0.5(N/r_n + r_n)$.

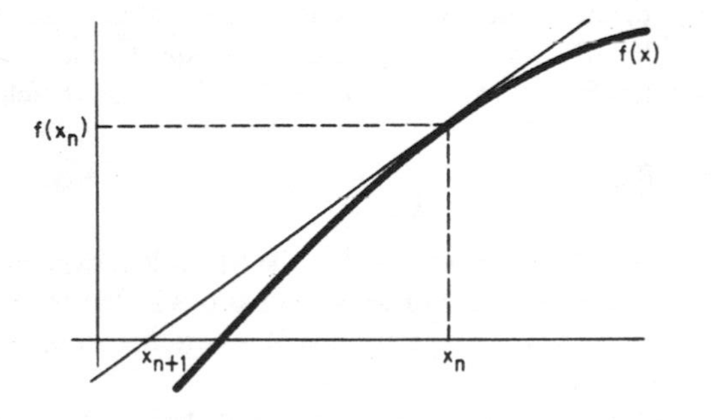

Fig. 6.15. Illustration of Newton's iteration.

Another method for deriving Eq. (6.63) is based on Taylor's series. If the function y is expanded in a Taylor's series [Eq. (5.6)] and only the first two terms retained, then (6.61) will be the result and the rest of the derivation follows. This approach suggests the possibility of developing a method for finding a root of a function that is based on a three (or more) term Taylor's series expansion of the function. Since such a method is not the topic of this section, its derivation has been omitted here.

6.9.1 Convergence Criterion for Newton's Method

It is fairly simple to determine the conditions that must be satisfied for either Newton's method or the secant method to converge to a solution. It is not so easy to determine if these conditions are satisfied.

Equation (6.63) is already in the form $x = g(x)$ so we need only apply the condition (6.50), which was established in Sec. 6.8.1. Therefore, since

$$g(x) = x - f(x)/f'(x) \tag{6.64}$$

then,

$$g'(x) = f(x)f''(x)/(f'(x))^2 \tag{6.65}$$

Therefore,

$$\frac{|f(x)f''(x)|}{(f'(x))^2} < 1 \tag{6.66}$$

Since $f(x) = 0$ at the solution, the above inequality always holds, at the solution, providing that $f'(x) \neq 0$ and $f''(x)$ is not infinite. This is usually true. If $f'(x) = 0$ at the solution, it means there is a multiple root. If $f''(x)$ is infinite, then $f(x)$ must also be infinite, which contradicts $f(x) = 0$. It appears then that Newton's method will converge except possibly for multiple roots. This is usually the case, but occasionally the condition (6.66) does not hold for the initial guess and the process may not converge because the next computed guess may exceed the capacity of the computer. Also, occasionally an iteration will oscillate between two values.

6.9.2 Convergence Criterion for the Secant Method

The convergence of the secant method is not as amenable to analysis as Newton's method. However, the similarities of the methods indicate that the conditions for convergence are similar. Equation (6.46) is of the form $x = g(x)$; therefore we may write

$$g(x_n) = x_{n-1} - f(x_{n-1})\left(\frac{x_n - x_{n-1}}{f(x_n) - f(x_{n-1})}\right) \tag{6.67}$$

Note that this is very similar to (6.64), which is the equivalent expression for Newton's method. The only difference between these two equations is that the slope of the secant line $(f(x_n) - f(x_{n-1}))/(x_n - x_{n-1})$ appears in (6.67), where the slope of the tangent line $f'(x)$ appears in (6.64).

In order to determine the derivative of $g(x_n)$ with respect to x_n, we assume x_{n-1} and f_{n-1} are constant. Thus,

$$g'(x_n) = -f(x_{n-1})(f(x_n) - f(x_{n-1}) - (x_n - x_{n-1})f'(x_n))/(f(x_n) - f(x_{n-1}))^2 \tag{6.68}$$

Dividing the numerator and denominator by $(x_n - x_{n-1})^2$ and rearranging,

$$g'(x_n) = f(x_{n-1})\frac{\dfrac{f'(x_n) - \dfrac{f(x_n) - f(x_{n-1})}{x_n - x_{n-1}}}{x_n - x_{n-1}}}{\left[\dfrac{f(x_n) - f(x_{n-1})}{x_n - x_{n-1}}\right]^2} \tag{6.69}$$

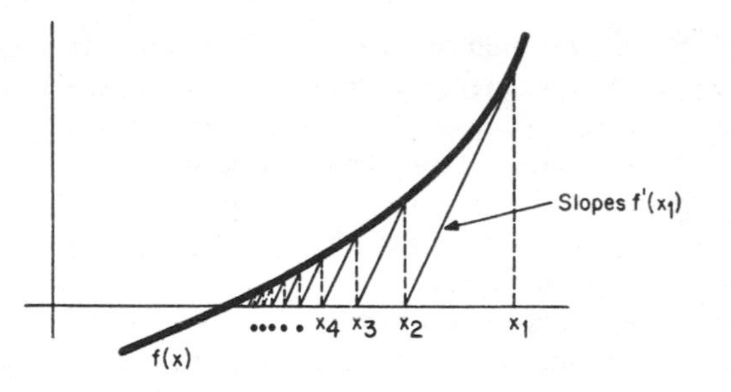

Fig. 6.16. Illustration of Newton's iteration with constant slope.

It is difficult to attach a physical significance to this form, but as we approach the solution, x_n approaches x_{n-1}. From fundamental definitions of differential calculus, we can write

$$f'(x_{n-1}) = \lim_{x_n \to x_{n-1}} \frac{f(x_n) - f(x_{n-1})}{x_n - x_{n-1}} \tag{6.70}$$

and

$$f''(x_{n-1}) = \lim_{x_n \to x_{n-1}} \frac{f'(x_n) - f'(x_{n-1})}{x_n - x_{n-1}} \tag{6.71}$$

Therefore,

$$\lim_{x_n \to x_{n-1}} g'(x_n) = \frac{f(x_{n-1})f''(x_{n-1})}{[f'(x_{n-1})]^2} \tag{6.72}$$

We see then that in the limit, the same condition (6.66) for convergence applies for the secant method and Newton's method. When it is either difficult or impossible to calculate the derivative $f'(x)$, it may be more advantageous or even mandatory to use the secant method instead of Newton's method.

An interesting fact that can be very useful is that Newton's iteration will usually converge even if the derivative is not known accurately. For some problems, this will greatly simplify the computations. We estimate the slope as $K = f'(x)$ and then write Eq. (6.63) in the form

$$x_{n+1} = x_n - f(x_n)/K \tag{6.73}$$

where K is held constant. At the solution, $f(x) = 0$ and hence, $x_{n+1} = x_n$. Figure 6.16 illustrates how the iteration converges to a solution.

6.10 COMPLEX ROOTS

In the previous sections, we have discussed how we could find real roots of equations numerically. In most cases, the problem of finding complex roots was deferred. In Sec. 6.4.3, the method for finding the roots of a quartic equation involved finding the roots of two quadratics. In that case, there was no restriction on the roots being real. In solving a cubic equation (Sec. 6.4.2), we can always find a real root. Then by dividing the given equation by the corresponding linear factor, the problem will be reduced to that of solving a quadratic equation.

In fact, once a root of a polynomial equation has been found, the order of the equation can be reduced by one by dividing by the linear factor $x - r$ (where r is the root already found). The quotient of this division will be a polynomial that can be solved by one of the given methods. This process is *not* recommended because sizable errors can accumulate in the coefficients of the reduced polynomials due to errors in the calculated roots. If all roots are real, other roots usually can be found by the iterative process by merely using a different first estimate. Finding roots individually by iteration is usually more accurate than reducing the polynomial by division. But suppose that we have a polynomial of even order greater than four or a transcendental equation that has only complex roots. Actually, the iterative methods discussed in this chapter also can be used to find these complex roots. The only difficulty is that complex arithmetic must be employed. Equation (6.74) is a third-order equation:

$$x^3 + 7x^2 + 19x + 13 = 0 \tag{6.74}$$

with roots

$$r_1 = -1.$$
$$r_2 = -3. + 2.i$$
$$r_3 = -3. - 2.i$$

Employing Newton's iteration using complex arithmetic with an initial guess of $r = -4. + i$ results in the series of estimates shown in Table 6.2. This iteration converges to the root identified as r_2. There is no need to attempt to find its complex conjugate by iteration, since we can write it directly from the previous result. However, it is interesting to see what will happen with a different initial guess. Using a first guess of $r = 1. + i$ leads to the series of estimates shown in Table 6.3.

TABLE 6.2

Real	*Imaginary*
−4.0000000	1.0000000
−2.9024390	1.1219512
−2.2579300	2.5752559
−2.6359526	1.9589392
−3.0345992	1.9195791
−2.9956572	2.0010436
−2.9999999	1.9999888
−3.0000000	2.0000000
−3.0000000	2.0000000

TABLE 6.3

Real	*Imaginary*
1.0000000	1.0000000
−0.1349899	0.6272666
−0.7966922	0.3116224
−1.0160387	0.0602258
−1.0017271	−0.0009221
−0.9999989	0.0000016
−1.0000000	0.0000000
−1.0000000	0.0000000

Note that this iteration converges to the real root r_1 even though the arithmetic used was complex.

6.10.1 Other Methods for Finding Complex Roots

Other methods of finding complex roots exist. They usually involve a procedure for iterating for quadratic factors of the form

$$x^2 + px + q \tag{6.75}$$

This means that two quantities p and q must be iterated for simultaneously just as we had to iterate for two quantities, the real and imaginary parts of the root, in the previous example.

These methods are useful only for finding quadratic factors (and hence pairs of roots) of polynomial equations. The following paragraphs give a presentation of two classic methods.

An nth-order polynomial can always be written in the form of Eq. (6.7); i.e., it can be normalized so that the coefficient of the highest order term is equal to unity. Furthermore, this polynomial can be represented by the product of a quadratic factor such as (6.75) and an (n − 2)th-order polynomial plus a linear remainder term. Thus,

$$x^n + a_{n-1}x^{n-1} + \cdots + a_1x + a_0 = (x^2 + px + q)(x^{n-2} + b_{n-1}x^{n-3} + \cdots + b_3x + b_2) + b_1x + b_0 \tag{6.76}$$

By multiplying out the right side of (6.76), the following relations are obtained:

$$\begin{aligned} a_{n-1} &= b_{n-1} + p \\ a_{n-2} &= b_{n-2} + pb_{n-1} + q && (6.77) \\ a_{n-j} &= b_{n-j} + pb_{n+1-j} + qb_{n+2-j} \qquad (j = 3, 4, \ldots, n-2) \\ a_1 &= b_1 + pb_2 + qb_3 && (6.78) \\ a_0 &= b_0 + qb_2 && (6.79) \end{aligned}$$

In order that the quadratic $x^2 + px + q$ be an exact factor of the given polynomial, the remainder term must be zero. Hence, $b_1 = b_0 = 0$. In *Lin's method*, Eqs. (6.77) are rewritten to solve for the b's:

$$\begin{aligned} b_{n-1} &= a_{n-1} - p \\ b_{n-2} &= a_{n-2} - pb_{n-1} - q && (6.80) \\ b_{n-j} &= a_{n-j} - pb_{n+1-j} - qb_{n+2-j} \qquad (j = 3, 4, \ldots, n-2)^{\dagger} \end{aligned}$$

and after setting $b_1 = b_0 = 0$, (6.78) and (6.79) are solved for p and q, respectively:

$$p = (a_1 - qb_3)/b_2 \tag{6.81}$$

$$q = a_0/b_2 \tag{6.82}$$

Then, starting with initial estimates for p and q, (6.80), (6.81), and (6.82) are solved iteratively for p and q.

In this chapter, we have been concerned with finding a root of a single equation. Lin's method, given here, requires the solution of two equations, (6.81) and (6.82), in two unknowns. Note that Eqs. (6.80) are only auxiliary equations needed to obtain values for b_3 and b_2. The solution of simultaneous equations by iteration is presented in Sec. 7.6 in Chap. 7.

Bairstow's method for finding quadratic factors of polynomials is also based on Eqs. (6.77), (6.78), and (6.79). However, Newton's iteration is employed in the solution.

† This formula also holds for n–1, but is not used in this form in Lin's method.

If the reader is not already familiar with Newton's iteration for simultaneous equations, which is the topic of Sec. 7.7, it is suggested that he read that section before continuing here.

The quantities that we wish to make zero are b_1 and b_0, which are functions of p and q. Equations (7.73) when specialized for this problem become

$$\begin{aligned} \Delta p \frac{\partial b_1}{\partial p} + \Delta q \frac{\partial b_1}{\partial q} &= -b_1 \\ \Delta p \frac{\partial b_0}{\partial p} + \Delta q \frac{\partial b_0}{\partial q} &= -b_0 \end{aligned} \tag{6.83}$$

where b_1 is found by the recursion relation (6.80) and b_0 must be found from

$$b_0 = a_0 - qb_2 \tag{6.84}$$

after finding b_2 from the general recursion relations. The partial derivatives of b_1 and b_0 can be found by differentiating the relations (6.80) *and* Eq. (6.84). Differentiating (6.80) and (6.84) with respect to p and substituting $c_k = \partial b_k/\partial p$ results in the relations

$$\begin{aligned} c_{n-1} &= -1 \\ c_{n-2} &= -b_{n-1} + p \\ c_{n-j} &= -b_{n+1-j} - pc_{n+1-j} - qc_{n+2-j} \qquad (j = 3, 4, \ldots, n-1) \\ c_0 &= -qc_2 \end{aligned} \tag{6.85}$$

Differentiating the same equations, but with respect to q, and letting $d_k = \partial b_k/\partial q$ results in the relations

$$d_{n-1} = 0 \tag{6.86a}$$

$$d_{n-2} = -1 \tag{6.86b}$$

$$d_{n-j} = -b_{n+2-j} - pd_{n+1-j} - qd_{n+2-j} \qquad (j = 3, 4, \ldots, n-1) \tag{6.86c}$$

$$d_0 = -b_2 - qd_2 \tag{6.86d}$$

Since $d_k = c_{k+1}$ (except for $k = 0$), only the relations (6.85) are needed plus the following modification to Eq. (6.86d):

$$d_0 = -b_2 - qc_3 \tag{6.87}$$

Using the new notation for derivatives, (6.83) can be written as

$$\begin{aligned} c_1\Delta p + c_2\Delta q &= -b_1 \\ c_0\Delta p + d_0\Delta q &= -b_0 \end{aligned} \tag{6.88}$$

The complete method then is as follows:

1. Select a p and q.
2. By Eqs. (6.80), find $b_{n-1}, b_{n-2}, \ldots, b_2$, and b_1.
3. By Eq. (6.84), find b_0.
4. By Eqs. (6.85), find $c_{n-2}, c_{n-1}, \ldots, c_1$, and c_0.
5. By Eq. (6.87), find d_0.
6. By Eqs. (6.88), find p and q.
7. Add the above increments to p and q and repeat as necessary.

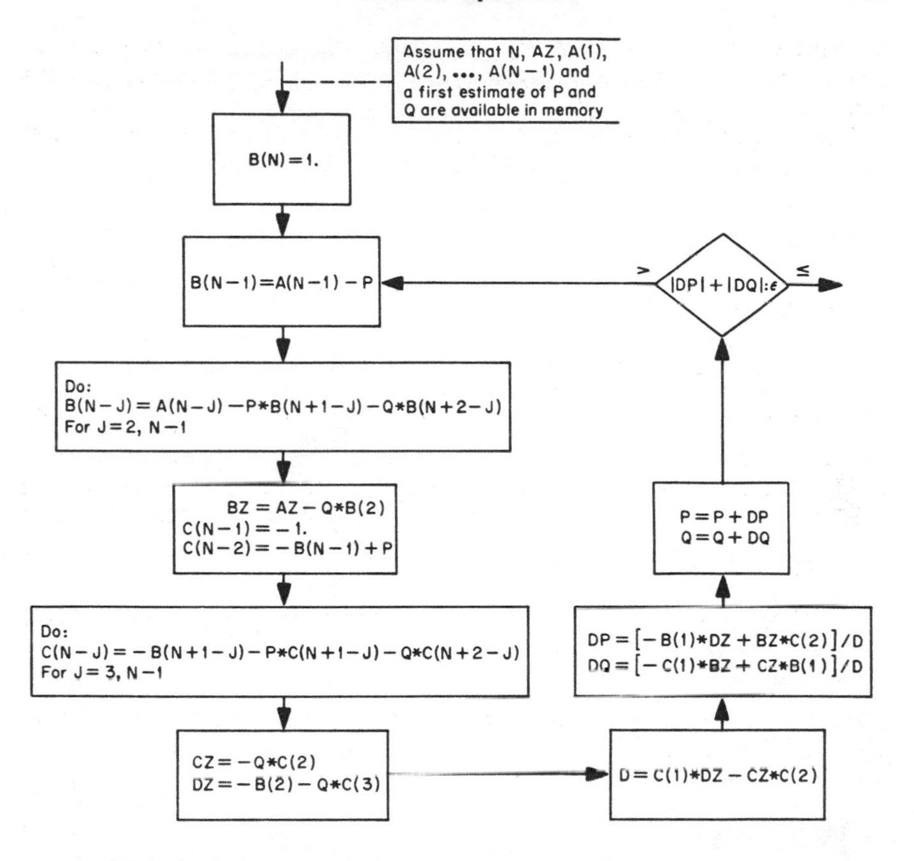

Fig.6.17. Bairstow's method for finding a quadratic factor of a fourth- or higher-degree polynomial.

The flowchart of Figure 6.17 depicts Bairstow's method in computer notation. Since FORTRAN does not allow a zero subscript, the symbols AZ, BZ, CZ, and DZ are used to represent the zero element of the arrays A, B, C, and D. No generality is lost, since these quantities must be calculated by special-purpose formulas.

6.11 DERIVATION OF CONVERGENCE CRITERIA AND THE RATES OF CONVERGENCE

In Sec. 6.8.1, it was established by demonstration that the condition required for convergence of the direct substitution method is[†]

$$|g'(x)| < 1 \tag{6.50}$$

The purpose of this section is to provide a more rigorous derivation of this criterion and to provide comparisons between the rates of convergence of various methods.

[†] The condition (6.50) must be satisfied at the solution and usually in some neighborhood of the solution.

Since the convergence criteria for other methods are based on the condition (6.50), it is only necessary to derive that relation. The condition required to assure convergence is that successive estimates become closer and closer together. This can be expressed mathematically as

$$|x_{n+1} - x_n| < |x_n - x_{n-1}| \tag{6.89}$$

or

$$\frac{|x_{n+1} - x_n|}{|x_n - x_{n-1}|} < 1 \tag{6.90}$$

or

$$\left|\frac{x_{n+1} - x_n}{x_n - x_{n-1}}\right| < 1 \tag{6.91}$$

The function $g(x_n)$ can be expanded in a Taylor's series about the point x_{n-1}.

$$g(x_n) = g(x_{n-1}) + (x_n - x_{n-1})\, g'(\xi) \tag{6.92}$$

The value ξ will lie somewhere between x_{n-1} and x_n. But Eq. (6.48) infers that we can write

$$x_n = g(x_{n-1}) \tag{6.93}$$

Then, by applying (6.93) and (6.48) to (6.92):

$$x_{n+1} = x_n + (x_n - x_{n-1})\, g'(\xi) \tag{6.94}$$

Solving this for $g'(\xi)$,

$$g'(\xi) = \frac{x_{n+1} - x_n}{x_n - x_{n-1}} \tag{6.95}$$

Therefore, if

$$|g'(\xi)| < 1 \tag{6.96}$$

the inequality (6.91) will be satisfied. Since the relation (6.96) is essentially the same as the relation (6.50), the latter is established.

6.11.1 Rate of Convergence for the False Position Method

The false position method uses a straight line passed through two points of the given function to estimate a zero of the function. This straight line is given by the equation

$$y = f(x_{n-1}) + (x - x_{n-1})\left(\frac{f(x_n) - f(x_{n-1})}{x_n - x_{n-1}}\right) \tag{6.97}$$

The function can be represented exactly by a Taylor's series:

$$f(x) = f(x_{n-1}) + (x - x_{n-1})f'(x_{n-1}) + (x - x_{n-1})^2 f''(\xi_1)/2 \tag{6.98}$$

where ξ_1 lies somewhere between x_{n-1} and x. Also, by Eq. (6.98), we can write

$$f(x_n) = f(x_{n-1}) + (x_n - x_{n-1})f'(x_{n-1}) + (x_n - x_{n-1})^2 f''(\xi_2)/2 \tag{6.99}$$

where ξ_2 lies somewhere between x_{n-1} and x_n. From (6.99) we find

$$\frac{f(x_n) - f(x_{n-1})}{x_n - x_{n-1}} = f'(x_{n-1}) + (x_n - x_{n-1})f''(\xi_2)/2 \tag{6.100}$$

Substituting this into (6.97) gives

$$y = f(x_{n-1}) + (x - x_{n-1})f'(x_{n-1}) + (x - x_{n-1})(x_n - x_{n-1})f''(\xi_2)/2 \tag{6.101}$$

The error then in the linear approximation (6.97) is given by

$$f(x) - y = (x - x_{n-1})^2 f''(\xi_1)/2 - (x - x_{n-1})(x_n - x_{n-1})f''(\xi_2)/2 \tag{6.102}$$

It can be shown that a ξ can be found such that when it is substituted for both ξ_1 and ξ_2, Eq. (6.102) is still valid. Making that substitution and rearranging, the error is

$$f(x) - y = (x - x_n)(x - x_{n-1})f''(\xi)/2 \tag{6.103}$$

Therefore, the exact function can be expressed as the sum of (6.97) and (6.103). When x is equal to one of the roots r, the resulting function will be zero.

$$0 = f(x_{n-1}) + (r - x_{n-1})\left(\frac{f(x_n) - f(x_{n-1})}{x_n - x_{n-1}}\right) + (r - x_n)(r - x_{n-1})f''(\xi)/2 \tag{6.104}$$

Since the error in the ith approximation is given by $\varepsilon_i = r - x_i$, Eq. (6.104) can be written as

$$0 = f(x_{n-1}) + (r - x_{n-1})\left(\frac{f(x_n) - f(x_{n-1})}{x_n - x_{n-1}}\right) + \varepsilon_n \varepsilon_{n-1} f''(\xi)/2 \tag{6.105}$$

Solving this for r gives

$$r = x_{n-1} - f(x_{n-1})\left(\frac{x_n - x_{n-1}}{f(x_n) - f(x_{n-1})}\right) - \frac{\varepsilon_n \varepsilon_{n-1} f''(\xi)}{2\left(\dfrac{f(x_n) - f(x_{n-1})}{x_n - x_{n-1}}\right)} \tag{6.106}$$

The general expression for obtaining the n + 1st approximation by means of the false position method is given by

$$x_{n+1} = x_{n-1} - f(x_{n-1})\left(\frac{x_n - x_{n-1}}{f(x_n) - f(x_{n-1})}\right) \tag{6.107}$$

[See Eq. (6.46).]

The *mean value theorem* states that for a function that is continuous between two points, the secant line joining those two points has the same slope as a tangent to the function at some point between the two given points. Therefore, we can write

$$\frac{f(x_n) - f(x_{n-1})}{x_n - x_{n-1}} = f'(\xi_3) \tag{6.108}$$

Substituting this in (6.106) and taking the difference between (6.106) and (6.107) to obtain the error,

$$\varepsilon_{n+1} \doteq -\varepsilon_n \varepsilon_{n-1} \frac{f''(\xi)}{2f'(\xi_3)} \tag{6.109}$$

This equation says that the error depends on the previous two errors and the first and second derivatives of the function, each evaluated in the neighborhood of the previous two guesses. In the neighborhood of the solution, we can assume that the above derivatives are fairly constant and therefore we can say approximately:

$$\varepsilon_{n+1} \doteq k\varepsilon_n \varepsilon_{n-1} \tag{6.110}$$

In the false position method, one point, say x_0, remains the same throughout the iteration. Therefore one of the errors, say ε_0, remains the same and the sequence of errors can be written as

$$\begin{aligned}
\varepsilon_2 &= k\varepsilon_1\varepsilon_0 \\
\varepsilon_3 &= k\varepsilon_2\varepsilon_0 = k^2\varepsilon_1\varepsilon_0^2 \\
\varepsilon_4 &= k\varepsilon_3\varepsilon_0 = k^3\varepsilon_1\varepsilon_0^3 \\
&\cdot \quad \cdot \quad \cdot \\
&\cdot \quad \cdot \quad \cdot \\
&\cdot \quad \cdot \quad \cdot \\
\varepsilon_i &= k\varepsilon_{i-1}\varepsilon_0 = (k\varepsilon_0)^{i-1}\varepsilon_1
\end{aligned} \tag{6.111}$$

or

$$k\varepsilon_i = (k\varepsilon_0)^{i-1}k\varepsilon_1 \tag{6.112}$$

Therefore, if $|k\varepsilon_0| < 1$ (i.e., if the initial error is sufficiently small), the iteration will converge.

6.11.2 Rate of Convergence for the Secant Method

The secant method uses the same linear approximation as the false position method, but the successive approximations are assigned in a different fashion. Equation (6.110) still applies for the error, but ε_n at one stage of the iteration becomes the ε_{n-1} for the next stage of the iteration.

The sequence of errors will be

$$\begin{aligned}
\varepsilon_2 &= k\varepsilon_1\varepsilon_0 \\
\varepsilon_3 &= k\varepsilon_2\varepsilon_1 = k^2\varepsilon_1^2\varepsilon_0 \\
\varepsilon_4 &= k\varepsilon_3\varepsilon_2 = k^4\varepsilon_1^3\varepsilon_0^2 \\
\varepsilon_5 &= k\varepsilon_4\varepsilon_3 = k^7\varepsilon_1^5\varepsilon_0^3 \\
&\cdot \quad \cdot \quad \cdot \\
&\cdot \quad \cdot \quad \cdot
\end{aligned} \tag{6.113}$$

It is not possible to write the general term in closed form, but an examination of the exponents of ε_1 and ε_0 shows that they are members of a Fibonacci sequence, i.e., each one is the sum of the previous two. We can say

$$k\varepsilon_i = (k\varepsilon_0)^p(k\varepsilon_1)^q \tag{6.114}$$

Therefore, if $|k\varepsilon_0| < 1$ and $|k\varepsilon_1| < 1$, the iteration will converge. For i sufficiently large, $p > i$ and $q > i$. Therefore a comparison of Eqs. (6.112) and (6.114) indicates that for the same starting estimates, the secant method will converge more rapidly than the false position method. (The false position algorithm given, Fig. 6.5, includes an initial search to bracket the solution before the actual method is started. The secant method algorithm given, Fig. 6.8, does not include such an initial search. Because of this, the former method sometimes converges more rapidly than the latter.)

6.11.3 Rate of Convergence for Newton's Method

The function f(x) can be expanded in a Taylor's series about the point x_n:

$$f(x) = f(x_n) + (x - x_n)f'(x_n) + (x - x_n)^2 f''(\xi)/2 \tag{6.115}$$

At a root r, this function will be zero. Therefore,

$$0 = f(x_n) + (r - x_n)f'(x_n) + (r - x_n)^2 f''(\xi)/2 \tag{6.116}$$

Letting $\varepsilon_n = r - x_n$ in the last term and solving for r,

$$r = x_n - f(x_n)/f'(x_n) - \frac{\varepsilon_n^2 f''(\xi)}{2f'(x_n)} \tag{6.117}$$

Subtracting Eq. (6.63) from (6.117) to obtain the error gives

$$\varepsilon_{n+1} = -\frac{f''(\xi)}{2f'(x_n)}\varepsilon_n^2 \tag{6.118}$$

Assuming that the derivatives do not vary markedly during the iteration, we have approximately:

$$\varepsilon_{n+1} \doteq k\varepsilon_n^2 \tag{6.119}$$

The sequence of errors will be

$$\begin{aligned}
\varepsilon_1 &= k\varepsilon_0^2 \\
\varepsilon_2 &= k\varepsilon_1^2 = k^3\varepsilon_0^4 \\
\varepsilon_3 &= k\varepsilon_2^2 = k^7\varepsilon_0^8 \\
\varepsilon_4 &= k\varepsilon_3^2 = k^{15}\varepsilon_0^{16} \\
&\;\;\vdots \\
\epsilon_i &= k\epsilon_{i-1}^2 = k^{2^i-1}\epsilon_0^{2^i}
\end{aligned} \tag{6.120}$$

or we can write

$$k\varepsilon_i = (k\varepsilon_0)^{2^i} \tag{6.121}$$

Therefore, if $|k\varepsilon_0| < 1$, the iteration will converge. Although no general term can be given for the sequence (6.113) (secant method), that sequence can be compared, line by line, with the sequence (6.120) (Newton's method) to determine which method converges more rapidly. Assuming that the initial estimates are about the same and k is constant and is the same for both methods, we can see that the exponent of ε_0 in Newton's method increases more rapidly than the sum of the exponents of ε_1 and ε_0 in the secant method. This means that if the sequences converge, Newton's method converges more rapidly than the secant method.

6.12 SUMMARY

This chapter presented methods for finding the roots of equations. Two types of methods have been defined: direct and indirect. Direct methods consist of a finite number of steps that, when completed, give the root(s) to an accuracy that depends only on the accuracy of the computational process. Indirect methods give an approximate answer in a finite number of steps. These steps (or a portion of them) can be repeated in an attempt to improve the accuracy of the result. This repetition can be continued indefinitely. However, it is usually terminated either when a test determines that the answer(s) is sufficiently accurate or when a test indicates that the process is diverging from the result and therefore further computations should be abandoned.

Two classes of equations were covered: polynomial and transcendental. General direct methods for finding roots of equations exist only for a small subclass of polynomial equations. This subclass is limited to first-, second-, third-, and fourth-degree polynomial equations. Although direct methods exist for the solution of some polynomial equations of other orders and for some transcendental equations, these methods are very specialized. The indirect methods can be applied to both classes of equations. For a given equation, one or more methods may fail to converge to a root; but if a root exists, there also exists some method for finding it to any degree of accuracy.

It was demonstrated in this chapter that the methods given for finding real roots of equations also can be used to find complex roots by using complex arithmetic in the computer program. Some specialized methods for finding complex roots of polynomial equations were also presented.

EXERCISES

1. Prepare a flowchart and write a FORTRAN program to find the roots of any quadratic equation of the form $Ax^2 + Bx + C = 0$ by the method of Sec. 6.4.1.
2. Execute the program in Exercise 1 for the following data:

a	*b*	*c*
14.7	−3.2	8.7
14.7	−3.2	−8.7
0.0	−5.6	4.3
− 9.8	0.0	6.3
0.0	0.0	8.5
−16.2	14.7	0.0

3. Prepare a flowchart and write a FORTRAN program to find the roots of any cubic equation of the form $x^3 + ax^2 + bx + c = 0$ by the method of Sec. 6.4.2.

4. Execute the program in Exercise 3 for the following data:

a	*b*	*c*
6.0	9.0	4.0
9.7	3.0	−29.7
4.6	6.85	−21.35

5. Prepare a flowchart and write a FORTRAN program to find the roots of any quartic equation of the form $x^4 + ax^3 + bx^2 + cx + d = 0$ by the method of Sec. 6.4.3.

6. Execute the program in Exercise 5 for the following data:

a	*b*	*c*	*d*
8.0	19.25	−17.5	−122.
−2.0	−13.0	14.0	24.0

7. Find one real root, by the binary search method, for each cubic equation whose coefficients are given in Exercise 4.

8. Repeat Exercise 7 using the false position method.

9. Repeat Exercise 7 using the secant method.

10. Find at least one root of the equation $x = 1/\sin x$ by the iteration method (Sec. 6.8).

11. Prepare a flowchart and a FORTRAN program to find a zero of any function (given the function and its first derivative) by Newton's iteration.

12. Execute the program in Exercise 11 for the function $F = 1/x - \sin x$.

13. Execute the program in Exercise 11 for the function $F = \sin x - e^{-x}$.

14. Modify the program in Exercise 11 so that complex arithmetic is used to obtain a complex root.

15. Execute the program in Exercise 14 for the cubic equations

$$x^3 + 4.6x^2 + 6.85x - 21.35 = 0$$

$$x^3 + x^2 + x + 1. = 0$$

16. Execute the program in Exercise 14 for the polynomials

$$x^4 + 8x^3 + 19.25x^2 - 17.5x - 122. = 0$$

$$x^4 + x^3 + x^2 + x + 1. = 0$$

17. Prepare a flowchart and write a FORTRAN program to find a quadratic factor of a polynomial equation of any degree by Lin's method. Include provisions for stopping the iteration if convergence is not achieved in a reasonable number of steps. Print out the coefficients of the quadratic factor and also the coefficients of the reduced polynomial.

18. Execute the program in Exercise 17 using the data given in Exercise 4.

19. Execute the program in Exercise 17 using the data given in Exercise 6.

20. Complete the flowchart of Fig. 6.17 and write a FORTRAN program to find a quadratic factor of a polynomial equation of degree greater than three by Bairstow's method. Include provisions for stopping the iteration if convergence is not achieved in a reasonable number of steps. Print out the coefficients of the quadratic factor and also the coefficients of the reduced polynomial.

21. Execute the program in Exercise 20 using the data given in Exercise 6.

22. Find all four quadratic factors by executing the program in Exercise 20 on the polynomial

$$x^8 + x^7 + x^6 + x^5 + x^4 + x^3 + x^2 + x + 1. = 0$$

and on each succeeding reduced polynomial.

7

Simultaneous Equations and Matrix Operations

7.1 INTRODUCTION

In Chap. 6, we discussed methods of finding the roots of equations. That discussion was limited to equations with a single unknown. In many physical problems, we have systems of equations involving several unknown quantities. In general, if we have n unknowns, we must have n *independent* equations in order to be able to arrive at a unique solution set. In this chapter, two types of systems of simultaneous equations will be considered: linear and nonlinear.

Linear equations are equations in which the variables (unknowns) exist in the first order, in multiplicative combination only with constants and in additive combination only with similar terms or constants. Expressing this algebraically, we can write

$$a_1x_1 + a_2x_2 + a_3x_3 + \cdots + a_nx_n = c \tag{7.1}$$

Equation (7.1) is a linear equation in the variables $x_1, x_2, x_3, \cdots, x_n$. The quantities $a_1, a_2, a_3, \ldots, a_n$, c are constants. An equation of the form

$$3x_1 + x_2/5 - 2x_3 = 7 \tag{7.2}$$

is also linear because the multiplicative and additive terms are to be interpreted in the broad sense, namely to include division and subtraction, respectively. Furthermore, the equations

$$\begin{aligned} 4x_1 - 2x_2^2 - 2\sin x_3 &= -5 \\ 2x_1 + x_2^2 \qquad\qquad &= 6 \\ x_1 \qquad + 2\sin x_3 &= 2 \end{aligned} \tag{7.3}$$

are also linear because the unknown x_2 always appears raised to the second power and x_3 only appears as an argument to the sine function. Therefore, by the simple substitutions $y_1 = x_2^2$ and $y_2 = \sin x_3$, Eqs. (7.3) can be rewritten as

$$\begin{aligned} 4x_1 - 2y_1 - 2y_2 &= -5 \\ 2x_1 + y_1 \qquad &= 6 \\ x_1 \qquad + 2y_2 &= 2 \end{aligned} \tag{7.4}$$

In this form, the equations are more easily recognized as being linear. Solution of (7.4) yields $x_1 = 1$, $y_1 = 4$, $y_2 = 1/2$. Therefore, $x_2 = \pm 2$ and $x_3 = \pi/6$ (principle value).

Nonlinear equations are those equations that do not qualify as linear equations.

Most of this chapter will be devoted to systems of linear equations. Linear equations are either the primary subject or one of the subjects of Secs. 7.2, 7.3, 7.4, 7.5, 7.6, 7.7, and 7.9. These sections include definitions concerning linear equations and solutions of systems of linear equations by substitution, determinants, iteration, and matrices. In addition, the solution of systems of nonlinear equations are discussed in Secs. 7.3, 7.6, and 7.7. Other topics presented are determinants, matrices, and eigenvalues and eigenvectors in Secs. 7.4, 7.8, and 7.10, respectively.

7.2 LINEAR EQUATIONS — DEFINITIONS

Equation (7.1) gives the general form for a linear equation. A more general form for a *set* of linear equations is

$$\begin{aligned} a_{11}x_1 + a_{12}x_2 + \cdots + a_{1n}x_n &= c_1 \\ a_{21}x_1 + a_{22}x_2 + \cdots + a_{2n}x_n &= c_2 \\ &\vdots \\ a_{m1}x_1 + a_{m2}x_2 + \cdots + a_{mn}x_n &= c_m \end{aligned} \tag{7.5}$$

Normally, $m = n$ because we must have as many equations as unknowns in order to be able to effect a unique solution. If $m < n$, the system is said to be *under determined*; if $m > n$, the system is said to be *over determined*. If the set is under determined, a unique solution for all unknowns is not possible. If the set is over determined, a solution may or may not exist (i.e., the set may be consistent or inconsistent), and if it exists, it can be unique. Over-determined sets resulting from observational data will usually not be consistent, but will be almost consistent. The techniques of Chap. 8 can be applied in such cases to obtain a "best" solution. In this chapter, only the case $m = n$ will be considered.

If the c_i's are all zero, the equations are said to be homogeneous. In this case, the only solution is all x_i equal zero (i.e., providing that the equations are all independent). There is a condition under which homogeneous equations can have a solution other than the trivial one cited. This condition will be discussed in Sec. 7.10.

If any equation of Eqs. (7.5) can be represented exactly by a linear combination of any other equations in the set, then the equations are *dependent*. In this case, the set can be reduced to an under determined set that we cannot solve (at least not for all unknowns).

7.2.1 Example of a Dependent Set of Equations

Consider the following set of equations:

$$3x_1 + 2x_2 - 4x_3 = -5 \tag{7.6}$$

$$-x_1 + x_2 + x_3 = 4 \tag{7.7}$$

$$7x_1 + 3x_2 - 9x_3 = -14 \tag{7.8}$$

This is not an independent set because (7.8) is a linear combination of (7.6) and 7.7). Namely,

Eq. 7.8 = (2 * Eq. 7.6) – Eq. 7.7

The above relationship is not obvious and is given to illustrate that a set of equations can be dependent without this fact being obvious. We will see in later sections that there are tests that can be performed to determine if the equations of a set are dependent.

7.3 SOLUTION BY SUBSTITUTION

Substitution is a method of solving simultaneous equations that can be used for solving linear equations and can also be applicable for some nonlinear equations. This method consists of solving algebraically for one unknown in terms of the other unknowns and substituting in the remaining equations to eliminate one unknown. By doing this repeatedly, expressions can be derived for each of the unknowns in terms of constants only.

7.3.1 Example of Two Linear Equations

Given

$$a_{11}x_1 + a_{12}x_2 = c_1 \tag{7.9}$$

$$a_{21}x_1 + a_{22}x_2 = c_2 \tag{7.10}$$

Equation (7.9) can be solved for x_1 giving

$$x_1 = (c_1 - a_{12}x_2)/a_{11} \tag{7.11}$$

Substitution of (7.11) into (7.10) gives

$$a_{21}c_1/a_{11} - a_{21}a_{12}x_2/a_{11} + a_{22}x_2 = c_2 \tag{7.12}$$

Solving for x_2

$$x_2 = \frac{c_2 - a_{21}c_1/a_{11}}{a_{22} - a_{21}a_{12}/a_{11}} \tag{7.13}$$

which can also be written as

$$x_2 = \frac{a_{11}c_2 - a_{21}c_1}{a_{11}a_{22} - a_{21}a_{12}} \tag{7.14}$$

Substitution of (7.14) into (7.11) gives, after simplifying,

$$x_1 = \frac{a_{22}c_1 - a_{12}c_2}{a_{11}a_{22} - a_{21}a_{12}} \tag{7.15}$$

Equations (7.15) and (7.14) give algebraic relations for evaluating x_1 and x_2, respectively. These relations can be used to solve any pair of linear equations in two unknowns, provided that the denominator $a_{11}a_{22} - a_{21}a_{12}$ is not zero. If this denominator is zero, then the equations are singular. If the denominator and both of the numerators are zero, then the right sides of Eqs. (7.14) and (7.15) are indeterminate (zero over zero) and this tells us that the original equations are *dependent*.

Expressions can also be derived in a similar manner for three or more equations in three or more unknowns, but the algebra rapidly becomes prohibitive. We will see in Sec. 7.4.1 that each unknown can be expressed in the form

$$x_i = N_i/D \tag{7.16}$$

The conditions regarding numerators N_i and denominators D, which were cited for the case of two equations, also apply in general.

7.3.2 Example of Two Nonlinear Equations

In some cases, nonlinear sets of equations can also be solved by direct substitution. Consider

$$Ax^2 + Bx + Cy = D \tag{7.17}$$

$$Ex^2 + Fx + Gy = H \tag{7.18}$$

Solving (7.17) for y gives

$$y = (D - Ax^2 - Bx)/C \tag{7.19}$$

Substituting in (7.18) and rearranging,

$$(E - AG/C)x^2 + (F - BG/C)x - H + GD/C = 0 \tag{7.20}$$

This equation is a quadratic in x only and can be solved by the method given in Sec. 6.4.1. Once x has been found, y can be found from (7.19).

7.4 DETERMINANTS

Determinants can be used to solve *linear* simultaneous equations. We will see that their use is not very practical for computer solution. The study of determinants, however, can be very useful in gaining a better understanding of simultaneous linear equations.

A determinant is a *square* array of numbers (elements) that has a single value. Depicted in literal form,

$$D = \begin{vmatrix} a_{11} & a_{12} & \cdots & a_{1n} \\ a_{21} & a_{22} & \cdots & a_{2n} \\ \cdot & \cdot & & \cdot \\ \cdot & \cdot & & \cdot \\ \cdot & \cdot & & \cdot \\ a_{n1} & a_{n2} & \cdots & a_{nn} \end{vmatrix} \tag{7.21}$$

The vertical bars denote that this array is a determinant. The determinant's single numerical value can be found by methods discussed in this section.

The methods for evaluating second- and third-order determinants are quite simple and are given in the following. A second-order determinant

$$D = \begin{vmatrix} a_{11} & a_{12} \\ a_{21} & a_{22} \end{vmatrix} \tag{7.22}$$

is evaluated by first multiplying the elements along the diagonal shown by the arrow in Fig. 7.1, namely $a_{11}a_{22}$; then multiplying the elements on the other diagonal as shown by Fig. 7.2, $a_{21}a_{12}$; and subtracting from the first product to obtain

$$D = a_{11}a_{22} - a_{21}a_{12} \tag{7.23}$$

We note that the right side of this equation is identical to the denominators of (7.14) and (7.15).

In the case of a third-order determinant, the procedure is somewhat more complex.

$$D = \begin{vmatrix} a_{11} & a_{12} & a_{13} \\ a_{21} & a_{22} & a_{23} \\ a_{31} & a_{32} & a_{33} \end{vmatrix} \tag{7.24}$$

To evaluate this determinant, the product of the elements along each arrow in Fig. 7.3 is found and the sum of the products is found: $a_{11}a_{22}a_{33} + a_{12}a_{23}a_{31} + a_{32}a_{21}a_{13}$. Then a similar operation is performed according to the arrows shown in Fig. 7.4 which results in the sum of products $a_{31}a_{22}a_{13} + a_{21}a_{12}a_{33} + a_{23}a_{32}a_{11}$. Subtracting the second sum from the first gives the value of the determinant.

$$D = a_{11}a_{22}a_{33} + a_{12}a_{23}a_{31} + a_{32}a_{21}a_{13} - a_{31}a_{22}a_{13} - a_{21}a_{12}a_{33} - a_{23}a_{32}a_{11} \tag{7.25}$$

The preceding simple methods for second- and third-order determinants *cannot* be extended to higher-order determinants.

The value of any determinant of order n can be found by applying the following rules:

1. Find all products of elements taken n at a time such that
   (a) Two or more elements are not taken from the same column.
   (b) Two or more elements are not taken from the same row.
2. Each product is assigned a sign by the following rules:
   (a) Write the literals of the elements in each product in order by column subscripts; then consider the row subscript. Call these $i_1, i_2, i_3, \ldots, i_n$. Count the number of i's that are *less* than i_1. This number is called the number of inversions in the sequence with respect to i_1. Count the number of i's (with subscripts greater than 2) that are less than i_2 to find the number of inversions with respect to i_2. Do this for each subscript.
   (b) Add the number of inversions for each subscript to obtain an index.
   (c) Minus one raised to this index gives the sign of the product. (In step 2(a), the words *row* and *column* can be interchanged and the results will be the same.)

To better understand this process, the following example for the third-order determinant Eq. (7.24), is given.

To find all products of terms taken 3 at a time and subject to preceding rules 1(a) and 1(b): Select a_{11} from column 1, a_{22} from column 2, and a_{33} from column 3 giving $a_{11}a_{22}a_{33}$; select a_{11} from column 1, a_{32} from column 2, and a_{23} from column 3 giving $a_{11}a_{32}a_{23}$. No other combinations are possible which include a_{11}. Continue using each of a_{21} and a_{31}. The resulting products are tabulated in Table 7.1, with the elements in each product written in order by column subscripts.

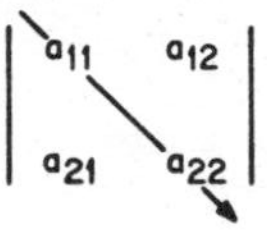

a11 a12
a21 a22

Fig. 7.1. Second-order determinant—positive term.

a11 a12
a21 a22

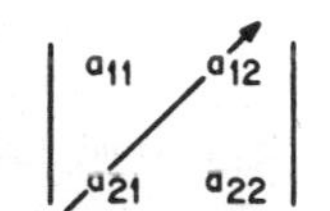

Fig. 7.2. Second-order determinant—negative term.

a11 a12 a13
a21 a22 a23
a31 a32 a33

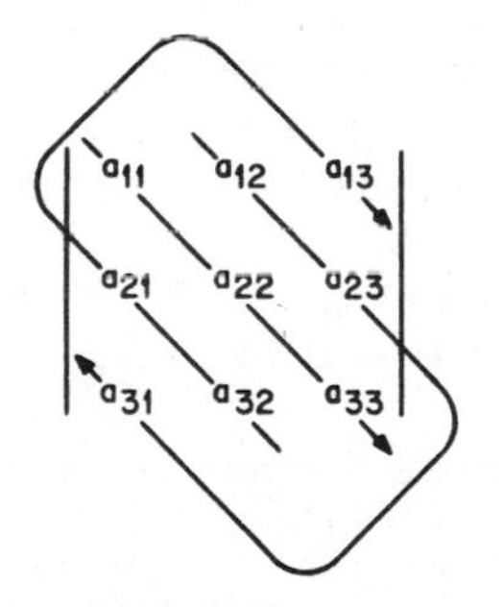

Fig. 7.3. Third-order determinant—positive term.

a11 a12 a13
a21 a22 a23
a31 a32 a33

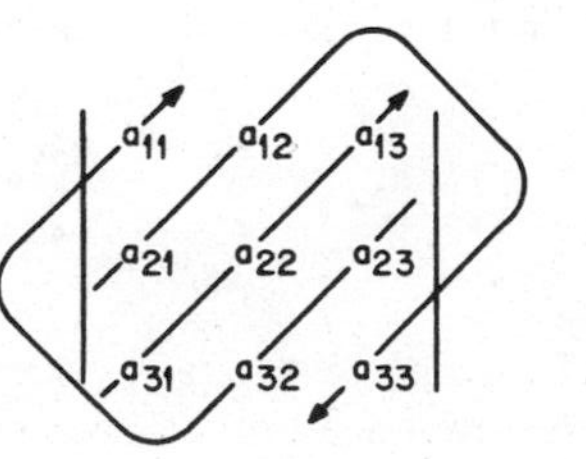

Fig. 7.4. Third-order determinant—negative term.

TABLE 7.1 PRODUCTS AND THEIR SIGNS FOR A THIRD-ORDER DETERMINANT

Product	*Row Subscripts*	*Inversions*	*Sign*
$a_{11}a_{22}a_{33}$	1 2 3	0	+
$a_{11}a_{32}a_{23}$	1 3 2	1	−
$a_{21}a_{12}a_{33}$	2 1 3	1	−
$a_{21}a_{32}a_{13}$	2 3 1	2	+
$a_{31}a_{12}a_{23}$	3 1 2	2	+
$a_{31}a_{22}a_{13}$	3 2 1	3	−

Also tabulated in Table 7.1 are the row subscripts, the total number of inversions, and the appropriate sign for each product. To clarify the counting of inversions, consider the first product; the row subscripts are 1, 2, 3. In this case, no subscript that follows the first one is less than the first one and no subscript that follows the second one is less than the second one. Hence, there are no inversions; $(-1)^0 = +1$, so the sign of the product is plus. Consider the last product; the row subscripts are 3, 2, 1. There are two subscripts that follow the first subscript that are less than the first subscript; there is one subscript following the second subscript that is less than the second subscript. Therefore, there is a total of three inversions; $(-1)^3 = -1$, so the sign of the product is minus.

If the products in Table 7.1 are added, using the appropriate signs, we will find the result is identical to that given in Eq. (7.25).

Although this process is relatively simple in this example, we can show that for a fourth-order determinant there would be four times as many products, and the work involved in determining each sign would be greater than for each of the products in the example. In general, an nth-order determinant will contain n! products of n terms each. The work involved in evaluating determinants increases with increasing order at a rate that exceeds the rate that n! increases. This conclusion makes determinants rather unattractive.

Another method of evaluating determinants is through the use of *cofactors*. A cofactor is defined in the following manner:

1. The *minor* or any element a_{ij} of an nth-order determinant is the $(n - 1)$th-order determinant that remains when row i and column j are removed from the original determinant.
2. The *cofactor* of an element a_{ij} is the minor of that element multiplied by $(-1)^{i+j}$.

Using these definitions, the value of a determinant can be found by selecting a row (or column) and multiplying each element in that row (or column) by its cofactor. The sum of these products will be the value of the determinant. Thus, Eq. (7.24) can be expanded:

$$D = a_{11} \begin{vmatrix} a_{22} & a_{23} \\ a_{32} & a_{33} \end{vmatrix} - a_{12} \begin{vmatrix} a_{21} & a_{23} \\ a_{31} & a_{33} \end{vmatrix} + a_{13} \begin{vmatrix} a_{21} & a_{22} \\ a_{31} & a_{32} \end{vmatrix} \tag{7.26}$$

Each of the determinants in (7.26) can be evaluated by one of the techniques given earlier (or by using cofactors) and the result will be the same as (7.25).

7.4.1 Cramer's Rule

So far, we have only considered how to evaluate determinants. In this section, we will see how determinants can be used to solve simultaneous linear equations. In 1750, the Swiss mathematician Gabriel Cramer discovered the following rule.

Consider the array of coefficients of Eq. (7.5) as a determinant.

$$D = \begin{vmatrix} a_{11} & a_{12} & \cdots & a_{1n} \\ a_{21} & a_{22} & \cdots & a_{2n} \\ \vdots & \vdots & & \vdots \\ a_{n1} & a_{n2} & \cdots & a_{nn} \end{vmatrix} \tag{7.27}$$

In the previous section, we stated that each unknown of a set of linear equations can be expressed in the form

$$x_i = N_i/D \tag{7.16}$$

Cramer proved this and showed that D is the determinant of the array of coefficients of Eqs. (7.5) as illustrated by (7.27). He also showed that the ith numerator can be found by substituting the column of constants $c_1, c_2, \ldots, c_n$ from Eqs. (7.5) into the ith column of the determinant of (7.27). This is illustrated for $i = 2$ by Eq. (7.28).

$$N_2 = \begin{vmatrix} a_{11} & c_1 & \cdots & a_{1n} \\ a_{21} & c_2 & \cdots & a_{2n} \\ \vdots & \vdots & & \vdots \\ a_{n1} & c_n & \cdots & a_{nn} \end{vmatrix} \tag{7.28}$$

This can be used to evaluate

$$x_2 = N_2/D \tag{7.29}$$

To solve n equations in n unknowns, $n + 1$ determinants must be evaluated.

7.4.2 Properties of Determinants

Even though we usually find that it is impractical to use determinants to solve equations, there are some useful properties of determinants that can aid in the solution of equations by other methods.

1. If the values of all elements in one row are zero, the value of the determinant is zero.
2. If the values of corresponding elements in two rows are proportional to each other by the same constant of proportionality, the value of the determinant is zero. (The special case of the constant being 1 means two rows are identical.)

3. If the values of each of the elements in one row are multiplied by a constant c, the value of the resulting determinant is equal to the value of the original determinant multiplied by c.
4. If the values of each of the elements in one row are multiplied by a constant c and then added to the values of corresponding elements in another row, the value of the determinant is not changed.
5. If the values of corresponding elements in any two rows are exchanged, the sign of the determinant is reversed.

Rules 1 to 5 can also be stated for columns by substituting column for row in each rule. In addition:

6. The values of all the elements in a row i can be exchanged with the values of all the elements in a column j without changing the value of the determinant. This exchange must be made so that the column order of the row becomes the row order of the column and vice versa, namely, $a_{ik} \leftrightarrows a_{kj}$.

These rules are primarily useful in determining if a set of equations is dependent.

7.5 SOLUTION BY ELIMINATION

We saw in Sec. 7.3 how equations could be solved by substitution. Elimination is basically the same technique, but the operations involved are easier to perform by computer. The method amounts to manipulating each equation in such a manner that a term containing one of the unknowns can be eliminated in all but one equation. Such an elimination then reduces the order of the equations by one, which if carried out repeatedly, simplifies the problem eventually to one equation in one unknown. The rules governing the allowable manipulation, which does not change the solution of the equations, are corollaries to the rules given for determinants.

1. The equations can be written in any order. This is obvious and can be proved by applying Rule 5 of Sec. 7.4.2 to both the numerator and denominator determinants.
2. Any equation can be multiplied by a constant. This can be proved by applying Rule 3 of Sec. 7.4.2.
3. One equation can be multiplied by a constant and added to any other equation. This can be proved by applying Rule 4 of Sec. 7.4.2.

To simplify the following discussion, the unknowns can be omitted from Eqs. (7.5) and the constant terms written as an array:

$$\begin{array}{ccccc} a_{11} & a_{12} & \cdots & a_{1n} & c_1 \\ a_{21} & a_{22} & \cdots & a_{2n} & c_2 \\ \cdot & \cdot & & \cdot & \cdot \\ \cdot & \cdot & & \cdot & \cdot \\ \cdot & \cdot & & \cdot & \cdot \\ a_{n1} & a_{n2} & \cdots & a_{nn} & c_n \end{array} \tag{7.30}$$

For storing the values of the array in a computer memory, it is more convenient to use the same name for the constant terms c by letting $a_{i\,n+1} = c_i$. Thus, the stored array would have the form

$$\begin{matrix} a_{11} & a_{12} & \cdots & a_{1n} & a_{1\,n+1} \\ a_{21} & a_{22} & \cdots & a_{2n} & a_{2\,n+1} \\ \vdots & \vdots & & \vdots & \vdots \\ a_{n1} & a_{n2} & \cdots & a_{nn} & a_{n\,n+1} \end{matrix} \tag{7.31}$$

Our objective is to modify array (7.31) such that a new array results of the following form:

$$\begin{matrix} b_{11} & b_{12} & \cdots & b_{1n} & b_{1\,n+1} \\ 0 & b_{22} & \cdots & b_{2n} & b_{2\,n+1} \\ \vdots & \vdots & & \vdots & \vdots \\ 0 & b_{n2} & \cdots & b_{nn} & b_{n\,n+1} \end{matrix} \tag{7.32}$$

This array represents a new set of equations having the same solution as the original set of equations, but with the variable x_1 eliminated from all but one equation.

7.5.1 Gaussian Elimination

To accomplish the foregoing, we can apply a technique attributed to Gauss. (See Sec. 1.3.) This method is sometimes called *triangularization*. The procedure consists of normalizing the first equation by dividing each coefficient by the lead coefficient a_{11} (Rule 2 of Sec. 7.5); then, in turn, multiplying the normalized equation by the lead coefficient a_{i1} of each other equation and subtracting from each corresponding equation (Rule 3 of Sec. 7.5). Algebraically, this means

$$b_{1j} = a_{1j}/a_{11} \tag{7.33}$$

and

$$b_{ij} = a_{ij} - a_{i1}b_{1j} \qquad (i > 1) \tag{7.34}$$

This can be generalized by substituting k where 1 appears in (7.33) and (7.34), where k may be $1, 2, \ldots, n$, depending on which variable is being eliminated [$i > k$ in Eq. (7.34)]. The b's of one stage become the a's of the next stage. An example will help to clarify this method. Solve the following equations by Gaussian elimination:

$$\begin{aligned} x_1 - x_2 + x_3 - x_4 &= 1 \\ -x_1 - x_2 + x_3 + x_4 &= -2 \\ 2x_1 + 4x_2 + 3x_3 + 5x_4 &= -2 \\ 3x_1 + x_2 + x_3 + x_4 &= -1 \end{aligned} \tag{7.35}$$

The array to be manipulated for this set of equations is

$$\begin{array}{lrrrrr} A_1 & 1 & -1 & 1 & -1 & 1 \\ A_2 & -1 & -1 & 1 & 1 & -2 \\ A_3 & 2 & 4 & 3 & 5 & -2 \\ A_4 & 3 & 1 & 1 & 1 & -1 \end{array} \tag{7.36}$$

Each of the rows in array (7.36) has been identified by a row number to help in denoting the operations performed to obtain subsequent arrays. Eliminating terms involving x_1,

$$\begin{array}{lrrrrr} B_1 = A_1/1 & 1 & -1 & 1 & -1 & 1 \\ B_2 = A_2 - (-1)B_1 & 0 & -2 & 2 & 0 & -1 \\ B_3 = A_3 - (2)B_1 & 0 & 6 & 1 & 7 & -4 \\ B_4 = A_4 - (3)B_1 & 0 & 4 & -2 & 4 & -4 \end{array} \tag{7.37}$$

Eliminating terms involving x_2,

$$\begin{array}{lrrrrr} B_1 & 1 & -1 & 1 & -1 & 1 \\ C_2 = B_2/-2 & 0 & 1 & -1 & 0 & 1/2 \\ C_3 = B_3 - (6)C_2 & 0 & 0 & 7 & 7 & -7 \\ C_4 = B_4 - (4)C_2 & 0 & 0 & 2 & 4 & -6 \end{array} \tag{7.38}$$

Eliminating terms involving x_3,

$$\begin{array}{lrrrrr} B_1 & 1 & -1 & 1 & -1 & 1 \\ C_2 & 0 & 1 & -1 & 0 & 1/2 \\ D_3 = C_3/7 & 0 & 0 & 1 & 1 & -1 \\ D_4 = C_4 - (2)D_3 & 0 & 0 & 0 & 2 & -4 \end{array} \tag{7.39}$$

Reducing the coefficient of x_4,

$$\begin{array}{lrrrrr} B_1 & 1 & -1 & 1 & -1 & 1 \\ C_2 & 0 & 1 & -1 & 0 & 1/2 \\ D_3 & 0 & 0 & 1 & 1 & -1 \\ E_4 = D_4/2 & 0 & 0 & 0 & 1 & -2 \end{array} \tag{7.40}$$

The array (7.40) can be rewritten in equation form as

$$x_1 - x_2 + x_3 - x_4 = 1 \tag{7.41}$$

$$x_2 - x_3 = 1/2 \tag{7.42}$$

$$x_3 + x_4 = -1 \tag{7.43}$$

$$x_4 = -2 \tag{7.44}$$

Now that the array has been triangularized, we have the value of only one unknown and a method for determining the other unknowns must be provided. This is called *back substitution* and is very simple since the equations are in triangular form.

Substituting Eq. (7.44) into (7.43) yields

$$x_3 = 1 \tag{7.45}$$

Further substitution of (7.45) and (7.44) into (7.42) yields

$$x_2 = 3/2 \tag{7.46}$$

And finally,

$$x_1 = -1/2 \tag{7.47}$$

can be obtained from (7.46), (7.45), (7.44), and (7.41).

The array (7.40) could have been operated on further in order to reduce all off-diagonal elements to zero. This is called *diagonalization*. In the following section, we will combine these procedures into one procedure.

7.5.2 Gauss–Jordan Elimination

In the Gauss–Jordan elimination method, both the forward and backward solutions are combined as suggested in the previous section. The only difference between this method and the Gaussian elimination method is that the condition on Eq. (7.34) becomes $i \neq k$ instead of $i > k$. The row k, which is normalized by (7.33), is called the *pivot* row. In the Gaussian elimination method, variables are only eliminated below the pivot row. In the Gauss–Jordan method, they are eliminated both above and below the pivot row.

The example equation (7.35) will be solved by this method.

$$\begin{array}{lrrrrr} A_1 & 1 & -1 & 1 & -1 & 1 \\ A_2 & -1 & -1 & 1 & 1 & -2 \\ A_3 & 2 & 4 & 3 & 5 & -2 \\ A_4 & 3 & 1 & 1 & 1 & -1 \end{array} \tag{7.48}$$

The array (7.48) is identical to array (7.36) since this is the initial array. On eliminating x_1, the array (7.49) is also identical to array (7.37).

$$\begin{array}{lrrrrr} B_1 = A_1/1 & 1 & -1 & 1 & -1 & 1 \\ B_2 = A_2 - (-1)B_1 & 0 & -2 & 2 & 0 & -1 \\ B_3 = A_3 - (2)B_1 & 0 & 6 & 1 & 7 & -4 \\ B_4 = A_4 - (3)B_1 & 0 & 4 & -2 & 4 & -4 \end{array} \tag{7.49}$$

At the next elimination, x_2 is eliminated in the first equation (first row) as well as in the third and fourth.

$$\begin{array}{lrrrrr} C_1 = B_1 - (-1)C_2 & 1 & 0 & 0 & -1 & 3/2 \\ C_2 = B_2/-2 & 0 & 1 & -1 & 0 & 1/2 \\ C_3 = B_3 - (6)C_2 & 0 & 0 & 7 & 7 & -7 \\ C_4 = B_4 - (4)C_2 & 0 & 0 & 2 & 4 & -6 \end{array} \tag{7.50}$$

Likewise in the next step, rows 1 and 2 are operated on in addition to row 4.

$$\begin{array}{lrrrrr} D_1 = C_1 - (0)D_3 & 1 & 0 & 0 & -1 & 3/2 \\ D_2 = C_2 - (-1)D_3 & 0 & 1 & 0 & 1 & -1/2 \\ D_3 = C_3/7 & 0 & 0 & 1 & 1 & -1 \\ D_4 = C_4 - (2)D_3 & 0 & 0 & 0 & 2 & -4 \end{array} \tag{7.51}$$

Finally, operations are performed on the first three rows in addition to the normalization of the pivot row.

$$
\begin{array}{llllll}
E_1 = D_1 - (-1)D_4 & 1 & 0 & 0 & 0 & -1/2 \\
E_2 = D_2 - (1)D_4 & 0 & 1 & 0 & 0 & 3/2 \\
E_3 = D_3 - (1)D_4 & 0 & 0 & 1 & 0 & 1 \\
E_4 = D_4/2 & 0 & 0 & 0 & 1 & -2
\end{array} \tag{7.52}
$$

Writing the array (7.52) in equation form immediately gives

$$
\begin{aligned}
x_1 \qquad\qquad\qquad &= -1/2 \\
x_2 \qquad\qquad &= 3/2 \\
x_3 \qquad &= 1 \\
x_4 &= -2
\end{aligned} \tag{7.53}
$$

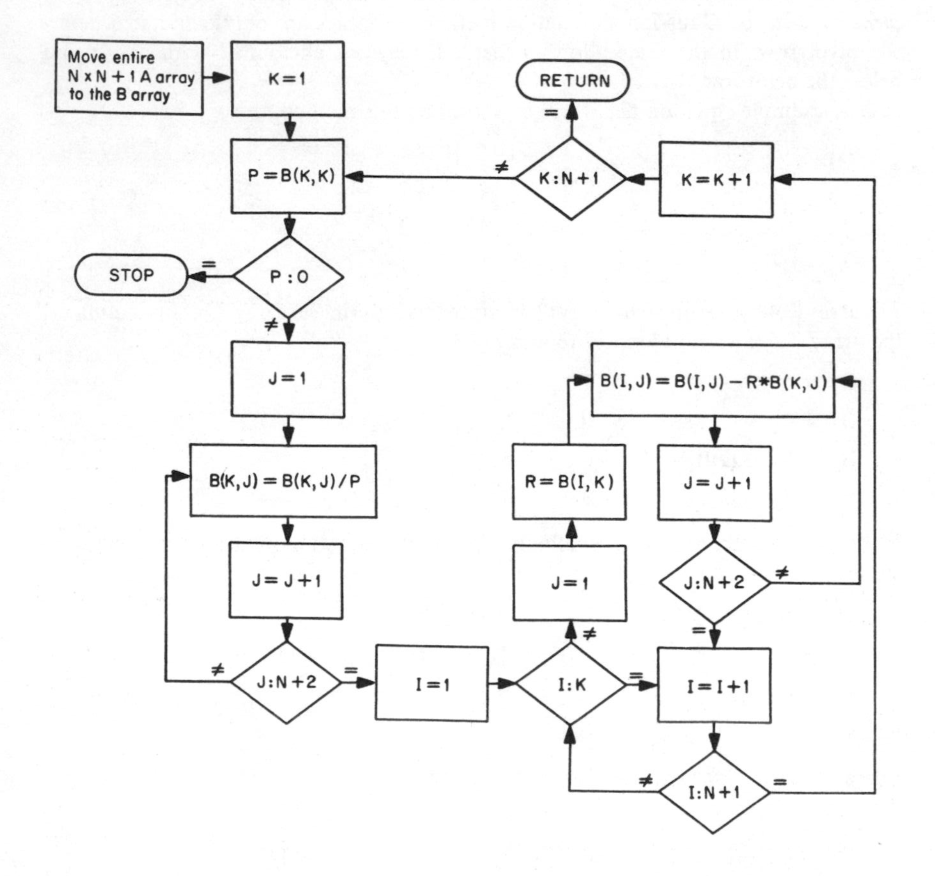

Fig. 7.5. Flowchart for the Gauss–Jordan elimination method for solving a set of linear equations.

Figure 7.5 gives a flowchart for the Gauss–Jordan elimination method. This flowchart calls for moving the entire original array A into a second array B. The purpose of this is to preserve the original array. The process destroys the array that is operated on. If the array need not be preserved, then this transfer is not required. A common situation is to require solving equations with the same coefficients, but each time with different constant terms. Figure 7.5 can be easily modified to do this. In this flowchart, the pivot element B(K, K) is called P and stored temporarily for a more efficient execution. Likewise, the row multiplier B(I, K) is called R for the same reason. The efficiency of the method (Fig. 7.5) can be further improved by eliminating steps that are known to result in 0 or 1. These steps are included here for the sake of clarity. Later methods will consider ways of improving program and storage efficiency.

7.5.3 Zero Pivot Element

It should also be noted that Fig. 7.5 checks to see if the pivot element is zero. Should this element be zero, the normalization of the pivot row cannot be performed. In such a case, some other action must be taken. One possibility is to type out an error message and either stop or read in new data.

By Rule 1 of Sec. 7.5, it is possible to change the order of the equations without changing the solution. At any stage (at the end of any row operation) of the elimination, the array in memory represents a set of equations that has the same solution as the original equations. Therefore, the array can be rearranged to avoid a zero pivot element. However, this cannot be done indiscriminately. Although several methods can be devised, a safe method is to interchange the row with zero pivot element with any row *below* it which does not have a zero element in that position.

This raises the question of whether there is any preference as to which row is exchanged with the one with the zero pivot element. It can be shown that greatest accuracy is attained when the pivot element has the greatest magnitude. Therefore, the row with a zero pivot element should be exchanged with the row below it which has the largest (in absolute value) element in that same column.

The preceding also suggests that such interchanges may improve accuracy even when the pivot element is not zero. A procedure to carry this out would involve searching all elements in the first column for the largest one and then exchanging the first row with the row containing that element. Then proceed with the elimination. When the second row becomes the pivot row, search the second column, from the second row down, for the greatest element and exchange the second row with that row, etc. Such a procedure will not only eliminate zero pivot elements, but will also probably increase overall computing accuracy.

7.6 SOLUTION OF SIMULTANEOUS EQUATIONS BY ITERATION

In Chap. 6, we saw that one method of finding the root of an equation was by the use of iteration. This method can also be applied for sets of simultaneous equations. Iteration can be used on both linear and nonlinear sets of equations. Iteration may be the *only* way that a set of nonlinear equations can be solved. Consider the equations

$$\begin{aligned} x_1 + \sin x_2 &= 1.5471976 \\ \cos x_1 - \quad x_2 &= -0.0235988 \end{aligned} \tag{7.54}$$

These equations cannot be solved by the means discussed in previous sections. They can be rewritten in the form

$$\begin{aligned} x_1 &= 1.5471976 - \sin x_2 \\ x_2 &= \cos x_1 + 0.0235988 \end{aligned} \tag{7.55}$$

and then by guessing an x_2 and x_1 solve, respectively, for x_1 and x_2; then these values can be used to get a new x_1 and x_2, etc. This sequence may converge to a set of answers of the desired accuracy. This method is called the *total step iteration*, since at each iteration, estimates for all unknowns are used to solve for a complete new set of values for the unknowns. This method is also known as *Jacobi iteration* or *simple iteration.*

Starting with the estimates $x_1 = 1.$ and $x_2 = 1.$, Eqs. (7.55) converge to the solutions $x_1 = 1.0471976$ and $x_2 = 0.5235988$ after 100 iterations by the total step iteration.

Since the values of the unknowns must be computed one at a time, it seems reasonable that the most recently calculated values should be used immediately in subsequent calculations. This version is called the *Gauss–Seidel iteration.* Applying this method to the foregoing example yields the same results in 48 iterations. Although the Gauss–Seidel iteration usually converges more rapidly than the total step iteration, it may not converge more rapidly and, in fact, in some cases diverges when total step converges.

In the case of a set of linear equations such as

$$\begin{aligned} a_{11}x_1 + a_{12}x_2 + a_{13}x_3 + \cdots + a_{1n}x_n &= c_1 \\ a_{21}x_1 + a_{22}x_2 + a_{23}x_3 + \cdots + a_{2n}x_n &= c_2 \\ a_{31}x_1 + a_{32}x_2 + a_{33}x_3 + \cdots + a_{3n}x_n &= c_3 \\ \vdots \qquad \vdots \qquad \qquad \vdots \quad &\quad \vdots \\ a_{n1}x_1 + a_{n2}x_2 + a_{n3}x_3 + \cdots + a_{nn}x_n &= c_n \end{aligned} \tag{7.56}$$

we can solve each equation for a different unknown. We will see in Sec. 7.6.2 that it is usually best to solve for the unknown that is on the diagonal (coefficient is of the form a_{kk}). Solving each equation for the unknown on the diagonal results in

$$\begin{aligned} x_1 &= (c_1 - a_{12}x_2 - a_{13}x_3 - \cdots - a_{1n}x_n)/a_{11} \\ x_2 &= (c_2 - a_{21}x_1 - a_{23}x_3 - \cdots - a_{2n}x_n)/a_{22} \\ x_3 &= (c_3 - a_{31}x_1 - a_{32}x_2 - \cdots - a_{3n}x_n)/a_{33} \\ &\vdots \\ x_n &= (c_n - a_{n1}x_1 - a_{n2}x_2 - a_{n3}x_3 - \cdots)/a_{nn} \end{aligned} \tag{7.57}$$

Rewriting these in iterative form using a superscript to indicate the number of the iteration, gives, for total step method,

$$\begin{aligned} x_1^{(m+1)} &= (c_1 - a_{12}x_2^{(m)} - a_{13}x_3^{(m)}) - \cdots - a_{1n}x_n^{(m)})/a_{11} \\ x_2^{(m+1)} &= (c_2 - a_{21}x_1^{(m)} - a_{23}x_3^{(m)} - \cdots - a_{2n}x_n^{(m)})/a_{22} \\ y_3^{(m+1)} &= (c_3 - a_{31}x_1^{(m)} - a_{32}x_2^{(m)} - \cdots - a_{3n}x_n^{(m)})/a_{33} \\ &\vdots \\ x_n^{(m+1)} &= (c_n - a_{n1}x_1^{(m)} - a_{n2}x_2^{(m)} - a_{n3}x_3^{(m)} - \cdots)/a_{nn} \end{aligned} \tag{7.58}$$

and for the Gauss–Seidel method,

$$
\begin{aligned}
x_1^{(m+1)} &= (c_1 - a_{12}x_2^{(m)} - a_{13}x_3^{(m)} - \cdots - a_{1n}x_n^{(m)})/a_{11} \\
x_2^{(m+1)} &= (c_2 - a_{21}x_1^{(m+1)} - a_{23}x_3^{(m)} - \cdots - a_{2n}x_n^{(m)})/a_{22} \\
x_3^{(m+1)} &= (c_3 - a_{31}x_1^{(m+1)} - a_{32}x_2^{(m+1)} - \cdots - a_{3n}x_n^{(m)})/a_{33} \\
&\vdots \\
x_n^{(m+1)} &= (c_n - a_{n1}x_1^{(m+1)} - a_{n2}x_2^{(m+1)} - a_{n3}x_3^{(m+1)} - \cdots)/a_{nn}
\end{aligned}
\tag{7.59}
$$

7.6.1 Convergence of the Iteration Method

Consider n equations of the form

$$x_i = F_i(x_1, x_2, \ldots, x_n) \tag{7.60}$$

To prove those conditions sufficient for this system to converge is not simple and will not be presented here. However, there are several conditions any one of which may be shown to be sufficient. These are as follows. All n inequalities of the form

$$\sum_{i=1}^{n} \left|\frac{\partial F_i}{\partial x_j}\right| < 1 \qquad j = 1, 2, \ldots, n \tag{7.61}$$

must be satisfied; or all n inequalities of the form

$$\sum_{j=1}^{n} \left|\frac{\partial F_i}{\partial x_j}\right| < 1 \qquad i = 1, 2, \ldots, n \tag{7.62}$$

must be satisfied; or

$$\sum_{i=1}^{n} \sum_{j=1}^{n} \left|\frac{\partial F_i}{\partial x_j}\right| < 1 \tag{7.63}$$

These conditions are analagous to condition (6.50) and also must be satisfied at the solution and in the neighborhood of the solution.

Each of the preceding conditions is probably more stringent than need be. In fact, we can see that if either conditions (7.61) or (7.62) are satisfied, then the left side of condition (7.63) will be less than n. Therefore, condition (7.63) imposes a more stringent requirement than either of the conditions (7.61) or (7.62).

We saw that Eqs. (7.55) converged to a solution. This could have been predicted by the following analysis. Taking partial derivatives,

$$
\begin{aligned}
&\frac{\partial F_1}{\partial x_1} = 0; \qquad \frac{\partial F_1}{\partial x_2} = -\cos x_2; \\
&\frac{\partial F_2}{\partial x_1} = -\sin x_1; \qquad \frac{\partial F_2}{\partial x_2} = 0
\end{aligned}
\tag{7.64}
$$

Substituting these values in the left side of conditions (7.61), we have

$$
\begin{aligned}
&\left|\frac{\partial F_1}{\partial x_1}\right| + \left|\frac{\partial F_2}{\partial x_1}\right| = |\sin x_1| \\
&\left|\frac{\partial F_1}{\partial x_2}\right| + \left|\frac{\partial F_2}{\partial x_2}\right| = |\cos x_2|
\end{aligned}
\tag{7.65}
$$

Likewise, substituting in the left side of conditions (7.62),

$$\left|\frac{\partial F_1}{\partial x_1}\right| + \left|\frac{\partial F_1}{\partial x_2}\right| = |\cos x_2|$$
$$\left|\frac{\partial F_2}{\partial x_1}\right| + \left|\frac{\partial F_2}{\partial x_2}\right| = |\sin x_1| \tag{7.66}$$

The right-hand sides of (7.65) and (7.66) can never exceed 1; and for the example given, $0.7 \leqq x_1 \leqq \pi/3$ and $\pi/6 \leqq x_2 \leqq 1$. Therefore, $\sin x_1 \leqq 0.866$ and $\cos x_2 \leqq 0.866$ and hence conditions (7.61) and (7.62) are satisfied.

The iteration of the example was relatively slow, which is consistent with the large value (0.866) of the sums of the partial derivatives. Although the conditions given are sufficient, the iteration will converge most rapidly if the sums of the partial derivatives are near zero.

7.6.2 Convergence Criteria for Linear Equations

If a set of linear equations such as (7.56) is rewritten in the form (7.57) and the partial derivatives are found, they will be of the form

$$\frac{\partial F_i}{\partial x_j} = \frac{-a_{ij}}{a_{ii}} \qquad i \neq j$$
$$\frac{\partial F_i}{\partial x_j} = 0 \qquad i = j \tag{7.67}$$

These expressions can be substituted in any of the conditions (7.61), (7.62), or (7.63) to find the special criteria for the convergence of the iteration method for linear equations. We will only do it for the most useful ones, namely, conditions (7.62). Substituting (7.67) into conditions (7.62) and simplifying,

$$\sum_{j=1}^{n} |a_{ij}| < |a_{ii}| \qquad \begin{matrix} i = 1, 2, \ldots, n \\ j \neq i \end{matrix} \tag{7.68}$$

This tells us that in a set of equations such as (7.56), if the absolute value of the coefficient of the diagonal element a_{ii} for each equation is greater than the sum of the absolute values of the coefficients of all other elements in that equation, the iteration method will converge to a solution.

Earlier, we had said that a system of equations must have a strong diagonal; the preceding statement expresses this more definitively.

7.7 NEWTON'S ITERATION FOR SIMULTANEOUS EQUATIONS

The methods discussed in Chap. 6 suggest that similar methods can be applied in the case of simultaneous equations. One of these methods is Newton's iteration. To see how it is applied to simultaneous equations, let us first consider only two equations in two unknowns.

$$F_1(x_1, x_2) = 0$$
$$F_2(x_1, x_2) = 0 \tag{7.69}$$

Now if x_1 and x_2 are the solutions and $x_1^{(1)}$ and $x_2^{(1)}$ are some initial estimates, then

$$\begin{aligned} x_1 &= x_1^{(1)} + \varepsilon_1 \\ y_2 &= x_2^{(1)} + \varepsilon_2 \end{aligned} \tag{7.70}$$

Substituting Eqs. (7.70) into (7.69),

$$\begin{aligned} F_1(x_1^{(1)} + \varepsilon_1, x_2^{(1)} + \varepsilon_2) &= 0 \\ F_2(x_1^{(1)} + \varepsilon_1, x_2^{(1)} + \varepsilon_2) &= 0 \end{aligned} \tag{7.71}$$

These equations can be expanded using Taylor's series. Ignoring terms that are of order greater than one, we obtain the approximate relations

$$\begin{aligned} F_1(x_1^{(1)}, x_2^{(1)}) + \varepsilon_1 \left(\frac{\partial F_1}{\partial x_1}\right)^{(1)} + \varepsilon_2 \left(\frac{\partial F_1}{\partial y_2}\right)^{(1)} &\doteq 0 \\ F_2(x_1^{(1)}, x_2^{(1)}) + \varepsilon_1 \left(\frac{\partial F_2}{\partial x_1}\right)^{(1)} + \varepsilon_2 \left(\frac{\partial F_2}{\partial x_2}\right)^{(1)} &\doteq 0 \end{aligned} \tag{7.72}$$

(Note that $(\partial F/\partial x)^{(m)}$ means the partial derivative evaluated with the mth approximants of the unknowns $x_1, x_2, \ldots, x_n$). There must exist some ε's, say $\varepsilon_1^{(2)}$ and $\varepsilon_2^{(2)}$ such that Eqs. (7.72) are no longer approximate.

$$\begin{aligned} F_1(x_1^{(1)}, x_2^{(1)}) + \varepsilon_1^{(2)} \left(\frac{\partial F_1}{\partial x_1}\right)^{(1)} + \varepsilon_2^{(2)} \left(\frac{\partial F_1}{\partial x_2}\right)^{(1)} &= 0 \\ F_2(x_1^{(1)}, x_2^{(1)}) + \varepsilon_1^{(2)} \left(\frac{\partial F_2}{\partial x_1}\right)^{(1)} + \varepsilon_2^{(2)} \left(\frac{\partial F_2}{\partial x_2}\right)^{(1)} &= 0 \end{aligned} \tag{7.73}$$

These equations are linear in $\varepsilon_1^{(2)}$ and $\varepsilon_2^{(2)}$ and can be solved by any of the previous methods given. However, the sums $x_1^{(1)} + \varepsilon_1^{(2)}$ and $x_2^{(1)} + \varepsilon_2^{(2)}$ will not necessarily be the solutions x_1 and x_2, but we hope they will be a better approximation than $x_1^{(1)}$ and $x_2^{(1)}$. Therefore, we write

$$\begin{aligned} x_1^{(2)} &= x_1^{(1)} + \varepsilon_1^{(2)} \\ x_2^{(2)} &= x_2^{(1)} + \varepsilon_2^{(2)} \end{aligned} \tag{7.74}$$

and these values can be substituted into equations similar to (7.73) to obtain yet a better approximation. This method can easily be extended to n equations, and the general form becomes

$$\sum_{j=1}^{n} \varepsilon_j^{(m+1)} \left(\frac{\partial F_i}{\partial x_j}\right)^{(m)} = -F_i(x_1^{(m)}, x_2^{(m)}, \ldots, x_n^{(m)}) \tag{7.75}$$

$$x_i^{(m+1)} = x_i^{(m)} + \varepsilon_i^{(m+1)} \tag{7.76}$$

where $i = 1, 2, \ldots, n$.

Since the foregoing method leads to a new set of simultaneous equations, Eq. (7.75), it may appear that there is no advantage to this method. However, there are two very distinct advantages.

1. If the original equations are nonlinear, this method provides a means for linearizing the solution.

2. Often, in the solution of systems of equations, the numbers and types of numerical operations involved are such that the direct methods of solution do not yield accurate results. The results obtained can usually be improved by iteration.

7.7.1 Example Using Newton's Iteration

Equations (7.75) and (7.76) can be applied to the example of Sec. 7.6 [Eqs. (7.54)]. Substituting and solving the system of two equations in ε_1 and ε_2:

$$\begin{aligned} x_1^{(m+1)} &= x_1^{(m)} + (-F(x_1^{(m)}, x_2^{(m)}) \\ &\quad + F_2(x_1^{(m)}, x_2^{(m)}) \cos x_2^{(m)})/(1 - \sin x_1^{(m)} \cos x_2^{(m)}) \\ x_2^{(m+1)} &= x_2^{(m)} + (-F_2(x_1^{(m)}, x_2^{(m)}) \\ &\quad + F_1(x_1^{(m)}, x_2^{(m)}) \sin x_1^{(m)})/(1 - \sin x_1^{(m)} \cos x_2^{(m)}) \end{aligned} \tag{7.77}$$

[Note that F_1 and F_2 are modifications of Eqs. (7.54) to the form of (7.69)]. Equations (7.77) are written for total step iteration, but can easily be modified for Gauss–Seidel iteration. Table 7.2 gives the results of the above iteration. It is apparent that this is a great improvement over the 48 iterations required previously.

TABLE 7.2 RESULTS OF AN EXAMPLE USING NEWTON'S ITERATION FOR TWO NONLINEAR EQUATIONS

x_1	x_2
1.0000000	1.0000000
0.8924580	0.6543945
1.0198031	0.5519459
1.0458418	0.5249474
1.0471942	0.5236021
1.0471976	0.5235987
1.0471976	0.5235987

7.8 MATRICES

Matrix operations can be employed in the solution of simultaneous linear equations. In this section, we will discuss some matrix fundamentals and their application to solving systems of equations. There are other numerical applications for matrices, but they will not be covered in this text.

7.8.1 Definitions

A *matrix* is a rectangular array of numbers. These numbers are called *elements* of the matrix and can be expressed literally using a name with two subscripts. The subscripts are used to specify the position of the element in the array. The first subscript is used to identify the *row* of the element and the second subscript, the *column* of the element. The array of elements is usually enclosed in square brackets, parentheses, or sometimes double bars to identify it as a matrix rather than as a determinant. A matrix can be referred to by a single name, but it must be remembered that this name refers to an array of numbers which usually does not have a single value. A name of an entire matrix is usually shown in bold face or is underlined to denote its special properties. Equation (7.78) is an example of an $m \times n$ matrix written literally.

$$\mathbf{A} = \begin{bmatrix} a_{11} & a_{12} & a_{13} & \cdots & a_{1n} \\ a_{21} & a_{22} & a_{23} & \cdots & a_{2n} \\ a_{31} & a_{32} & a_{33} & \cdots & a_{3n} \\ \vdots & \vdots & \vdots & & \vdots \\ a_{m1} & a_{m2} & a_{m3} & \cdots & a_{mn} \end{bmatrix} \tag{7.78}$$

In the special cases where either m or n is 1, the matrix is called a *vector.* If $m = 1$, the matrix is a *row* vector such as

$$\mathbf{B} = [b_{11} \quad b_{12} \quad b_{13} \quad \cdots \quad b_{1n}] \tag{7.79}$$

Sometimes the row subscript is omitted. If $n = 1$, the matrix is a *column* vector. Equation (7.80) is an example of a column vector.

$$\mathbf{C} = \begin{bmatrix} c_{11} \\ c_{21} \\ c_{31} \\ \vdots \\ c_{n1} \end{bmatrix} \tag{7.80}$$

In this case, the column subscript can be omitted.

There are other special matrices that are defined as follows:

1. *Square*—a matrix with the same number of rows and columns. In such a matrix we can define the *principal diagonal* as being those elements with equal subscripts (i.e., of the form a_{ii}) or those elements on the diagonal from the upper left corner to the lower right corner.
2. *Symmetric*—a square matrix in which there is symmetry about the principal diagonal (i.e., $a_{ij} = a_{ji}$).
3. *Skew Symmetric*—a square matrix with symmetry about the principal diagonal such that the elements on one side of the diagonal are the negatives of those on the other side of the diagonal (i.e., $a_{ij} = -a_{ji}$ for all $i \neq j$).
4. *Triangular*—a square matrix such that all elements on one side of the principal diagonal are equal to zero (i.e., either $a_{ij} = 0$ for all elements with $i < j$ or $a_{ij} = 0$ for all elements with $i > j$).
5. *Diagonal*—a square matrix with all elements not on the principal diagonal equal to zero (i.e., $a_{ij} = 0$ for all $i \neq j$).
6. *Scalar*—a diagonal matrix with all diagonal elements equal to the same constant (i.e., $a_{ij} = 0$ for all $i \neq j$ *and* $a_{ij} = k$ for $i = j$).
7. *Identity*—a scalar matrix with all diagonal elements equal to one. It is also called the *unit* matrix. This special matrix is usually identified with the letter **I**.
8. *Unit Vector*—a column or row matrix with one element equal to 1 and all other elements equal to zero.

Examples of these special matrices are shown in Fig. 7.6.

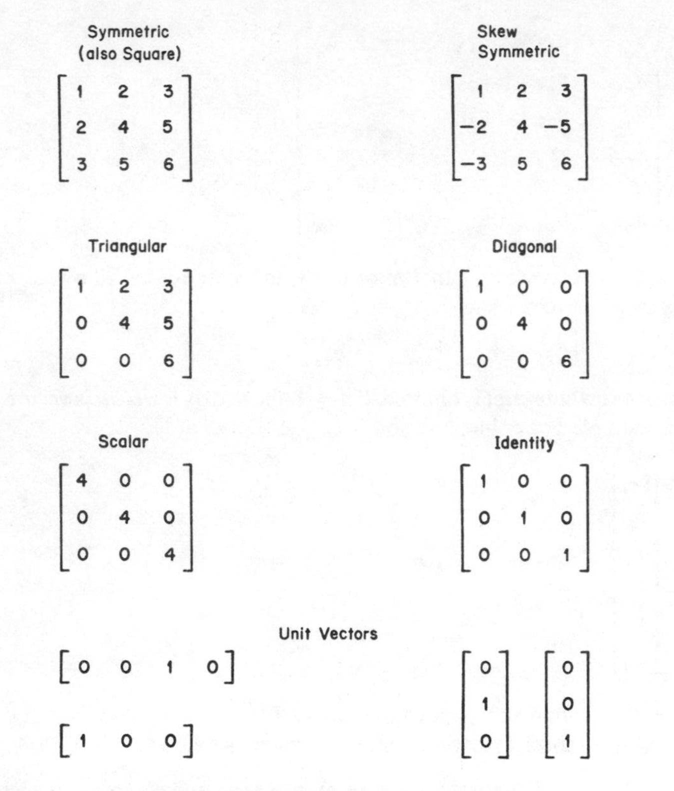

Fig. 7.6. Examples of special matrices.

7.8.2 Matrix Operations

Subsequent sections will discuss the elementary matrix operations of addition (and subtraction), multiplication, and inversion. Each of these operations can be represented algebraically in short form as follows:

$$\begin{aligned} \mathbf{A} &= \mathbf{B} + \mathbf{C} \\ \mathbf{A} &= \mathbf{B} \times \mathbf{C} \\ \mathbf{A} &= \mathbf{B}^{-1} \end{aligned} \tag{7.81}$$

However, it must be understood that Eqs. (7.81) cannot be interpreted in the same way as ordinary algebraic equations, since each of the letters represents an array of numbers. These equations imply rather that some sequence of operations are to be performed on the elements of **B** (and in the first two equations also **C**) to obtain the elements for **A**.

7.8.3 Matrix Addition

In order to add two matrices, both matrices must have the same dimensions. The resulting matrix will also have these same dimensions. The rules for matrix addition are simple. To perform the addition $\mathbf{A} = \mathbf{B} + \mathbf{C}$, the individual elements of **A** are found from the elements of **B** and **C** by the relation

$$a_{ij} = b_{ij} + c_{ij} \tag{7.82}$$

It is easy to see from this relation that for each element b, there must be a corresponding element c and vice versa. Also it can be shown by Eq. (7.82) that matrix addition is *associative*

$$(\mathbf{A} + \mathbf{B}) + \mathbf{C} = \mathbf{A} + (\mathbf{B} + \mathbf{C})$$

and *commutative*

$$\mathbf{A} + \mathbf{B} = \mathbf{B} + \mathbf{A}$$

Matrix subtraction is just a special case of matrix addition. For the operation $\mathbf{A} = \mathbf{B} - \mathbf{C}$, the following operation on the elements is performed.

$$a_{ij} = b_{ij} - c_{ij}$$

Matrix subtraction is also associative in the same sense as in ordinary algebra.

$$(\mathbf{A} - \mathbf{B}) + \mathbf{C} = \mathbf{A} + (\mathbf{C} - \mathbf{B})$$

However, it is *not* commutative.

$$\mathbf{A} - \mathbf{B} \neq \mathbf{B} - \mathbf{A}$$

7.8.4 Matrix Multiplication

Matrix multiplication is not quite as simple as addition, but it is still a straightforward operation. The equation $\mathbf{A} = \mathbf{B} \times \mathbf{C}$ implies the following operation on the elements of **B** and **C** to yield **A**.

$$a_{ij} = \sum_{k=1}^{n} b_{ik}c_{kj} \tag{7.83}$$

In the case of multiplication, the dimensions of the matrices **A**, **B**, and **C** need not be the same, but **B** must have the same number of columns as **C** has rows. The resulting matrix **A** will have the same number of rows as **B** and the same number of columns as **C**. Stating this in another way, if **B** is an $m \times n$ matrix and **C** is an $n \times p$ matrix, then **A** will be an $m \times p$ matrix.

The rule for matrix multiplication can be verbally stated as follows:

To obtain the ijth element of the result matrix, multiply each element in the ith row of the first matrix by the corresponding element in the jth column of the second matrix and sum all of these products.

Matrix multiplication is associative.

$$\mathbf{A} \times (\mathbf{B} \times \mathbf{C}) = (\mathbf{A} \times \mathbf{B}) \times \mathbf{C}$$

However, matrix multiplication is not commutative.

$$\mathbf{A} \times \mathbf{B} \neq \mathbf{B} \times \mathbf{A}$$

$$\begin{bmatrix} -1 & -2 \\ 1 & -1 \\ 2 & 3 \end{bmatrix} \times \begin{bmatrix} 2 & 1 & -1 & -2 \\ 1 & 2 & 3 & -4 \end{bmatrix} = \begin{bmatrix} -4 & -5 & -5 & 10 \\ 1 & -1 & -4 & 2 \\ 7 & 8 & 7 & -16 \end{bmatrix}$$

Fig. 7.7. Example of matrix multiplication.

To differentiate between the preceding two operations, we say that $\mathbf{A} \times \mathbf{B}$ means either matrix **A** is *post-multiplied* by **B** or matrix **B** is *pre-multiplied* by **A**; while $\mathbf{B} \times \mathbf{A}$ means either matrix **B** is post-multiplied by **A** or matrix **A** is pre-multiplied by **B**. An example of matrix multiplication is shown in Fig. 7.7.

For the special case of the product **A** of a scalar matrix **B** multiplied (either pre- or post-) by another matrix **C**, it may be shown that

$$a_{ij} = bc_{ij} \tag{7.84}$$

(Note that in the case of a scalar matrix, $b_{ij} = 0$ for $i \neq j$ and $b_{ij} = b$ (a constant) for $i = j$). It has been implied here that multiplication by a scalar matrix *is* commutative. That is,

$$\mathbf{B} \times \mathbf{C} = \mathbf{C} \times \mathbf{B}$$

if either **B** or **C** is a scalar matrix. For clarity, scalar matrices are usually indicated by the use of lower-case letters so the commutative property is shown by

$$b\mathbf{C} = \mathbf{C}b$$

For the special scalar matrix **I**, we can also write

$$\mathbf{I} \times \mathbf{C} = \mathbf{C} \times \mathbf{I} = \mathbf{C}$$

7.8.5 Matrix Inversion

Matrix division is not defined in the same sense as algebraic division. However, the matrix counterpart of division can be achieved through the use of matrix multiplication by the inverse of a matrix. The inverse of a matrix is indicated by the use of an exponent of -1. The relation

$$\mathbf{A} = \mathbf{B}^{-1}$$

means **A** is the inverse of **B**.

The operations on the elements of a matrix to find its inverse cannot be stated explicitly as was done for matrix addition and multiplication. Rather, we must define these operations implicitly by the fundamental definition

$$\mathbf{B} \times \mathbf{B}^{-1} = \mathbf{I}$$

In order to be able to perform this multiplication, **B** must be a square matrix. Also, we can show that the multiplication of a matrix by its own inverse is commutative.

$$\mathbf{B} \times \mathbf{B}^{-1} = \mathbf{B}^{-1} \times \mathbf{B} = \mathbf{I}$$

$$\begin{bmatrix} a_{11} & a_{12} & a_{13} \\ a_{21} & a_{22} & a_{23} \\ a_{31} & a_{32} & a_{33} \end{bmatrix} \times \begin{bmatrix} b_{11} & b_{12} & b_{13} \\ b_{21} & b_{22} & b_{23} \\ b_{31} & b_{32} & b_{33} \end{bmatrix} = \begin{bmatrix} c_{11} & c_{12} & c_{13} \\ c_{21} & c_{22} & c_{23} \\ c_{31} & c_{32} & c_{33} \end{bmatrix}$$

Fig. 7.8. Literal example of matrix multiplication.

$$\begin{bmatrix} a_{11} & a_{12} & a_{13} \\ a_{21} & a_{22} & a_{23} \\ a_{31} & a_{32} & a_{33} \end{bmatrix} \times \begin{bmatrix} b_{11} \\ b_{21} \\ b_{31} \end{bmatrix} = \begin{bmatrix} c_{11} \\ c_{21} \\ c_{31} \end{bmatrix}$$

$$\begin{bmatrix} a_{11} & a_{12} & a_{13} \\ a_{21} & a_{22} & a_{23} \\ a_{31} & a_{32} & a_{33} \end{bmatrix} \times \begin{bmatrix} b_{12} \\ b_{22} \\ b_{32} \end{bmatrix} = \begin{bmatrix} c_{12} \\ c_{22} \\ c_{32} \end{bmatrix}$$

$$\begin{bmatrix} a_{11} & a_{12} & a_{13} \\ a_{21} & a_{22} & a_{23} \\ a_{31} & a_{32} & a_{33} \end{bmatrix} \times \begin{bmatrix} b_{13} \\ b_{23} \\ b_{33} \end{bmatrix} = \begin{bmatrix} c_{13} \\ c_{23} \\ c_{33} \end{bmatrix}$$

Fig. 7.9. Matrix multiplication shown as a set of matrix–vector multiplications.

$$\begin{bmatrix} a_{11} & a_{12} & a_{13} \\ a_{21} & a_{22} & a_{23} \\ a_{31} & a_{32} & a_{33} \end{bmatrix} \times \begin{bmatrix} b_{11} & b_{12} & b_{13} \\ b_{21} & b_{22} & b_{23} \\ b_{31} & b_{32} & b_{33} \end{bmatrix} = \begin{bmatrix} 1 & 0 & 0 \\ 0 & 1 & 0 \\ 0 & 0 & 1 \end{bmatrix}$$

Fig. 7.10. Product of matrix and its inverse.

To understand the operations necessary to find the inverse of a matrix, we must examine more closely the process of matrix multiplication. The product of two matrices can be found directly, as shown in Fig. 7.8, or by three separate matrix–vector multiplications, as shown in Fig. 7.9. If in the example of Fig. 7.8 we know the a's and the c's and wish to find the b's, this can be accomplished by writing the multiplication as in Fig. 7.9. The multiplications shown in Fig. 7.9 can be interpreted as three sets of three equations in three unknowns. The first multiplication, for example, can be written as

$$\begin{aligned} a_{11}b_{11} + a_{12}b_{21} + a_{13}b_{31} &= c_{11} \\ a_{21}b_{11} + a_{22}b_{21} + a_{23}b_{31} &= c_{21} \\ a_{31}b_{11} + a_{32}b_{21} + a_{33}b_{31} &= c_{31} \end{aligned} \tag{7.85}$$

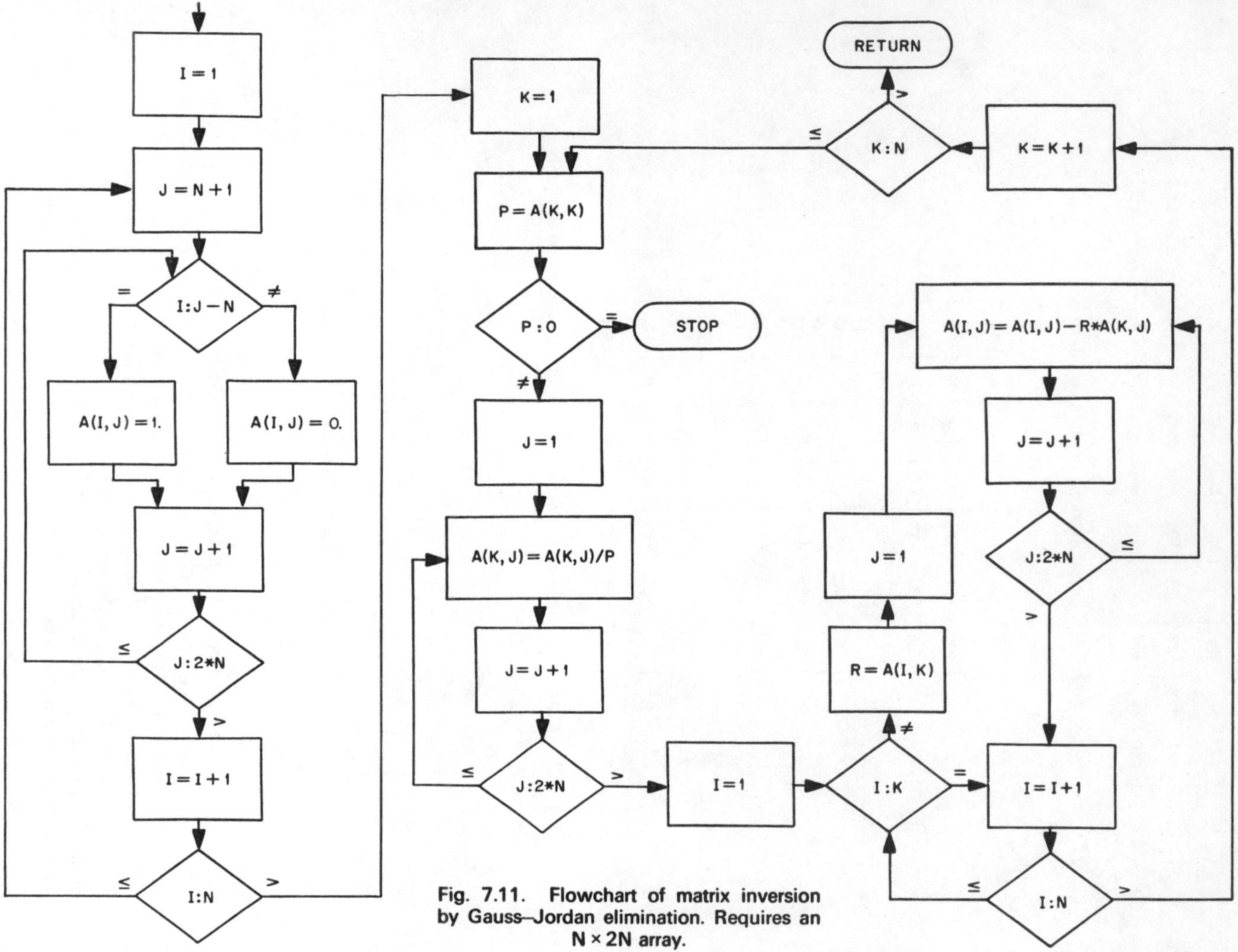

Fig. 7.11. Flowchart of matrix inversion by Gauss–Jordan elimination. Requires an N × 2N array.

If the a's and the c's are known, then Eqs. (7.85) can be solved by any of the previously given methods to obtain b_{11}, b_{21}, and b_{31}. By similar procedures, the other b's can also be found.

The point of the foregoing discussion is that when we wish to find the inverse of a matrix, the problem is similar to the one just described. We know the matrix **A** and we know the product of it and its inverse, namely **I**. We therefore can find the inverse by solving a set of linear equations. For a third-order matrix where $\mathbf{B} = \mathbf{A}^{-1}$, we may write the product shown in Fig. 7.10. The b's of Fig. 7.10 can be found in the same manner (write three sets of three equations in three unknowns) as shown in the example of Fig. 7.8.

To illustrate how this may be done by the Gauss–Jordan elimination method, let us consider that method. In Fig. 7.5 we assumed that an N X(N + 1) array A was stored in computer memory, where A(I, N + 1) represented the constant vector of our original equations. In the elimination process, the original array of coefficients is modified to an array containing zeros, except along the diagonal where the elements all become 1. The constant terms, A(I, N + 1) for I = 1, 2, . . . , N, eventually are modified until they become the sought for solutions. There is no reason that the method cannot be expanded to contain several constant vectors so that several solutions can be worked on concurrently. Figure 7.11 is a modification of Fig. 7.5 for a matrix inversion by Gauss–Jordan elimination.

Original matrix with auxiliary identity matrix:

$$\begin{bmatrix} 60 & 30 & 20 & 1 & 0 & 0 \\ 30 & 20 & 15 & 0 & 1 & 0 \\ 20 & 15 & 12 & 0 & 0 & 1 \end{bmatrix}$$

Using row 1 of the above matrix as pivot row gives:

$$\begin{bmatrix} 1 & 1/2 & 1/3 & 1/60 & 0 & 0 \\ 0 & 5 & 5 & -1/2 & 1 & 0 \\ 0 & 5 & 16/3 & -1/3 & 0 & 1 \end{bmatrix}$$

Using row 2 of the above matrix as pivot row gives:

$$\begin{bmatrix} 1 & 0 & -1/6 & 1/15 & -1/10 & 0 \\ 0 & 1 & 1 & -1/10 & 1/5 & 0 \\ 0 & 0 & 1/3 & 1/6 & -1 & 1 \end{bmatrix}$$

Using row 3 of the above matrix as pivot row yields the final result

$$\begin{bmatrix} 1 & 0 & 0 & 3/20 & -3/5 & 1/2 \\ 0 & 1 & 0 & -3/5 & 16/5 & -3 \\ 0 & 0 & 1 & 1/2 & -3 & 3 \end{bmatrix}$$

where the increase is contained in the last three columns.

Fig. 7.12. Example of matrix inversion by Gauss–Jordan elimination.

In Fig. 7.11, it is assumed that the matrix A is already in memory and it is permissible to destroy it in the process of inversion. The first two loops of the flowchart generate an $N \times N$ identity matrix in an extension of the A array. This extension means that an $N \times 2N$ array must be reserved for this process. The rest of the flowchart is essentially the same as Fig. 7.5.

Figure 7.12 is an example of the steps taken in the inversion of a 3×3 matrix by the Gauss–Jordan elimination method.

7.8.6 Economization of Matrix Inversion

It was noted earlier that the Gauss–Jordan elimination method for solving simultaneous linear equations could be made more efficient. In this section, we will present some of the techniques that are used to improve the Gauss–Jordan matrix inversion method.

It was noted in the previous section that the method of Fig. 7.11 requires $2N^2$ storage locations just for the array that is being manipulated. Furthermore, the procedure performs calculations that are unnecessary because we already know the answers, namely 1. or 0.; and furthermore, we will not use these answers. However, the method given is important in presenting a clear picture of how this method functions. It is possible and desirable to eliminate superfluous computations and decrease the amount of storage. Figure 7.13 is a flowchart of the Gauss–Jordan matrix inversion method that primarily eliminates superfluous computations, but also decreases storage requirements to $N \times (N + 1)$.

The method of Fig. 7.13 carried only one column vector in addition to the matrix being inverted. Referring to Fig. 7.12, we see that at the first stage of the elimination, the last two columns were not altered; therefore, we did not need them. In this method, the superfluous columns of the identity matrix are not carried in the computation. Each column of the identity matrix is generated as needed and appended to the matrix being manipulated. Furthermore, rather than alter the leading columns to ones and zeros, those computations are omitted and the matrix is moved over one column at each stage of the elimination.

Since we also know at each stage of the elimination the values of the $(N + 1)$th column of our array, it is possible to modify the method of Fig. 7.13 to eliminate even that column. Figure 7.14 gives this modification. The improvement, if any, in computational speed and program storage of Fig. 7.14 over Fig. 7.13 is not significant; the primary gain is in saving N units of storage previously required for data.

7.8.7 Zero Pivot Element in Matrix Inversion

Some attention was given to the possibility of having a zero pivot element in solving simultaneous linear equations by elimination. The same difficulty can arise in matrix inversion. Essentially the same techniques suggested in Sec. 7.5.3 can also be used to avoid zero pivot elements and also to improve the accuracy of the results.

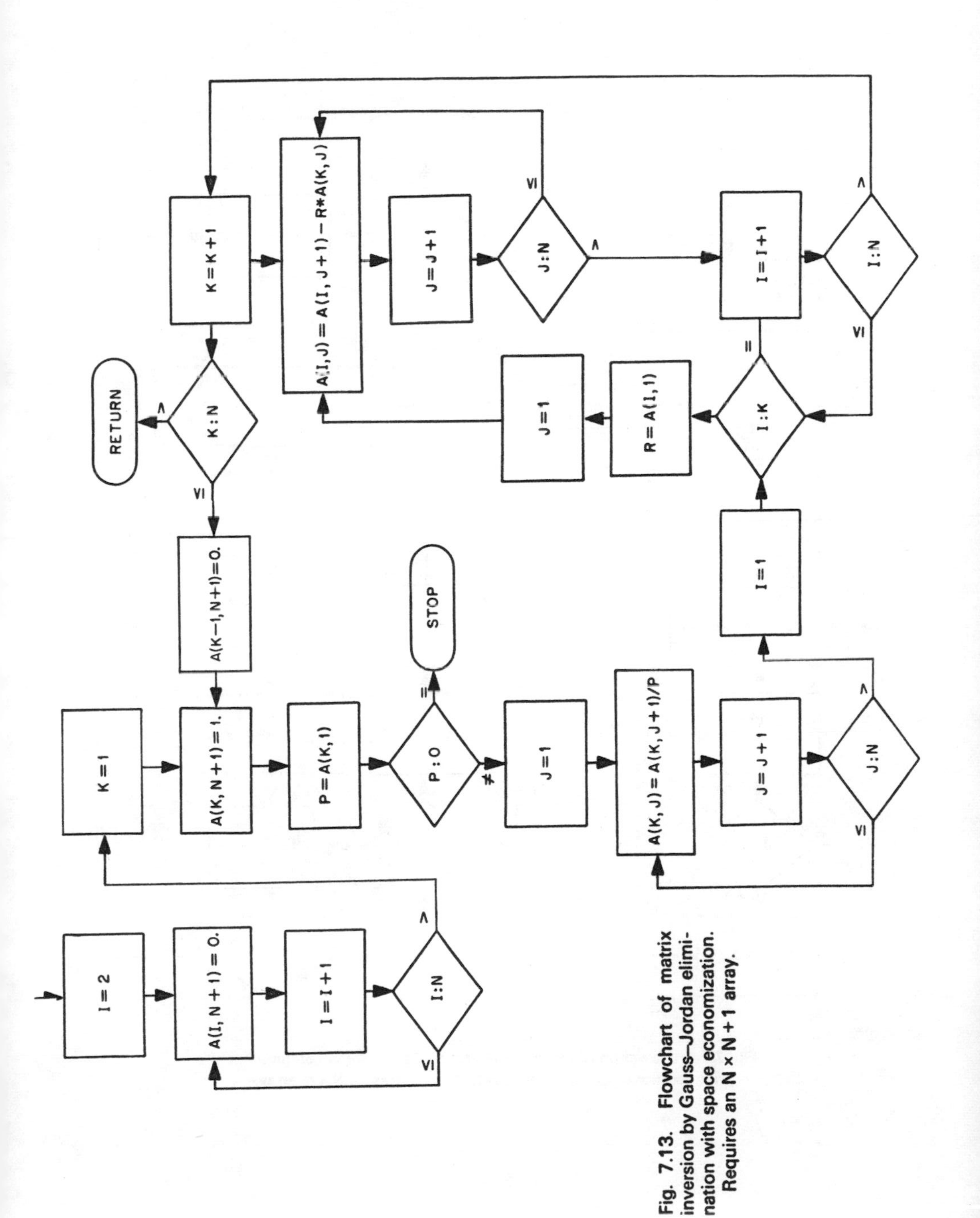

Fig. 7.13. Flowchart of matrix inversion by Gauss–Jordan elimination with space economization. Requires an N × N + 1 array.

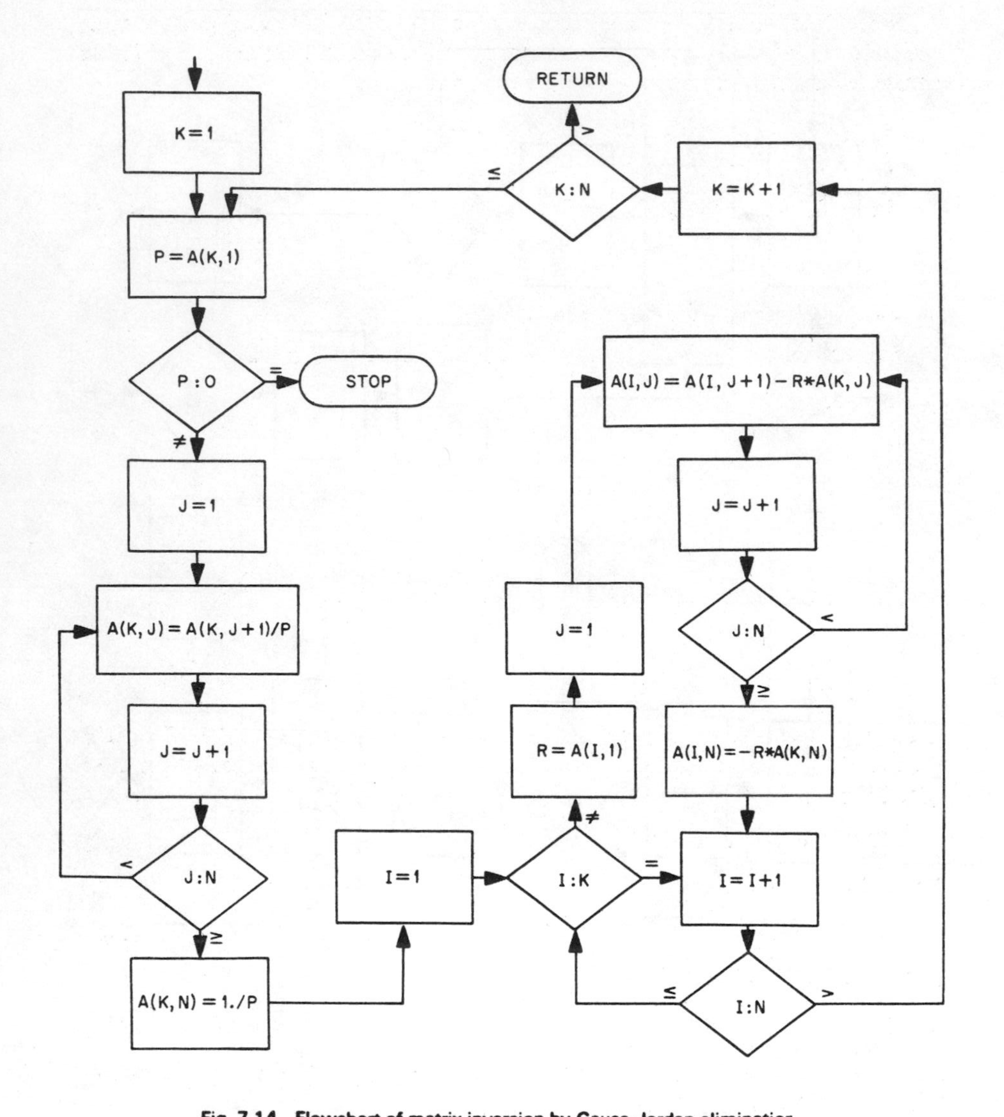

Fig. 7.14. Flowchart of matrix inversion by Gauss-Jordan eliminatior with additional space economization. Requires an N x N array.

7.9 MATRIX METHOD FOR SOLVING SIMULTANEOUS LINEAR EQUATIONS

In Sec. 7.8.5, we saw how an equation-solving method could be employed to invert a matrix. However, the point in inverting a matrix is usually to solve a set of linear equations. Consider the following set of equations:

$$\begin{aligned} a_{11}x_1 + a_{12}x_2 + a_{13}x_3 + \cdots + a_{1n}x_n &= c_1 \\ a_{21}x_1 + a_{22}x_2 + a_{23}x_3 + \cdots + a_{2n}x_n &= c_2 \\ a_{31}x_1 + a_{32}x_2 + a_{33}x_3 + \cdots + a_{3n}x_n &= c_3 \\ \vdots \quad\quad \vdots \quad\quad \vdots \quad\quad\quad\quad \vdots \quad &\quad \vdots \\ a_{n1}x_1 + a_{n2}x_2 + a_{n3}x_3 + \cdots + a_{nn}x_n &= c_n \end{aligned} \tag{7.86}$$

These equations can be represented more concisely in matrix form as

$$\mathbf{A} \times \mathbf{X} = \mathbf{C} \tag{7.87}$$

where

$$\mathbf{A} = \begin{bmatrix} a_{11} & a_{12} & a_{13} & \cdots & a_{1n} \\ a_{21} & a_{22} & a_{23} & \cdots & a_{2n} \\ a_{31} & a_{32} & a_{33} & \cdots & a_{3n} \\ \vdots & \vdots & \vdots & & \vdots \\ a_{n1} & a_{n2} & a_{n3} & \cdots & a_{nn} \end{bmatrix} \qquad \mathbf{X} = \begin{bmatrix} x_1 \\ x_2 \\ x_3 \\ \vdots \\ x_n \end{bmatrix} \qquad \mathbf{C} = \begin{bmatrix} c_1 \\ c_2 \\ c_3 \\ \vdots \\ c_n \end{bmatrix}$$

Our actual task is to find the column vector **X**. Since the quantity **AX** in Eq. (7.87) is a matrix, it can be multiplied by a matrix. Therefore, we can pre-multiply both sides of (7.87) by $\mathbf{A}^{-1}$.

$$\mathbf{A}^{-1}\mathbf{A}\mathbf{X} = \mathbf{A}^{-1}\mathbf{C} \tag{7.88}$$

But $\mathbf{A}^{-1}\mathbf{A} = \mathbf{I}$, and therefore,

$$\mathbf{I}\mathbf{X} = \mathbf{A}^{-1}\mathbf{C} \tag{7.89}$$

and $\mathbf{IX} = \mathbf{X}$ and we can write

$$\mathbf{X} = \mathbf{A}^{-1}\mathbf{C} \tag{7.90}$$

Equation (7.90) then tells us that if we know the inverse of **A**, we need only post-multiply it by **C** to obtain **X**.

Since, in order to invert a matrix, we must use a procedure similar to solving a set of equations, it would seem more efficient to solve the equations directly rather than to go through the matrix inversion. A particular advantage of solving the equations directly is that sometimes it is necessary to solve several sets of equations that have the same coefficient matrix **A**, but different constant vectors **C**. Therefore, the sets of equations

$$\begin{aligned} \mathbf{AX}_1 &= \mathbf{C}_1 \\ \mathbf{AX}_2 &= \mathbf{C}_2 \\ \mathbf{AX}_3 &= \mathbf{C}_3 \\ \vdots\;\; &\;\;\; \vdots \end{aligned} \tag{7.91}$$

can be solved by first finding $\mathbf{A}^{-1}$. Then the solutions for the individual vectors reduce to the matrix multiplications

$$\begin{aligned} \mathbf{X}_1 &= \mathbf{A}^{-1}\mathbf{C}_1 \\ \mathbf{X}_2 &= \mathbf{A}^{-1}\mathbf{C}_2 \\ \mathbf{X}_3 &= \mathbf{A}^{-1}\mathbf{C}_3 \\ \vdots\;\; &\;\;\; \vdots \end{aligned} \tag{7.92}$$

Equations (7.91) can be solved directly by the elimination method by carrying all constant vectors in an auxiliary array and operating on all of them in the elimination process. In fact, such a procedure usually takes fewer arithmetic operations than the matrix inversion method just described. However, the matrix inversion method has certain advantages over the direct elimination method:

1. Fewer memory locations are required for data.
2. The inverse can be found at one time and the constant vectors supplied individually at other times.
3. The inverse of the matrix is sometimes of interest in itself.

In the next section we will see an application where the inverse of a matrix is needed to solve a problem.

7.10 EIGENVALUES AND EIGENVECTORS

In some types of physical problems, primarily in problems involving free oscillation, simultaneous *homogeneous* equations of the form

$$\begin{aligned} a_{11}x_1 + a_{12}x_2 + \cdots + a_{1n}x_n &= \lambda x_1 \\ a_{21}x_1 + a_{22}x_2 + \cdots + a_{2n}x_n &= \lambda x_2 \\ \vdots \qquad \vdots \qquad\quad & \quad \vdots \qquad \vdots \\ a_{n1}x_1 + a_{n2}x_2 + \cdots + a_{nn}x_n &= \lambda x_n \end{aligned} \tag{7.93}$$

may arise. These equations can be written in matrix notation as

$$\mathbf{AX} = \lambda\mathbf{X} \tag{7.94}$$

or

$$(\mathbf{A} - \lambda\mathbf{I})\mathbf{X} = 0 \tag{7.95}$$

It was noted in Sec. 7.2 that such equations may have a nontrivial solution. The equations will have a nontrivial solution if and only if the determinant of $\mathbf{A} - \lambda\mathbf{I}$ is identically equal to zero. That is,

$$\begin{vmatrix} a_{11} - \lambda & a_{12} & \cdots & a_{1n} \\ a_{12} & a_{22} - \lambda & \cdots & a_{2n} \\ \vdots & \vdots & & \vdots \\ a_{n1} & a_{n2} & \cdots & a_{nn} - \lambda \end{vmatrix} = 0 \tag{7.96}$$

By the techniques of Sec. 7.4, the determinant of (7.96) can be evaluated and will, in general, result in an nth-order polynomial in λ. The roots of this polynomial equation are called the *characteristic roots*, or *eigenvalues*, of the matrix **A**. For example, consider the matrix

$$\mathbf{A} = \begin{bmatrix} 2 & -1 & 2 \\ -1 & -1 & 1 \\ 2 & 1 & 2 \end{bmatrix} \tag{7.97}$$

The determinant of $\mathbf{A} - \lambda\,\mathbf{I}$ is

$$|\mathbf{A} - \lambda\mathbf{I}| = \begin{vmatrix} 2 - \lambda & -1 & 2 \\ -1 & -1 - \lambda & 1 \\ 2 & 1 & 2 - \lambda \end{vmatrix} \tag{7.98}$$

Equating this to zero and evaluating, leads to the polynomial equation,

$$-8 + 6\lambda + 3\lambda^2 - \lambda^3 = 0 \tag{7.99}$$

This equation has roots $\lambda_1 = 4$, $\lambda_2 = 1$, and $\lambda_3 = -2$. Therefore, the characteristic roots, or eigenvalues, of matrix (7.97) are 4, 1, and -2.

For each eigenvalue, there is a vector **X** that satisfies Eq. (7.94). These are called the *characteristic vectors*, or *eigenvectors*, of the matrix **A**. The values of the individual components of an eigenvector cannot be determined uniquely. This is because both sides of Eq. (7.95) can be multiplied by a nonzero scalar c, which still satisfies the equation. This proves that any scalar multiple of **X**, namely c**X**, also satisfies (7.95). Therefore, only the ratio of each component to some other component can be found. For example, if the components are $x_1, x_2, \cdots, x_{n-1}, x_n$, we can only determine $x_1/x_n, x_2/x_n, \ldots, x_{n-1}/x_n$.

The eigenvectors of matrix (7.97) can be found by substituting the matrix **A** and each of the eigenvalues, found earlier, into (7.96). Thus for $\lambda = 4$,

$$\begin{aligned} -2x_1 - x_2 + 2x_3 &= 0 \\ -x_1 - 5x_2 + x_3 &= 0 \\ 2x_1 + x_2 - 2x_3 &= 0 \end{aligned} \tag{7.100}$$

Letting $u_i = x_i/x_3$, the Eqs. (7.100) reduce to

$$\begin{aligned} -2u_1 - u_2 &= -2 \\ -u_1 - 5u_2 &= -1 \\ 2u_1 + u_2 &= 2 \end{aligned} \tag{7.101}$$

Solving yields $u_1 = 1$, $u_2 = 0$. Although the first and third equations are equivalent, there is a unique solution.

For $\lambda = 1$, we can write

$$\begin{aligned} u_1 - u_2 &= -2 \\ -u_1 - 2u_2 &= -1 \\ 2u_1 + u_2 &= -1 \end{aligned} \tag{7.102}$$

which have the solutions $u_1 = -1$, $u_2 = 1$.

For $\lambda = -2$,

$$\begin{aligned} 4u_1 - u_2 &= -2 \\ -u_1 + u_2 &= -1 \\ 2u_1 + u_2 &= -4 \end{aligned} \tag{7.103}$$

which have the solutions $u_1 = -1$, $u_2 = -2$. Note that both Eqs. (7.102) and (7.103) are overdetermined, yet have unique solutions. It should be understood that in all three cases, we have chosen to set $u_3 = 1$. Therefore, the eigenvectors of matrix (7.97) are

$$\begin{bmatrix} 1 \\ 0 \\ 1 \end{bmatrix} \quad \begin{bmatrix} -1 \\ 1 \\ 1 \end{bmatrix} \quad \begin{bmatrix} -1 \\ -2 \\ 1 \end{bmatrix} \tag{7.104}$$

As noted earlier, any scalar multiples of these vectors are also eigenvectors of the example matrix.

The preceding description and analysis of eigenvalues and eigenvectors has been facilitated by the use of a third-order example. Second- and third-order determinants can be evaluated quite easily. However, the evaluation of higher-order determinants must use the general technique given in Sec. 7.4. This technique is not very amenable to computer implementation. In the case of the eigenvalue problem, the evaluation of the determinant is further complicated by the result being a polynomial rather than a single numerical value. Therefore, when the order of the matrix is large (large may be only four or five), some other technique must be employed to find the eigenvalues and eigenvectors of the matrix.

7.10.1 Iterative Method of Finding an Eigenvalue of a Matrix

We will now consider an iterative technique that can be used to find the largest (or smallest) eigenvalue of a matrix and also the corresponding eigenvector. Fortunately, in the physical problems in which eigenvalues are to be found, it is usually the largest and smallest eigenvalues that are of the most interest. Techniques for finding *all* eigenvalues and eigenvectors of a matrix will not be treated here. The reader, if interested, will find such techniques presented in a number of the references given in the bibliography.

The iterative technique to be described here is essentially the iteration method of Sec. 6.8, but applied to the matrix equation (7.94). Writing that equation in iterative form gives

$$\lambda_{k+1}\mathbf{X}_{k+1} = \mathbf{A}\mathbf{X}_k \tag{7.105}$$

First, we select an initial approximation for $\mathbf{X}$, say,

$$\mathbf{X}_0 = \begin{bmatrix} 0 \\ 0 \\ 0 \\ \vdots \\ 0 \\ 1 \end{bmatrix} \tag{7.106}$$

Then we perform the multiplication

$$\mathbf{Y}_{k+1} = \mathbf{A}\mathbf{X}_k \tag{7.107}$$

The vector $\mathbf{Y}$ is then normalized by dividing each of its components by one of the components, say y_n, giving $\mathbf{X}_{k+1}$:

$$\mathbf{X}_{k+1} = \frac{1}{(y_n)_{k+1}} \mathbf{Y}_{k+1} \tag{7.108}$$

Repeating the operations indicated by (7.107) and (7.108) usually will lead to a solution. If so, the solution will be the largest eigenvalues

$$\lambda = y_n \tag{7.109}$$

and the corresponding eigenvector will be $\mathbf{X}$.

This iterative method will converge if the largest eigenvalue is real and is not a multiple root. Convergence is most rapid when the ratio of the largest eigenvalue to the next largest eigenvalue is large. The flowchart of Fig. 7.15 illustrates this process.

In order to find the smallest eigenvalue of a matrix, we apply the principle that the reciprocals of the eigenvalues of a matrix are the eigenvalues of the inverse of the matrix. That is, if λ is an eigenvalue of $\mathbf{A}$, then

$$\mathbf{A}^{-1}\mathbf{X} = \frac{1}{\lambda}\mathbf{X} \tag{7.110}$$

Therefore, taking the inverse of $\mathbf{A}$ and then using the iteration we have just described will give the largest eigenvalue of the inverse of $\mathbf{A}$. The reciprocal of this value will then be the smallest eigenvalue of $\mathbf{A}$.

7.11 SUMMARY

The solution of sets of equations is a common type of problem that arises in engineering and science. This chapter presented various methods for solving sets of equations. Which of these methods to use depends primarily on the nature of the equations to be solved and to some extent on the type of computing equipment to be used. Two classes of equations have been discussed: linear and nonlinear. Methods of solution by substitution, determinants, elimination, iteration, and matrices were given for systems of linear equations. Of the methods covered, only the substitution and iteration methods were found to be also applicable to systems of nonlinear equations. Certain topics that are related to systems of linear equations, namely, determinants, matrices, eigenvalues, and eigenvectors, were also presented.

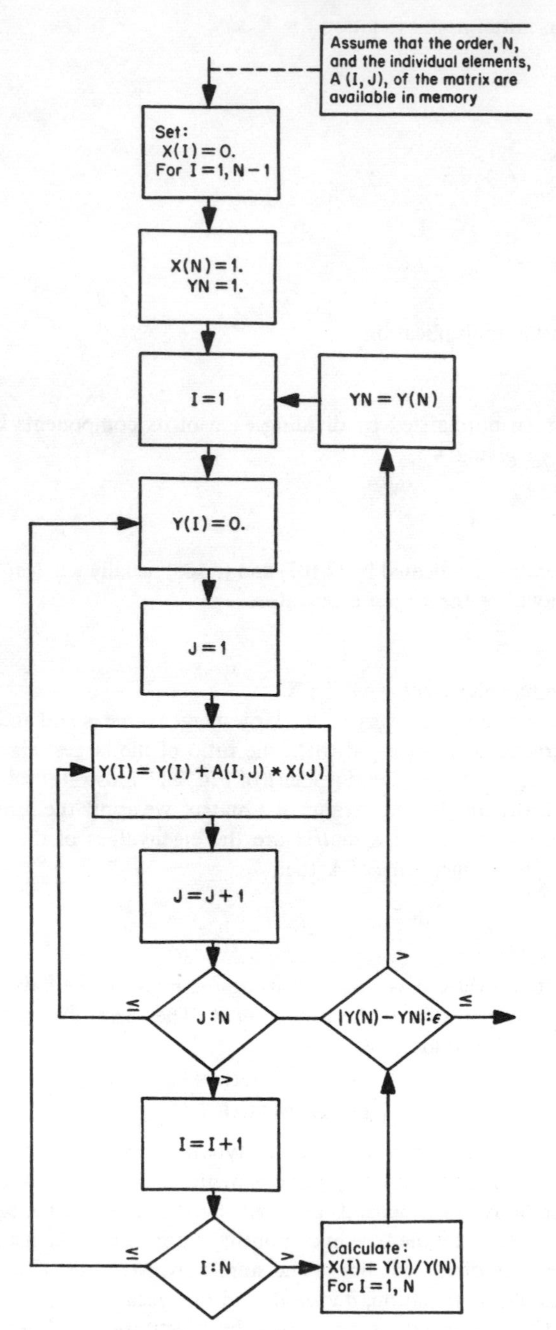

Fig. 7.15. Flowchart of iterative method for finding largest eigenvalue of a matrix and corresponding eigenvector.

EXERCISES

1. Write a FORTRAN subroutine for evaluating a fourth-order determinant.
2. Prepare a flowchart and write a FORTRAN program for solving a system of four equations in four unknowns by Cramer's rule. Use the subroutine of Exercise 1.
3. Execute Exercise 2 for the following sets of equations:

(a)
$$\begin{aligned} x_1 - x_2 + 2x_3 - x_4 &= 12 \\ -3x_1 - 4x_2 + x_3 + x_4 &= 1 \\ 5x_1 + 6x_2 - x_3 + 3x_4 &= -3 \\ x_1 + x_2 + x_3 + 3x_4 &= 0 \end{aligned}$$

(b)
$$\begin{aligned} x_1 + x_2 + x_3 + x_4 &= 0 \\ 2x_1 + 3x_2 + 2x_3 + x_4 &= 1 \\ 4x_1 + 5x_2 + 4x_3 + 3x_4 &= 1 \\ -x_1 + 3x_2 + 2x_3 + x_4 &= -2 \end{aligned}$$

(c)
$$\begin{aligned} x_1 + x_2 + x_3 + x_4 &= 0 \\ 2x_1 + x_2 + x_3 - x_4 &= \sqrt{3} \\ 2x_1 + 2x_2 + x_3 + x_4 &= \sqrt{5} \\ 3x_1 - x_2 - x_3 - 5x_4 &= 0 \end{aligned}$$

(d)
$$\begin{aligned} x_1 + x_2 + x_3 + x_4 &= 1 \\ 2x_1 + x_2 + x_3 - x_4 &= 2 \\ 2x_1 + 2x_2 + x_3 + x_4 &= 3 \\ 3x_1 - x_2 - x_3 - 5x_4 &= 4 \end{aligned}$$

(e)
$$\begin{aligned} x_1 - x_2 + x_3 &= 1 \\ 3x_1 + 2x_2 - x_3 &= -2 \\ -x_1 - 3x_2 + 2x_3 &= 1 \end{aligned}$$

4. Prepare a flowchart and write a FORTRAN program to solve a set of equations by the Gaussian elimination procedure. Allow for as many as eight equations in eight unknowns.
5. Execute the program in Exercise 4 with the data from Exercise 3.
6. Extend the flowchart for the Gauss–Jordan elimination method (Fig. 7.5) to include input, output, and an error stop and message for $P = 0$. Write a FORTRAN program from this flowchart. Allow for as many as eight equations in eight unknowns.
7. Execute the program in Exercise 6 with the data from Exercise 3.
8. Execute the program in Exercise 6 for the following equations:

$$\begin{aligned} x_1 + 2x_2 + x_3 - x_4 - x_5 + 2x_6 + x_7 - x_8 &= 2 \\ 2x_1 + 4x_2 + 3x_3 - x_4 - x_5 + 5x_6 + 3x_7 - x_8 &= 1 \\ -x_1 - x_2 + 2x_4 + 2x_5 - x_6 + 2x_8 &= -3 \\ x_2 + 2x_3 + 3x_4 - 4x_5 - 3x_6 - 2x_7 - x_8 &= 0 \\ x_1 + x_2 + 3x_3 - 6x_4 + 2x_5 + x_6 + 5x_7 + 2x_8 &= 3 \\ -2x_1 - 2x_2 + 2x_3 + 5x_5 - 2x_6 - 3x_7 &= 0 \\ 5x_1 + 7x_2 + 5x_3 - 3x_4 + x_5 + 4x_6 + 11x_7 + 2x_8 &= 12 \\ 4x_1 + 6x_2 - 4x_4 + 10x_6 + 5x_7 - 4x_8 &= 11 \end{aligned}$$

9. Prepare a flowchart and write a FORTRAN program to solve the following equations by Gauss–Seidel iteration:

$$
\begin{aligned}
x + \cos y &= 0.90710678 \\
x^2 + \tan y - e^z &= 0.67212056 \\
y + 5z &= -4.21460184
\end{aligned}
$$

Execute this program starting with the estimates $y = z = 0$.

10. Repeat Exercise 9 using Newton's iteration. Use starting estimates of $x = y = z = 0$.

11. Prepare a flowchart and write a FORTRAN subroutine to multiply a matrix and a vector. Allow for an order of eight for each.

12. Write a FORTRAN program to invert a matrix using the flowchart of Figure 7.11. Include provision for reading in the matrix to be inverted and for printing out the inverse.

13. Execute the program in Exercise 12 for the matrices defined by the coefficients of the left sides of the equations in Exercise 3.

14. Write a FORTRAN program to solve a set of linear equations by matrix inversion and matrix–vector multiplication. Use the inversion method of Fig. 7.14 and the subroutine of Exercise 11.

15. Execute the program in Exercise 14 for the equations in Exercises 3 and 8.

16. Prepare a flowchart and write a FORTRAN program to find *all* eigenvalues of a 3×3 matrix. Note that this program will require the solution of a cubic equation.

17. Prepare a flowchart and write a FORTRAN subroutine that will find the eigenvectors of a 3×3 matrix if the eigenvalues are given.

18. Combine the programs in Exercises 16 and 17 to obtain all eigenvalues and the corresponding eigenvectors for a 3×3 matrix. Execute this program for the matrix

$$
\begin{bmatrix}
1.7 & 2. & 1. \\
2. & -3.5 & -2. \\
1. & -2. & 1.
\end{bmatrix}
$$

19. Complete the flowchart of Fig. 7.15 to include input and output, and write a FORTRAN program to find the largest eigenvalue and corresponding eigenvector of any matrix of any size up to 10×10.

20. Execute the program in Exercise 19 for the matrix

$$
\begin{bmatrix}
1 & 1 & 1 & 1 \\
2 & 3 & 2 & 1 \\
4 & 5 & 4 & 3 \\
-1 & 3 & 2 & 1
\end{bmatrix}
$$

8
Curve Fitting

8.1 INTRODUCTION

In the physical sciences and engineering, it is often necessary to represent some tabular data with a mathematically expressible function (or express a complicated mathematical expression with a simpler one). This process is known as *curve fitting* and is the subject of this chapter. Curve fitting can be performed for any of the following purposes:

1. *To verify a theory and evaluate the associated parameters*—For example, the orbits of two point masses moving only under the influence of their common attraction are ellipses. From observations of the positions of the two masses, the shapes, the orientation, and the dimensions of the ellipses can be determined. In actual practice, the motions of two masses are affected by the presence of other masses and by the fact that their own masses are not concentrated at a point. Therefore, the orbits will not be true ellipses. However, the data can be fitted to the formula of an ellipse. This procedure provides a measure of how much the actual case differs from the ideal case. It also provides approximate orbital elements that may be sufficiently accurate to be used in predicting future positions.
2. *To measure the accuracy of observation*—Even though the mathematical theory of a physical phenomenon is known accurately, it is not possible to make exact observations. Fitting the actual data to the theory gives a measure of the accuracy of observation.
3. *To aid in computing*—For example, the characteristics of an electronic device (transistor, diode, vacuum tube, etc.) may be available in graphical or tabular form only. In order to use this data in a computer program for the analysis of a circuit containing such a device, it is often advantageous to express these characteristics in a mathematical form.

One method of curve fitting, which the term itself implies, is graphical curve fitting. In this method, observed data points are plotted and then, in the judgment of the drawer, a "best" curve is drawn through the data so that intermediate values can be read from the curve. This method has limited value. It can be no more accurate than the accuracy of plotting, and it gives no information about the mathematical properties of the curve. Therefore, this chapter will be devoted to more analytical forms of curve fitting. The term *curve fitting* will be used with the meaning: The process of finding a mathematical formula for approximating a set of data values or for approximating another mathematical formula.

In the following two subsections, the generalities of curve fitting are presented along with a brief discussion of some simple methods. These methods are the method of collocation, the method of averages, and other related but unnamed methods. These methods are usually used only for fitting a formula to data. In Sec. 8.2, the method of least squares is discussed in detail. This method is used for fitting both data and known functions. The latter application is the subject of a separate section. In addition, two methods that employ the method of least squares, but which use standard approximating formulas, are also presented. These are Chebyshev polynomials (Sec. 8.4) and Fourier series (Sec. 8.5). The Chebyshev polynomials can be used to reduce the number of of terms in a powers series and hence are the basis of another approximating technique, the economization of power series. This technique is given in Sec. 8.6.

8.1.1 General Considerations

Given a table of data where each given (or observed) quantity y_i is dependent on one or more variables $x_1, x_2, \ldots, x_n$, we can write the functional relationship

$$y_i = f(x_{1i}, x_{2i}, \ldots, x_{ni}) \tag{8.1}$$

It is desirable to find some mathematically expressible function of the independent variables, say,

$$u = g(x_1, x_2, \cdots, x_n) \tag{8.2}$$

such that when g is evaluated for each set of values of the independent variables, all errors

$$\varepsilon_i = y_i - u_i \tag{8.3}$$

are sufficiently small. There are two major parts to this problem:

1. finding a mathematical form for the function g,
2. evaluating the parameters of the function g.

The first part of the problem can be very easy or very difficult to do depending on the nature of the physical phenomenon under consideration. Ordinarily we have a theory to predict the behavior of the phenomenon being observed. Therefore, the mathematical form of the function is known. If the mathematical form is not known (or is a very complex one), it may be very difficult to develop a suitable function. There is an infinite number of function forms that can be selected to represent the given data. For this reason, it is often necessary to limit our investigations to certain standard forms such as polynomials and Fourier series. The second phase of the problem is usually more straightforward. That is, general computing methods exist for determining approximate values of the parameters of the function. An example will help clarify this concept. Suppose the selected function has the form

$$u = A + Bx_1 + Cx_2 \tag{8.4}$$

Then we must determine values for A, B, and C that provide us with the "best" set of values u_i to fit the data y_i. Because of the variety of functional forms for the function g, it is difficult to illustrate general methods for evaluating the parameters. The methods given in this chapter are given in narrative form and the user must specialize them to fit his own needs. However, flowcharts are given in Sec. 8.2 for specialized functions to illustrate the application of two methods.

8.1.2 Elementary Curve Fitting Methods

A number of methods are available for determining values for the parameters of the selected function. A few elementary ones are outlined in the following discussion.

If the selected function has K parameters, in order to determine values for all of these parameters, we must have at least K sets of data values $y_i, x_{1i}, x_{2i}, \ldots, x_{ni}$. That is, $i_{max} \geqq K$. If there are exactly the same number of data points as parameters, the values of the parameters may be determined by substituting the data values, one set at a time, into the selected function, thus creating a set of K simultaneous equations. These equations can then be solved for the parameters by one of the methods of Chap. 7. However, even if there are more data points available than parameters, we can select from that set just K of them and use only these to determine the parameters. Selection of the set can be based on various criteria such as equal spacing of independent variable values, random choice, precision of the data, etc. This method is sometimes known as the *method of collocation.*

A second method is based on the method of collocation. We select K data points, determine the parameters as before, and then examine the errors at the unused data points that result from the use of the computed parameters. This procedure is repeated using other sets of K data points until a set of parameters is found that satisfies some criterion for the errors at the unused data points. One possible criterion is to find the parameters that cause the least difference between the most positive error and the most negative error. Another criterion is to find those parameters with the least largest error.

A third method is also based on the first one given. We select several sets of K data points and for each set determine the parameter values. Then by some process, we average the values thus found to give the final parameter values. A variety of averaging processes can be used; for example, the arithmetic mean, the geometric mean, a weighted arithmetic mean, etc. Furthermore, a variety of methods can be employed to select the sets of data points. Some of the possible methods of grouping are 1) using all combinations of data points taken K at a time; 2) selecting the first K data points for one set, the second K points for another set, etc.; or 3) selecting the 1st through Kth points for one set, the 2nd through (K + 1)th points for a second set, etc.

A fourth method involves determining the parameters using all data points to minimize the sum of the errors. This may be done by the *method of averages.* In this method, the data points are divided into K groups. The data values are substituted into the selected function, such as Eq. (8.4), and then an average equation is determined for each group. The result of this process is K equations that can be solved for the parameters.

Another method is determining the parameters in a manner that minimizes the sum of the absolute values of the errors. This cannot be done analytically, but a trial and error method can be devised to accomplish this.

Only the rudiments of the foregoing methods have been given. The details of these methods are dependent on too many factors to be described generally. For example, it is possible to select data points in such a manner that when their values are substituted in the selected function, inconsistent redundant equations will result. Any general method must include provisions for avoiding this situation. Additional information on three of the preceding methods is provided in Sec. 8.5 in the form of an example problem. A method that is used much more widely than any of those described so far is discussed in Sec. 8.2.

8.2 LEAST SQUARES

The criterion of best fit of the *method of least squares* is that the sum of the squares of the errors be a minimum.

$$\Sigma \varepsilon_i^2 = \text{minimum} \tag{8.5}$$

The term *error*, in this application, means the difference between the observed and computed values, Eq. (8.3). In some discussions, the quantity expressed mathematically by (8.3) is called a *residual* or a *deviation*. The primary advantage of the method of least squares is that we can employ the methods of calculus to find the minimum sum of squares of errors. This is done in the following manner.

Suppose that a set of values y_i, which depend on certain other variables $x_{1i}, x_{2i}, \ldots, x_{ni}$, are to be represented by some function such as Eq. (8.2). This function will also contain certain parameters a_j that must be determined. Therefore, it is appropriate to write (8.2) in a manner that would illustrate this.

$$u = g(a_1, a_2, \ldots, a_k, x_1, x_2, \ldots, x_n) \tag{8.6}$$

This equation is called the *equation of condition*. Let S be the sum of the squares of all errors.

$$S = \sum_{i=1}^{N} \varepsilon_i^2 \tag{8.7}$$

where N is the total number of cases. Substituting (8.6), evaluated for the ith set of independent variables, into (8.3) and the result into (8.7) yields

$$S = \sum_{i=1}^{N} (y_i - g(a_1, a_2, \ldots, a_k, x_{1i}, x_{2i}, \ldots, x_{ni}))^2 \tag{8.8}$$

A *necessary* condition that a minimum for the function S exists is that the partial derivatives with respect to each of the parameters $a_1, a_2, \ldots, a_k$ be zero. We will not explore the *sufficiency* of these conditions, but we will assume that the equations

$$\begin{aligned} \frac{\partial S}{\partial a_1} &= 0 \\ \frac{\partial S}{\partial a_2} &= 0 \\ &\vdots \\ \frac{\partial S}{\partial a_k} &= 0 \end{aligned} \tag{8.9}$$

are satisfied only when S is minimum. Equations (8.9) give us k equations in k unknowns: $a_1, a_2, \ldots, a_k$. These equations are called *normal equations.*

8.2.1 Nonlinear Functions

In this section, we consider a function to be *nonlinear* if it is a nonlinear function of the parameters $a_1, a_2, \ldots, a_k$. We will see in Sec. 8.2.2 that we may have functions that are nonlinear functions of the independent variables $x_1, x_2, \ldots, x_n$ and are still linear functions of the parameters.

In Sec. 8.2, no restriction was placed on the form of Eq. (8.6). To better understand how this method is used, consider the following example.

The Perfect Gas Law states that for an adiabatic expansion or compression, the pressure p is related to the volume v by the formula

$$pv^k = C \tag{8.10}$$

Solving for p, this can be written

$$p = a_1 v^{a_2} \tag{8.11}$$

This then becomes the equation of condition. If we observe values of p for various values of v, we can, by the method of least squares, determine the parameters a_1 and a_2. The function S becomes

$$S = \sum_{i=1}^{N} (p_i - a_1 v_i^{a_2})^2 \tag{8.12}$$

and the normal equations are

$$\frac{\partial S}{\partial a_1} = -2 \sum_{i=1}^{N} \left[(p_i - a_i v_i^{a_2}) v_i^{a_2} \right] = 0$$
$$\frac{\partial S}{\partial a_2} = -2a \sum_{i=1}^{N} \left[(p_i - a_1 v_i^{a_2}) v_i^{a_2} \ln v_i \right] = 0 \tag{8.13}$$

Equations (8.13) can be simplified to

$$\sum_{i=1}^{N} p_i v_i^{a_2} = a_1 \sum_{i=1}^{N} v_i^{2a_2}$$
$$\sum_{i=1}^{N} p_i v_i^{a_2} \ln v_i = a_1 \sum_{i=1}^{N} v_i^{2a_2} \ln v_i \tag{8.14}$$

These equations can be solved for a_1 and a_2, but since a_2 appears under the summation sign, an iterative technique must be employed to solve for the unknowns.

We will discuss in Secs. 8.2.3 and 8.2.4 how the foregoing example can be simplified using linearizing techniques.

8.2.2 Linear Functions

In this section, we will discuss techniques for simplifying the least squares method for those functions that are linear functions of the parameters. The following examples of equations of condition, although nonlinear functions of the independent variables, *are* linear functions of the parameters.

$$\begin{aligned} y &= a_1 + a_2x_1 + a_3x_2 + a_4x_1x_2 \\ y &= a_1 + a_2x + a_3x^2 + a_4x^3 \\ y &= a_1 \sin x + a_2 \end{aligned} \tag{8.15}$$

These equations can be reduced to the general form

$$y = a_1 + a_2x_1 + a_3x_2 + a_4x_3 + \cdots \tag{8.16}$$

by making suitable substitutions such as $x_3 = x_1x_2$, $x_2 = x^2$, $x_3 = x^3$, and $x_1 = \sin x$. The function S then becomes

$$S = \sum_{i=1}^{N} (y_i - a_1 - a_2x_{1i} - a_3x_{2i} - \cdots - a_{n+1}x_{ni})^2 \tag{8.17}$$

By taking partial derivatives of Eq. (8.17), the following normal equations can be formed:

$$\begin{aligned} \frac{\partial S}{\partial a_1} &= -2 \sum_{i=1}^{N} (y_i - a_1 - a_2x_{1i} - a_3x_{2i} - \cdots - a_{n+1}x_{ni}) = 0 \\ \frac{\partial S}{\partial a_2} &= -2 \sum_{i=1}^{N} x_{1i}(y_i - a_1 - a_2x_{1i} - a_3x_{2i} - \cdots - a_{n+1}x_{ni}) = 0 \\ \frac{\partial S}{\partial a_3} &= -2 \sum_{i=1}^{N} x_{2i}(y_i - a_1 - a_2x_{1i} - a_3x_{2i} - \cdots - a_{n+1}x_{ni}) = 0 \\ &\vdots \\ \frac{\partial S}{\partial a_{n+1}} &= -2 \sum_{i=1}^{N} x_{ni}(y_i - a_1 - a_2x_{1i} - a_3x_{2i} - \cdots - a_{n+1}x_{ni}) = 0 \end{aligned} \tag{8.18}$$

Equations (8.18) can be simplified to the following form. (*Note:* The limits $i=1$ to $i=N$ have been omitted from the summations and the subscripts on the variables have also been omitted for simplicity in writing.)

$$\begin{aligned} a_1\Sigma 1 + a_2\Sigma x_1 + a_3\Sigma x_2 + \cdots + a_{n+1}\Sigma x_n &= \Sigma y \\ a_1\Sigma x_1 + a_2\Sigma x_1^2 + a_3\Sigma x_1x_2 + \cdots + a_{n+1}\Sigma x_1x_n &= \Sigma x_1y \\ a_1\Sigma x_2 + a_2\Sigma x_1x_2 + a_3\Sigma x_2^2 + \cdots + a_{n+1}\Sigma x_2x_n &= \Sigma x_2y \\ &\vdots \\ a_1\Sigma x_n + a_2\Sigma x_1x_n + a_3\Sigma x_2x_n + \cdots + a_{n+1}\Sigma x_n^2 &= \Sigma x_ny \end{aligned} \tag{8.19}$$

The notation $\Sigma 1$ means "sum of the ones," and its value is N.

An inspection of Eqs. (8.19) reveals the following interesting facts about linear equations that will permit a "cook book" approach to the method of least squares:

1. The coefficient matrix is symmetric. Note that the a's are the unknowns and the coefficient matrix consists of the sums. This symmetry will simplify the computing of the coefficient matrix and will also aid in solving the system.
2. The first equation is just the equation of condition summed over all i.

3. The second equation is just the equation of condition multiplied by x_1 and summed over all i.
4. Observations 2 and 3 above can be generalized to: The jth normal equation is formed by multiplying the equation of condition by the multiplier of a_j, from the equation of condition, and summing over all i.

As an example of the use of rule 4, consider the polynomial equation of condition, (8.20). This type of equation is often used to approximate functions of a single variable whose true form is not known.

$$y = a_0 + a_1 x + a_2 x^2 + \cdots + a_n x^n \tag{8.20}$$

The normal equations become

$$\begin{aligned} \Sigma y &= a_0 N + a_1 \Sigma x + a_2 \Sigma x^2 + \cdots + a_n \Sigma x^n \\ \Sigma xy &= a_0 \Sigma x + a_1 \Sigma x^2 + a_2 \Sigma x^3 + \cdots + a_n \Sigma x^{n+1} \\ \Sigma x^2 y &= a_0 \Sigma x^2 + a_1 \Sigma x^3 + a_2 \Sigma x^4 + \cdots + a_n \Sigma x^{n+2} \\ &\vdots \\ \Sigma x^n y &= a_0 \Sigma x^n + a_1 \Sigma x^{n+1} + a_2 \Sigma x^{n+2} + \cdots + a_n \Sigma x^{2n} \end{aligned} \tag{8.21}$$

Since there are n+1 linear equations in the n+1 unknowns a_j, the methods of Chap. 7 can be used to solve for the unknowns.

As an actual numerical example, the data in Table 8.1 is to be fit by a straight line.

TABLE 8.1 EXPERIMENTAL DATA

x	0	1	2	3	4
y	1.05	1.45	2.01	2.47	3.02

The equation of condition is

$$y = a_0 + a_1 x \tag{8.22}$$

The normal equations are

$$\begin{aligned} \Sigma y &= a_0 N + a_1 \Sigma x \\ \Sigma xy &= a_0 \Sigma x + a_1 \Sigma x^2 \end{aligned} \tag{8.23}$$

Evaluating the sums for the data in Table 8.1, the normal equations become

$$\begin{aligned} 10.00 &= 5a_0 + 10a_1 \\ 24.96 &= 10a_0 + 30a_1 \end{aligned} \tag{8.24}$$

Solution of (8.24) yields

$$a_0 = 1.008; \quad a_1 = 0.496$$

Substituting these values in (8.22) and evaluating for the given x's gives the computed values and errors (observed minus computed), shown in Table 8.2. The sum of the squares of the errors is 0.006240, and we can be assured that this is the least value that can be obtained.

TABLE 8.2 COMPUTED VALUES AND ERRORS FOR A STRAIGHT LINE FIT TO EXPERIMENTAL DATA

x	0	1	2	3	4
y_c	1.008	1.504	2.000	2.496	2.992
ε	0.042	−0.054	0.010	−0.026	0.028

A word of caution is appropriate at this point. Although the method of least squares is normally used when there are many more observations than unknown parameters, it will even work when the number of observations equals the number of unknown parameters. However, *no* solution can be obtained if the number of observations is less than the number of parameters.

It was noted in Sec. 8.1.1 that because of the variety of functional forms that can be used, it is difficult to illustrate general methods for curve fitting. However, by specializing the form of the selected function, some generality can be illustrated. In this section, we have discussed the use of the method of least squares to find the parameters of functions that are linear functions of the parameters. Suppose that we consider the specialized case of a function of two variables of the form

$$y = AF_1(x_1) + BF_2(x_2) \tag{8.25}$$

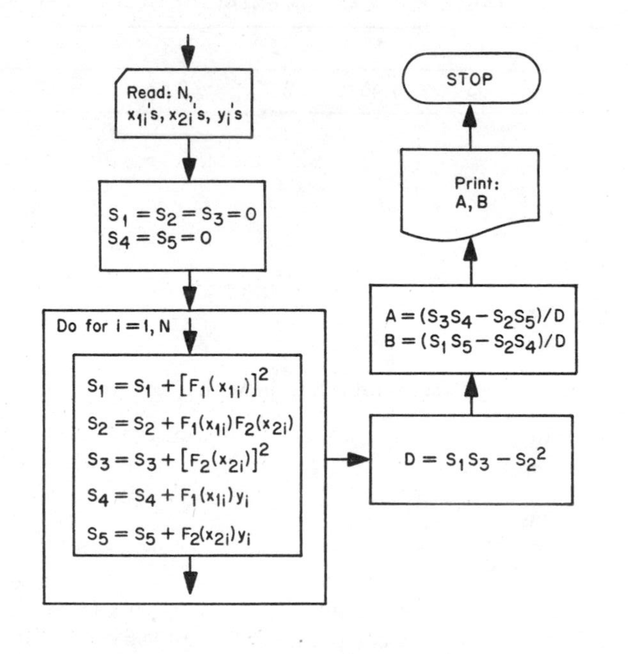

Fig. 8.1. Least squares solution for parameters of a function that is a linear function of two parameters.

Although this function is somewhat specialized, it still has considerable generality since the functions F_1 and F_2 can be almost any type of function. This form is linear in the parameters A and B, and hence the method of this subsection applies. Fig. 8.1 is a flowchart illustrating one computer approach for the determination of A and B. Note that by selecting $F_1(x) = 1$ and $F_2(x) = x$, the flowchart of Fig. 8.1 can be used to solve the example given earlier in this subsection.

8.2.3 Linearizing Nonlinear Functions

In Sec. 8.2.1 we saw an example of a difficulty encountered in applying the method of least squares to functions that are nonlinear functions of the parameters. In some cases, certain techniques can be applied that will either eliminate the nonlinearity or will simplify the problem.

In that example, we wished to fit a function

$$pv^k = C \tag{8.10}$$

to some data p_i, v_i. This function can be rewritten as

$$\ln p + k \ln v = \ln C \tag{8.26}$$

Letting $y = \ln p$, $x = \ln v$, $a_0 = \ln C$, and $a_1 = -k$, Eq. (8.26) becomes

$$y = a_0 + a_1 x \tag{8.27}$$

which is a linear form that can be treated by the methods of Sec. 8.2.2. We must remember, however, that in so doing, we minimize the sum of the squares of the errors in ln p rather than the sum of the squares of the errors in p.

A function of the form

$$y = A \sin(x + B) \tag{8.28}$$

will lead to normal equations that have one of the unknowns B under the summation signs. Equation (8.28) can be expanded to give

$$y = A \sin x \cos B + A \cos x \sin B \tag{8.29}$$

Letting $x_1 = \sin x$, $x_2 = \cos x$, $a_1 = A \cos B$, and $a_2 = A \sin B$, Eq. (8.29) becomes

$$y = a_1 x_1 + a_2 x_2 \tag{8.30}$$

which is in linear form.

No general rules can be given for linearizing nonlinear functions. Each such function must be treated individually. In many cases, the foregoing reductions may be useful; in other cases, it may be possible to linearize the function, but the method is unknown; and in still other cases, it may not be possible to linearize the function.

8.2.4 Method of Differential Corrections

Another method of linearizing a nonlinear form is to use Newton's iteration (Sec. 7.7). This method does not linearize the equation of condition as in Sec. 8.2.3, but approximates it with a linear form that is convenient to use for an iterative solution.

Suppose that the equation of condition is given by Eq. (8.6). Writing this with the x's not shown, we have

$$u = g(a_1, a_2, \ldots, a_n) \tag{8.31}$$

We wish to find a set of a's such that $\Sigma(y_i - g_i(a_1, a_2, \ldots, a_n))^2$ is a minimum. If we select a set of approximate values, $a_1^{(0)}, a_2^{(0)}, \ldots, a_n^{(0)}$, for the parameters, we can expand the function in a Taylor's series, and retaining only first-order terms of the series approximately have

$$g(a_1, a_2, \ldots, a_n) \doteq g(a_1^{(0)}, a_2^{(0)}, \ldots, a_n^{(0)}) + \Delta a_1 \left(\frac{\partial g}{\partial a_1}\right)^{(0)} + \Delta a_2 \left(\frac{\partial g}{\partial a_2}\right)^{(0)} + \cdots + \Delta a_n \left(\frac{\partial g}{\partial a_n}\right)^{(0)} \tag{8.32}$$

The function g and its partial derivatives, evaluated with the approximate values of the parameters, are just constants in (8.32). Therefore, that equation is a linear function of $\Delta a_1, \Delta a_2, \ldots, \Delta a_n$. Using the right-hand side of (8.32) for the function g, we can write the equation of condition for each set of variables as

$$y_i = g_i(a_1^{(0)}, a_2^{(0)}, \ldots, a_n^{(0)}) + \Delta a_1 \left(\frac{\partial g}{\partial a_1}\right)_i^{(0)} + \Delta a_2 \left(\frac{\partial g}{\partial a_2}\right)_i^{(0)} + \cdots + \Delta a_n \left(\frac{\partial g}{\partial a_n}\right)_i^{(0)} \tag{8.33}$$

The methods of Sec. 8.2.2 can be applied to solve for $\Delta a_1, \Delta a_2, \ldots, \Delta a_n$. These increments can then be added to the original approximations of the a's to get new approximations.

$$\begin{aligned} a_1^{(1)} &= a_1^{(0)} + \Delta a_1 \\ a_2^{(1)} &= a_2^{(0)} + \Delta a_2 \\ &\vdots \\ a_n^{(1)} &= a_n^{(0)} + \Delta a_n \end{aligned} \tag{8.34}$$

These approximations can then be used to find new values for the function and its partial derivatives, to form new sums, to find the normal equations evaluated for new corrections, etc.

Applying this method to the example of Sec. 8.2.1, we have

$$\begin{aligned} g &= a_1 v^{a_2} \\ \frac{\partial g}{\partial a_1} &= v^{a_2} \\ \frac{\partial g}{\partial a_2} &= a_1 v^{a_2} \ln v \end{aligned} \tag{8.35}$$

The equation of condition becomes

$$p = a_1^{(0)} v^{a_2^{(0)}} + \Delta a_1 v^{a_2^{(0)}} + \Delta a_2 a_1^{(0)} v^{a_2^{(0)}} \ln v \tag{8.36}$$

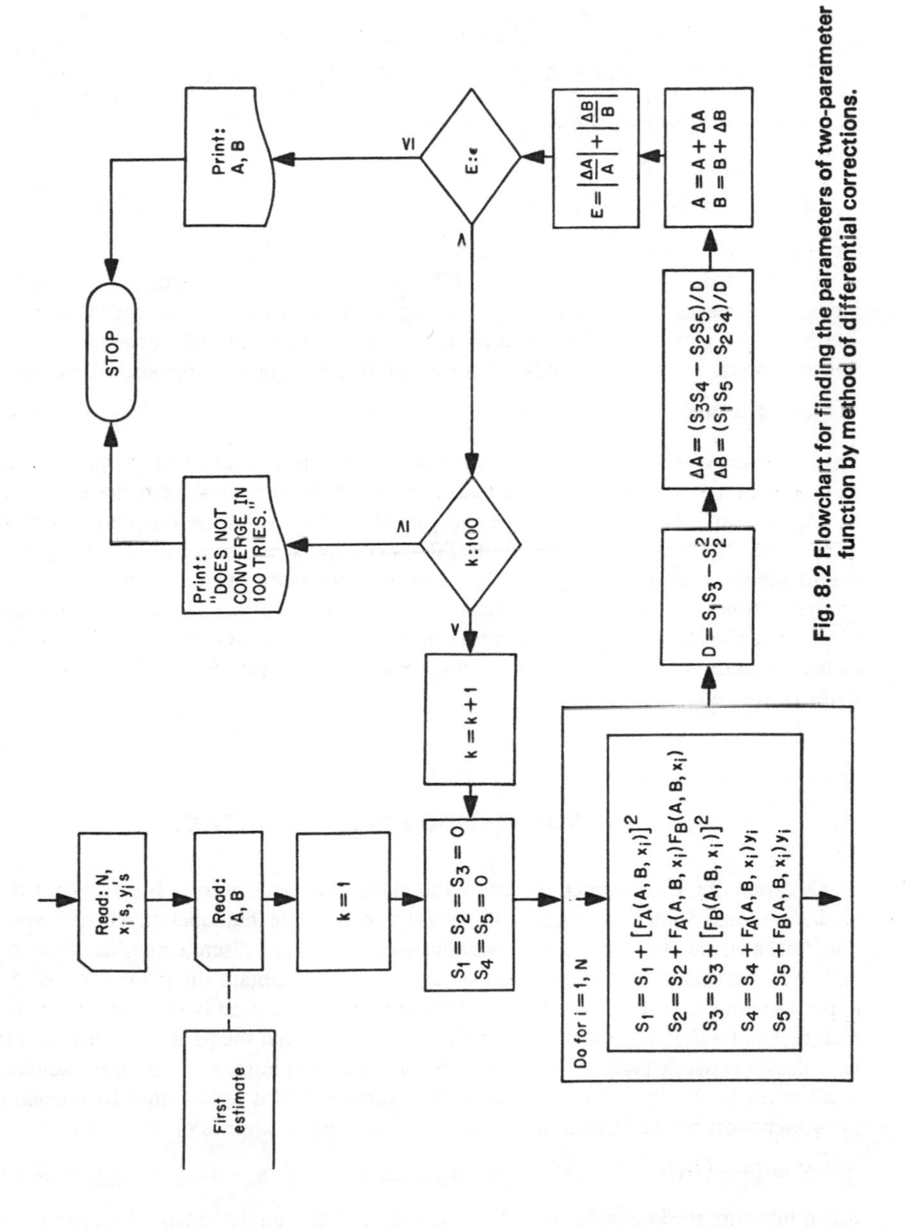

Fig. 8.2 Flowchart for finding the parameters of two-parameter function by method of differential corrections.

and the normal equations are

$$\begin{aligned} \Sigma(p - a_1^{(0)}v^{a_2^{(0)}})v \quad &= \Delta a_1 \Sigma v^{2a_2^{(0)}} + \Delta a_2 \Sigma a_1^{(0)} v^{2a_2^{(0)}} \ln v \\ \Sigma(p - a_1^{(0)}v^{a_2^{(0)}})v \quad \ln v &= \Delta a_1 \Sigma v^{2a_2^{(0)}} \ln v + \Delta a_2 \Sigma a_1^{(0)} v^{2a_2^{(0)}} (\ln v)^2 \end{aligned} \tag{8.37}$$

With the auxiliary equations

$$\begin{aligned} a_1^{(n)} &= a_1^{(n-1)} + \Delta a_1 \\ a_2^{(n)} &= a_2^{(n-1)} + \Delta a_2 \end{aligned} \tag{8.38}$$

the system (8.37) can be solved iteratively to obtain a_1 and a_2.

It was noted in Sec. 8.1.1 that it is difficult to illustrate, in general, a curve fitting method. However, by judicious specializing, a method can be illustrated without sacrificing all generality. To illustrate the method of differential corrections for a function which is *not* a linear function of its parameters, consider the specialized form

$$y = F(A, B, x) \tag{8.39}$$

The flowchart of Fig. 8.2 describes a method for finding the values of the parameters A and B by the method of differential corrections. In this flowchart, the notations F_A (A, B, x) and F_B (A, B, x) denote the partial derivatives of the function A and B, respectively. By means of suitable FORTRAN coding, a program can be written that would require only the changing of three arithmetic statement functions to alter the program from one function of the form of Eq. (8.39) to another function of the same form. Note that the example given earlier in this subsection could be solved by the use of the program of Fig. 8.2 and by setting $F(A, B, x) = Ax^B$, $F_A(A, B, x) = x^B$, and $F_B(A, B, x) = Ax^B \ln x$.

8.3 APPROXIMATING KNOWN FUNCTIONS

Occasionally, even though we know the mathematical form of a function and the values of the associated parameters, we need to approximate that function with a simpler one. One method of doing this is to evaluate the function at a discrete number of points, and then use this data in the method of least squares to obtain the parameters of the approximating function. Since the selection of points to be used is arbitrary, the parameters found will not be unique. The only way to assure that the parameters are unique is to use *all* possible points. This means that we must minimize the sum of the squares of all errors in the domain of the independent variables. This can be done by replacing the summation by the integral in the fundamental sum of squares of the errors.

$$S = \int \cdots \int\int (y(x_1, x_2, \ldots, x_n) - g(a_1, a_2, \ldots, a_k, x_1, x_2, \ldots, x_n))^2 dx_1 dx_2 \ldots dx_n \tag{8.40}$$

From this sum, the known forms of y and g, and the limits on the independent variables, the normal equations can be derived providing all needed antiderivatives are known. It is of some interest to present the details of this method for the case where the approximating function is a polynomial of a single variable and the function being approximated is a function of a single variable. For this case, if the equation of condition is

$$y = a_1 + a_2 x + a_3 x^2 + \cdots + a_k x^{k-1} \tag{8.41}$$

then the normal equations are

$$
\begin{aligned}
\int y dx &= a_1 \int dx + a_2 \int x dx + a_3 \int x^2 dx + \cdots + a_k \int x^{k-1} dx \\
\int xy dx &= a_1 \int x dx + a_2 \int x^2 dx + a_3 \int x^3 dx + \cdots + a_k \int x^k dx \\
\int x^2 y dx &= a_1 \int x^2 dx + a_2 \int x^3 dx + a_3 \int x^4 dx + \cdots + a_k \int x^{k+1} dx \\
&\vdots \\
\int x^{k-1} y dx &= a_1 \int x^{k-1} dx + a_2 \int x^k dx + a_3 \int x^{k+1} dx + \cdots + a_k \int x^{2k-2} dx
\end{aligned}
\tag{8.42}
$$

where all the integrals are taken over the limits $x = x_I$ to $x = x_F$. By the simple transformation of the independent variable,

$$x = x_I + (x_F - x_I)z \tag{8.43}$$

the limits of integration are changed to 0 and 1. When (8.43) is substituted in (8.41) and simplified, we still have a polynomial, but in z rather than in x. The resulting individual integrals on the right-hand side of the normal equations can then be evaluated for all cases.

$$\int_0^1 z^m dz = 1/(m+1) \tag{8.44}$$

Then the normal equations become

$$
\begin{aligned}
\int_0^1 y dz &= A_1 + A_2\, 1/2 + A_3\, 1/3 + \cdots + A_k\, 1/k \\
\int_0^1 zy dz &= A_1\, 1/2 + A_2\, 1/3 + A_3\, 1/4 + \cdots + A_k\, 1/(k+1) \\
\int_0^1 z^2 y dz &= A_1\, 1/3 + A_2\, 1/4 + A_3\, 1/5 + \cdots + A_k\, 1/(k+2) \\
&\vdots \\
\int_0^1 z^{k-1} y dz &= A_1\, 1/k + A_2\, 1/k+1) + A_3\, 1/(k+2) + \cdots + A_k\, 1/(2k-1)
\end{aligned}
\tag{8.45}
$$

The coefficient matrix is

$$
\begin{bmatrix}
1 & 1/2 & 1/3 & \cdots & 1/k \\
1/2 & 1/3 & 1/4 & \cdots & 1/(k+1) \\
1/3 & 1/4 & 1/5 & \cdots & 1/(k+2) \\
\vdots & \vdots & \vdots & & \vdots \\
1/k & 1/(k+1) & 1/(k+2) & \cdots & 1/(2k-1)
\end{bmatrix}
$$

and is a special case of a *Hilbert matrix*. Other powers series will lead to other coefficient matrices. This type matrix is difficult to invert because as k increases, the value of the determinant of the matrix rapidly approaches zero, and therefore the system of equations is near-singular. However, the inverse of this type matrix can be computed without going through the matrix inversion process, thus improving the accuracy. The elements b of the inverse can be found by the following relations. If the elements of the coefficient matrix can be expressed as

$$a_{ij} = \frac{1}{p(i + j) - q}$$

then

$$b_{11} = a_{11}\left[\frac{\prod_{\nu=1}^{k} (\nu p + p - q)}{(k - 1)!\, p^{k-1}}\right]^2 \tag{8.46}$$

$$b_{ij} = b_{ji}$$

$$b_{ij} = -\frac{(pi + pj - q - p)(k - j + 1)(pk + pj - q)}{(pi + pj - q)(j - 1)(pj - q)}\, b_{i,j-1}$$

To use this in an example, suppose that we want to approximate the sine function over the interval 0 to $\pi/2$ with the polynomial

$$\sin x = a_1 x + a_2 x^3 + a_3 x^5 \tag{8.47}$$

Making the substitution $x = (\pi/2)z$, Eq. (8.47) becomes

$$\sin (\pi/2)z = A_1 z + A_2 z^3 + A_3 z^5 \tag{8.48}$$

In this example, $p = 2$, $q = 1$, and $k = 3$. The use of Eqs. (8.46) gives

$$\begin{bmatrix} 3675/64 & -13230/64 & 10395/64 \\ -13230/64 & 56700/64 & -48510/64 \\ 10395/64 & -48510/64 & 43659/64 \end{bmatrix} \tag{8.49}$$

as the inverse of the coefficient matrix. The constant vector is

$$\begin{aligned} \int_0^1 z \sin (\pi/2)z\, dz &= (2/\pi)^2 &&= 0.40528473 \\ \int_0^1 z^3 \sin (\pi/2)z\, dz &= 3(2/\pi)^2 - 6(2/\pi)^4 &&= 0.23031991 \\ \int_0^1 z^5 \sin (\pi/2)z\, dz &= 5(2/\pi)^2 - 60(2/\pi)^4 + 120(2/\pi)^6 &&= 0.15952082 \end{aligned} \tag{8.50}$$

Multiplying the matrix (8.49) by the constant vector given by Eqs. (8.50) results in

$$A_1 = 1.57043683 \qquad A_2 = -0.64270786 \qquad A_3 = 0.07243169$$

and

$$a_1 = 0.99977114 \qquad a_2 = -0.16582652 \qquad a_3 = 0.00757407$$

The error curve for the resulting approximation is shown in Fig. 8.5 in Sec. 8.6.

8.4 CHEBYSHEV (TCHEBICHEFF) POLYNOMIALS‡

In the previous sections, we have discussed some specialized forms of the approximating function Eq. (8.6). Another specialized form is

$$u = a_1\phi_1(x_1, x_2, \ldots, x_n) + a_2\phi_2(x_1, x_2, \ldots, x_n) + \cdots + a_k\phi_k(x_1, x_2, \ldots, x_n) \tag{8.51}$$

If the ϕ's contain no unknown parameters, then (8.51) is a linear function of the parameters. Using this function to approximate a set of function values y_i, we can immediately write the normal equations

$$\begin{aligned}
\Sigma y_i\phi_{1i} &= a_1\Sigma\phi_{1i}^2 + a_2\Sigma\phi_{1i}\phi_{2i} + \cdots + a_k\Sigma\phi_{1i}\phi_{ki} \\
\Sigma y_i\phi_{2i} &= a_1\Sigma\phi_{1i}\phi_{2i} + a_2\Sigma\phi_{2i}^2 + \cdots + a_k\Sigma\phi_{2i}\phi_{ki} \\
&\;\;\vdots \\
\Sigma y_i\phi_{ki} &= a_1\Sigma\phi_{1i}\phi_{ki} + a_2\Sigma\phi_{2i}\phi_{ki} + \cdots + a_k\Sigma\phi_{ki}^2
\end{aligned} \tag{8.52}$$

where ϕ_{ji} is the abbreviated notation for $\phi_j(x_{1i}, x_{2i}, \ldots, x_{ki})$. For a given set of functions ϕ_j, Eqs. (8.52) can be solved and the a's determined. However, the purpose of putting the equation of condition in the form of Eqs. (8.51) is to simplify the solution of Eqs. (8.52). It can be shown that functions can be determined such that

$$\begin{aligned}
\sum_{i=1}^{N} \phi_{pi}\phi_{qi} &= 0 \qquad p \neq q \\
\sum_{i=1}^{N} \phi_{pi}\phi_{qi} &= K \qquad p = q
\end{aligned} \tag{8.53}$$

Such functions are called an *orthogonal set of functions.*

Although there can be many sets of orthogonal functions, we will confine our discussion to those introduced by Chebyshev. These are the Chebyshev polynomials, which are defined only on the interval $-1 \leqq x \leqq 1$ and are identified by the notation $T_j(x)$. Some Chebyshev polynomials are given in Table 8.3.

TABLE 8.3 EXAMPLES OF CHEBYSHEV POLYNOMINALS

$T_0(x) = 1$
$T_1(x) = x$
$T_2(x) = 2x^2 - 1$
$T_3(x) = 4x^3 - 3x$

‡ The letter T is used to denote the Chebyshev polynomials because the first transliteration of the name from Russian was to French, giving the spelling Tchebicheff. An analysis of the spelling of Chebyshev appears in "Chebyshev Series for Mathematical Functions" by C. W. Clenshaw. In *Mathematical Tables,* vol. 5. London: Her Majesty's Stationery Office, 1962

Additional polynomials can be calculated from the recursion relation

$$T_{m+1}(x) = 2xT_m(x) - T_{m-1}(x) \tag{8.54}$$

where x is given by

$$x = \cos\theta \tag{8.55}$$

The $T_j(x)$ of Table 8.3 and the recursion relation (8.54) are derived from the definition

$$T_j(\cos\theta) = \cos j\theta \tag{8.56}$$

Also from this definition it can be shown that

$$\begin{aligned}
&\sum_{i=1}^{N} T_p(x_i)T_q(x_i) = 0 \qquad p \neq q \\
&\sum_{i=1}^{N} T_p(x_i)T_q(x_i) = N/2 \qquad p = q \neq 0 \\
&\sum_{i=1}^{N} [T_0(x_i)]^2 = N
\end{aligned} \tag{8.57}$$

If our approximating function is

$$u = a_1T_0 + a_2T_1 + \cdots + a_kT_{k-1} \tag{8.58}$$

then the normal equations become

$$\begin{aligned}
\Sigma y_iT_0(x_i) &= a_i\Sigma\,[T_0(x_i)]^2 + a_2\Sigma\,T_0(x_i)T_1(x_i) + \cdots + a_k\Sigma\,T_0(x_i)T_{k-1}(x_i) \\
\Sigma y_iT_1(x_i) &= a_1\Sigma\,T_0(x_i)T_1(x_i) + a_2\Sigma\,[T_1(x_i)]^2 + \cdots + a_k\Sigma\,T_1(x_i)T_{k-1}(x_i) \\
&\vdots \\
\Sigma y_iT_{k-1}(x_i) &= a_i\Sigma\,T_0(x_i)T_{k-1}(x_i) + a_2\Sigma\,T_1(x_i)T_{k-1}(x_i) + \cdots + a_k\Sigma\,[T_{k-1}(x_i)]^2
\end{aligned} \tag{8.59}$$

By the relations (8.57), Eqs. (8.59) can be solved to obtain

$$\begin{aligned}
a_1 &= 2\Sigma y_iT_0(x_i)/N = \Sigma y_i/N \\
a_2 &= 2\Sigma y_iT_1(x_i)/N \\
&\vdots \\
a_k &= 2\Sigma y_iT_{k-1}(x_i)/N
\end{aligned} \tag{8.60}$$

The x_i and hence the y_i must be determined at equally spaced values of θ from 0 to π. These can be found by applying

$$\theta_i = \frac{\pi}{2N}(2i-1) \qquad 1 \leqq i \leqq N \tag{8.61}$$

where N is the total number of points to be used in the summations. In evaluating the preceding sums, only those x_i given by (8.55) may be used.

If the independent variable, say z, varies over the interval $z_I \leqq z \leqq z_F$, then the transformation

$$x = (2z - z_I - z_F)/(z_F - z_I) \tag{8.62}$$

can be used to obtain the new variable x that conforms to the conditions $-1 \leqq x \leqq 1$.

In an actual problem, a set of N values for x_i would be selected using Eqs. (8.61) and (8.55), the corresponding values of the polynomials $T_j(x_i)$ would be computed, and then a set of z's would be calculated to determine the corresponding y's. The z's can be found by solving (8.62).

$$z = (z_I + z_F)/2 + x(z_F - z_I)/2 \tag{8.63}$$

For known functions, we can replace the summations in Eqs. (8.60) by integrations, but the variable of integration must be θ in order to preserve the orthogonality properties. After making appropriate substitutions and solving, we can write

$$\begin{aligned} a_1 &= \frac{1}{\pi}\int_{-1}^{1} \frac{y}{\sqrt{1-x^2}}\,dx \\ a_2 &= \frac{2}{\pi}\int_{-1}^{1} \frac{y\,T_1(x)\,dx}{\sqrt{1-x^2}} \\ &\;\;\vdots \\ a_k &= \frac{2}{\pi}\int_{-1}^{1} \frac{y\,T_{k-1}(x)\,dx}{\sqrt{1-x^2}} \end{aligned} \tag{8.64}$$

In practice it may not be possible or convenient to find anti-derivatives for Eqs. (8.64), so it may be necessary to evaluate them by numerical integration.

8.5 FOURIER SERIES

Another type of useful function for approximating known functions is the Fourier series. This series is defined to be the infinite series

$$g(x) = a_0/2 + \sum_{i=1}^{\infty} (a_i \cos ix + b_i \sin ix) \tag{8.65}$$

This series can be used to represent exactly any single-valued function on the interval $c \leqq x \leqq c + 2\pi$ that has no infinite discontinuities. Since the Fourier series has periodicity of 2π, it can also be used to represent exactly those functions that also have a periodicity of 2π, for all x. It can be shown that to determine the series to represent a function f(x), the coefficients are found by the following formulas

$$\begin{aligned} a_0 &= \int_c^{c+2\pi} f(x)dx/\pi \\ a_i &= \int_c^{c+2\pi} f(x)\cos ixdx/\pi \qquad i \neq 0 \\ b_i &= \int_c^{c+2\pi} f(x)\sin ixdx/\pi \end{aligned} \tag{8.66}$$

These formulas result from applying the least squares method (Sec. 8.2) to Eq. (8.65). The resulting set of normal equations reduces to Eqs. (8.66), because the terms on the left sides of the normal equations satisfy the orthogonality conditions (8.53). In actual practice, it is necessary to use a finite number of terms of the series. Although the truncated series cannot represent the given function exactly, the coefficients determined are independent of the number of terms retained in the approximation.

By a simple translation of the argument

$$z = x - c - \pi \tag{8.67}$$

we can always change our interval to $-\pi \leqq z \leqq \pi$. So without loss of generality, Eqs. (8.66) can be written as

$$a_0 = \int_{-\pi}^{\pi} y dz/\pi$$

$$a_i = \int_{-\pi}^{\pi} y \cos iz dz/\pi \qquad i \neq 0 \tag{8.68}$$

$$b_i = \int_{-\pi}^{\pi} y \sin iz dz/\pi$$

where y is the function f(x) appropriately modified for the new argument z.

If a function is an *odd* function, i.e.,

$$y(z) = -y(-z) \tag{8.69}$$

it can be shown that $a_i = 0$ for all i, and the relation for b_i simplifies to

$$b_i = 2\int_0^{\pi} y \sin iz dz/\pi \tag{8.70}$$

Similarly, for an *even* function, i.e.,

$$y(z) = y(-z) \tag{8.71}$$

it can be shown that $b_i = 0$ for all i, and the expressions for a_i simplify to

$$a_0 = 2\int_0^{\pi} y dz/\pi$$

$$a_i = 2\int_0^{\pi} y \cos iz dz/\pi \tag{8.72}$$

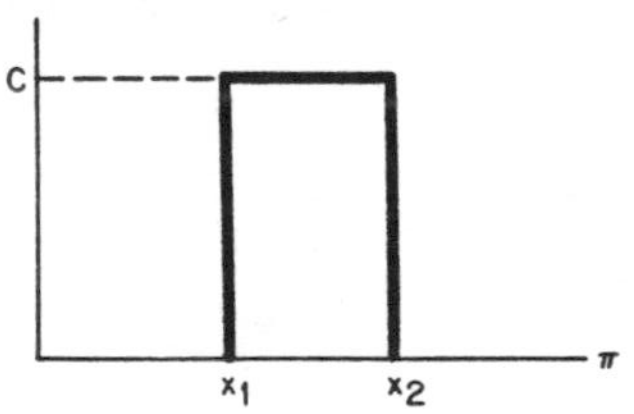

Fig. 8.3. Plot of a square wave.

Even if we do not have an odd or even function, we can approximate a function on the interval $0 \leqq z \leqq \pi$ by use of either the relations (8.70) or (8.72). The resultant approximating function will either be an odd or even function, respectively, with periodicity of 2π. The approximating functions found by use of Eqs. (8.70) or (8.72) are called *half-range series.*

If we wish to approximate a function over some other interval, say $-L \leqq z \leqq L$, then Eqs. (8.68) can be written as

$$
\begin{aligned}
a_0 &= \int_{-L}^{L} y\,dz/L \\
a_i &= \int_{-L}^{L} y \cos(i\pi z/L)\,dz/L \\
b_i &= \int_{-L}^{L} y \sin(i\pi z/L)\,dz/L
\end{aligned}
\tag{8.73}
$$

And similarly, the coefficients for the half-range series can be found by

$$
b_i = 2\int_0^L y \sin(i\pi z/L)\,dz/L \tag{8.74}
$$

for the half-range sine series, and

$$
\begin{aligned}
a_0 &= 2\int_0^L y\,dz/L \\
a_i &= 2\int_0^L y \cos(i\pi z/L)\,dz/L
\end{aligned}
\tag{8.75}
$$

for the half-range cosine series.

The Fourier series can be used to approximate tabular data by using a numerical integration method to evaluate the coefficients. The integrals can be replaced by simple sums, but we would have to define the argument values z_i in such a manner that the orthogonality relations (8.53) still hold.

One of the most important uses of Fourier series is to approximate functions with finite discontinuities. Consider the function

$$
\begin{aligned}
f(x) &= 0 \qquad 0 \leqq x \leqq x_1 \\
f(x) &= C \qquad x_1 \leqq x \leqq x_2 \\
f(x) &= 0 \qquad x_2 \leqq x \leqq \pi
\end{aligned}
\tag{8.76}
$$

This function is illustrated in Fig. 8.3. To approximate this with a half-range cosine series, we would evaluate the formulas

$$
\begin{aligned}
a_0 &= 2\int_0^{\pi} f(x)\,dx/\pi \\
a_i &= 2\int_0^{\pi} f(x) \cos ix\,dx/\pi
\end{aligned}
\tag{8.77}
$$

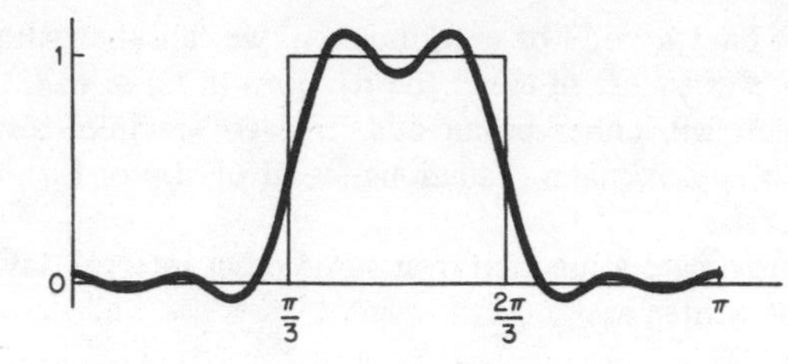

Fig. 8.4. Plot of an example square wave with a Fourier approximation.

Substituting the function given by Eqs. (8.76) into (8.77) gives

$$
\begin{aligned}
a_0 &= 2\int_{x_1}^{x_2} dx/\pi = 2(x_2 - x_1)/\pi \\
a_i &= 2\int_{x_1}^{x_2} \cos ix dx/\pi = 2(\sin ix_2 - \sin ix_1)/(i\pi)
\end{aligned}
\tag{8.78}
$$

Given values for x_1 and x_2, (8.78) can be evaluated to obtain the coefficients a_i. For example, suppose that $x_1 = \pi/3$ and $x_2 = 2\pi/3$. Then all coefficients with odd subscripts are zero and

$$
\begin{aligned}
a_0 &= 2/3 \\
a_2 &= -\sqrt{3}/\pi \\
a_4 &= \sqrt{3}/2\pi \\
a_6 &= 0 \\
a_8 &= -\sqrt{3}/4\pi \\
a_{10} &= \sqrt{3}/5\pi \\
a_{12} &= 0 \\
&\vdots
\end{aligned}
\tag{8.79}
$$

Applying these coefficients to (8.65) and truncating all terms where $i > 13$ gives

$$
g(x) = \frac{1}{3} + \frac{\sqrt{3}}{\pi}\left(-\cos 2x + \frac{1}{2}\cos 4x \frac{1}{4} - \cos 8x + \frac{1}{5}\cos 10x\right) \tag{8.80}
$$

This function is illustrated in Fig. 8.4 (dashed curve) with the function being approximated (solid lines).

8.6 ECONOMIZATION OF POWER SERIES

It has been stated in Sec. 8.1.2 that various criteria can be employed as the basis for curve fitting methods. No one of these criteria will be satisfactory for all problems.

The least squares criterion, which is widely used, may be less acceptable than some other criterion for certain problems.

If the method of least squares is applied to fit a function y to a polynomial such as

$$u = Ax + Bx^3 + Cx^5 \tag{8.81}$$

the resulting sum of the squares of the errors will indeed be less than (or equal to) the sum of the squares of the errors obtained by any other method. However, the errors could be very poorly distributed. If y is zero when x is zero, then the error will also be zero when x is zero. The error will usually be largest when x has its maximum value. The error curve will typically look like the one given in Fig. 8.5. This is the actual error curve for the least squares approximation to the sine function, which was given as an example in Sec. 8.3.

A function that can be expressed as a powers series can be approximated directly by use of Chebyshev polynomials. This method has the property of distributing the error more uniformly over the interval of approximation than the method of least squares. It also allows for a reduction (economization) of terms from that required by a Taylor's series approximation. It can be shown that the error whose absolute value is largest on the interval is least by this method. That is, the method to be described satisfies the criterion

$$|\varepsilon|_{max} = \text{minimum} \tag{8.82}$$

Table 8.3 gave a few examples of the Chebyshev polynomials. These equations can be solved for the powers of x as functions of the Chebyshev polynomials. Table 8.4 lists the first nine powers of x represented in terms of Chebyshev polynomials. The table setup emphasizes the fact that the odd powers of x are functions of the odd Chebyshev polynomials only, and that the even powers are functions of the even Chebyshev polynomials only.

The values in Table 8.4 can be substituted directly in a given powers series to obtain a new series that is a function of Chebyshev polynomials. Since the independent variable of the Chebyshev polynomials is defined only on the interval (−1, 1), the substitutions in Table 8.4 can be made only when the independent variable of the given powers series is limited to the same range.

TABLE 8.4 POWERS OF X AS A FUNCTION OF CHEBYSHEV POLYNOMIALS

Odd	Even
$x = T_1$	$x^2 = (T_0 + T_2)/2$
$x^3 = (3T_1 + T_3)/4$	$x^4 = (3T_0 + 4T_2 + T_4)/8$
$x^5 = (10T_1 + 5T_3 + T_5)/16$	$x^6 = (10T_0 + 15T_2 + 6T_4 + T_6)/32$
$x^7 = (35T_1 + 21T_3 + 7T_5 + T_7)/64$	$x^8 = (35T_0 + 56T_2 + 28T_4 + 8T_6 + T_8)/128$
$x^9 = (126T_1 + 84T_3 + 36T_5 + 9T_7 + T_9)/256$	

To fit a given function f(z), first transform the independent variable z using relations (8.63). Substitute this in the given function so that it is a function of x that varies only over the interval (−1, 1). If the function is not already a powers series, expand it by use of Taylor's series and truncate the series, retaining sufficient terms to attain the

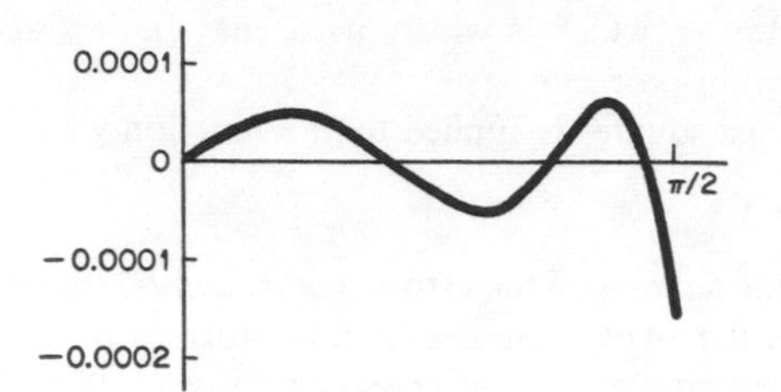

Fig. 8.5. Error curve for least squares fit to sine function.

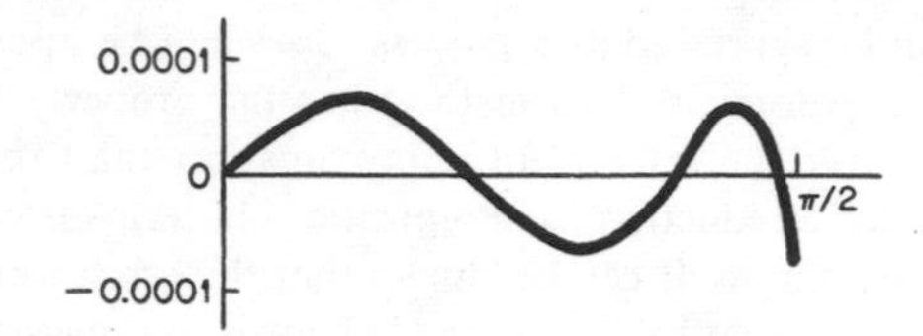

Fig. 8.6. Error curve for a Chebyshev approximation to sine function.

desired accuracy. Next substitute the expressions for the powers of x from Table 8.4 into the series. Collect terms giving a function of the form

$$u = A_0T_0 + A_1T_1 + A_2T_2 + \cdots \tag{8.83}$$

Since each Chebyshev polynomial satisfies the condition

$$-1 \leqq T_n \leqq 1 \tag{8.84}$$

neglecting a term A_nT_n will result in an error whose absolute value is less than $|A_n|$.

A second advantage of this method of curve fitting is that usually a number of terms can be discarded from the resulting function, Eq. (8.83), because the coefficients A_n of some terms may be sufficiently small. This is how economization is achieved. Before discarding more than one term, we must be certain that the sum of the absolute values of the coefficients of the terms to be discarded is less than the acceptable error. Once the function is truncated, it can be transformed back into a powers series in x by use of the relations in Table 8.3. Then, if necessary, that powers series can be transformed back into a powers series in the original variable using Eq. (8.62).

The following example illustrates this method. Suppose we want to approximate the sine function on the interval $(-\pi/2, \pi/2)$, without the absolute value of the error exceeding 0.0001. These conditions are satisfied by a Taylor's series truncated after the term in the ninth power of the independent variable.

$$\sin z = z - z^3/3! + z^5/5! - z^7/7! + z^9/9! \tag{8.85}$$

In order to obtain a powers series in which the independent variable is in the range $(-1, 1)$, we must make the substitution

$$z = (\pi/2)x \tag{8.86}$$

which gives

$$\sin(\pi/2)x = (\pi/2)x - (\pi/2)^3x^3/3! + (\pi/2)^5x^5/5! - (\pi/2)^7x^7/7! + (\pi/2)^9x^9/9! \qquad (8.87)$$

Substituting from Table 8.4 and collecting terms gives

$$\sin(\pi/2)x = 1.13364978\, T_1 - 0.13807064\, T_3 + 0.00449128\, T_5 \qquad (8.88)$$

The terms containing T_7 and T_9 have been omitted, since they are less than the allowable error. Substituting for the T's from Table 8.3 gives

$$\sin(\pi/2)x = 1.57031810\, x - 0.64210816\, x^3 + 0.07186048\, x^5 \qquad (8.89)$$

Thus

$$\sin z = 0.99969555\, z - 0.16567179\, z^3 + 0.00751434\, z^5 \qquad (8.90)$$

Fig. 8.6 gives the error curve for this approximation. Compare this error curve with that of Fig. 8.5. Note how the error is more evenly distributed and is confined to a smaller range. The largest error by the method of least squares is 0.00016, while the largest error for the Chebyshev economized series is 0.00007.

8.7 SUMMARY

This chapter discussed various methods of approximating tabular data or continuous functions. We will now summarize these methods and give some comparative examples.

In Sec. 8.1 the general problem of curve fitting was presented along with a brief description of a number of curve fitting methods. In Sec. 8.2, the method of least squares was presented. An example problem will help to illustrate some of the methods of Sec. 8.1.2 and will serve to compare these methods with the method of least squares.

Suppose that values of a dependent variable y have been determined by experimental means for various combinations of two independent variables x_1 and x_2. The results of this experiment are given in Table 8.5. Further suppose that the data in Table 8.5 should obey the function given by Eq. (8.4). Since there are (for $K = 3$) three unknown parameters A, B, and C, then method 1 of Sec. 8.1.2 says that these parameter values can be determined from three of the data points. This is done by substituting each of three values of y_i, x_{1i}, and x_{2i} into Eq. (8.4) for u, x_1, and x_2, respectively, thus forming three equations in three unknowns. For the sake of later analysis, all nine data points have been used to create the following equations:

TABLE 8.5 EXPERIMENTAL DATA

i	x_{1i}	x_{2i}	y_i
1	−1	−1	0.01
2	−1	0	1.87
3	−1	1	4.03
4	0	−1	−1.10
5	0	0	1.05
6	0	1	2.94
7	1	−1	−2.02
8	1	0	−0.08
9	1	1	1.94

$$0.01 = A - B - C \tag{8.91a}$$
$$1.87 = A - B \tag{8.91b}$$
$$4.03 = A - B + C \tag{8.91c}$$
$$-1.10 = A - C \tag{8.91d}$$
$$1.05 = A \tag{8.91e}$$
$$2.94 = A + C \tag{8.91f}$$
$$-2.02 = A + B - C \tag{8.91g}$$
$$-0.08 = A + B \tag{8.91h}$$
$$1.94 = A + B + C \tag{8.91i}$$

In choosing the three equations to use, we may select (8.91a), (8.91e), and (8.91i). If we do, elimination of A from (8.91a) and (8.91i) using (8.91e) will yield

$$\begin{aligned} 1.04 &= B + C \\ 0.89 &= B + C \end{aligned} \tag{8.92}$$

which are redundant and inconsistent. However, Eqs. (8.91a), (8.91e), and (8.91g) can be solved to yield $A = 1.05$, $B = -1.015$, and $C = 2.055$. After rounding these to two decimal places and substituting in Eq. (8.4), we obtain an equation that can be evaluated for various values of x_1 and x_2. Substituting the values from Table 8.5, we can determine u_i for each data point. Then, using (8.3), the error ε_i can be determined for each data point. These errors are given in column I of Table 8.6.

Method 3 of Sec. 8.1.2 suggests finding the parameter values from several sets of equations and then averaging all of the corresponding parameter values. Use of Eqs. (8.91b), (8.91f), and (8.91h) gives the values $A = 0.895$, $B = -0.975$, and $C = 2.045$. Also, using Eqs. (8.91c), (8.91d), and (8.91i) gives $A = 0.9425$, $B = -1.045$, and $C = 2.0425$. Averaging these two sets of values along with the set determined earlier and rounding off gives $A = 0.96$, $B = -1.01$, and $C = 2.05$. Use of these parameters gives the errors tabulated in column II of Table 8.6.

Method 4 requires that the equations be divided into K groups (in this case $K = 3$) and then averaging the equations in each group. Again we must be careful that our grouping does not lead to an inconsistent set of equations. Actually, the equations within a group need only be added, since averaging would only divide this sum by a constant. Using the groupings (a, b, g), (c, d, e), and (f, h, i) of Eqs. (8.91) and summing within each group results in

$$\begin{aligned} 3A - B - 2C &= -0.14 \\ 3A - B &= 3.98 \\ 3A + 2B + 2C &= 4.80 \end{aligned} \tag{8.93}$$

The solution of these equations gives $A = 0.96$, $B = -1.10$, and $C = 2.06$. The resulting errors from the use of these parameters in the original equation are given in column III of Table 8.6.

Finally, the rounded off results from the least squares method gives the parameter values $A = 0.96$, $B = -1.01$, and $C = 2.03$. The corresponding error values are shown in column IV of Table 8.6.

TABLE 8.6 COMPARATIVE ERRORS ($\varepsilon = y - A - Bx_1 - Cx_2$)

	I	*II*	*III*	*IV*
i	*Method 1*	*Method 3*	*Method 4*	*Least Squares*
1	0.00	0.09	0.01	0.07
2	−0.20	−0.10	−0.19	−0.10
3	−0.10	0.01	−0.09	0.03
4	−0.09	−0.01	0.00	−0.03
5	0.00	0.09	0.09	0.09
6	−0.17	−0.07	−0.08	−0.05
7	0.01	0.08	0.18	0.06
8	−0.11	−0.03	0.06	−0.03
9	−0.15	−0.06	0.02	−0.04
Sum of errors	−0.81	0.00	0.00	0.00
Sum of $(\text{errors})^2$	0.1217	0.0422	0.0952	0.0334

No conclusions should be drawn from Table 8.6, since it only illustrates the results of selected methods for one problem. However, it does show that, of the methods illustrated, the least squares method gives the smallest sum of squares of the errors.

The method of least squares is based on a criterion that allows for the method to be developed analytically. It is a general method, having many applications. Certain specialized forms, namely Chebyshev polynomials and Fourier series, are of particular interest and have been treated in Sec. 8.4 and 8.5, respectively.

It was demonstrated that least squares can be applied to mathematical forms that are both linear and nonlinear functions of the parameters. However, it was shown that the least squares method for those forms that are linear functions of the parameters can be reduced to a simple "cook book" procedure. Illustrations of how nonlinear forms could be linearized were also given.

Another important phase of curve fitting is to approximate continuous functions with simpler ones. Sections 8.3, 8.4, and 8.5 gave various methods of accomplishing this. It is important to note that these methods also permit the approximation of functions that are *sectionally continuous*; i.e., these functions may have *finite* discontinuities.

The methods given in Sec. 8.3, 8.4, and 8.5 are all based on the method of least squares. Another method of approximating functions was given in Sec. 8.6. Although that method is based on Chebyshev polynomials, it does not employ the method of least squares. The use of this method may be simpler to apply than the method of least squares and the results obtained, for some problems, may be more desirable than those obtained by least squares.

EXERCISES

1. Prepare a flowchart and write a FORTRAN program to fit a set of data points with a second-degree polynomial by the method of averages. Assume that inconsistent equations will *not* occur when consecutive data points are grouped together. Allow for as many as 100 data points.

2. Execute the program in Exercise 1 for the following data:

x	y	x	y	x	y	x	y	x	y
0.0	439.6	2.5	647.5	5.0	810.8	7.5	902.8	10.0	934.2
0.5	477.5	3.0	686.4	5.5	836.4	8.0	916.1	10.5	931.3
1.0	515.5	3.5	718.7	6.0	855.4	8.5	924.7	11.0	927.5
1.5	563.9	4.0	754.7	6.5	874.4	9.0	930.4	11.5	918.0
2.0	603.8	4.5	785.1	7.0	889.6	9.5	932.3	12.0	908.5

3. Prepare a flowchart and write a FORTRAN program to fit a set of data points with a second-degree polynomial by the method of least squares. Allow for as many as 100 data points.

4. Execute the program in Exercise 3 for the data of Exercise 2.

5. Prepare a flowchart and write a FORTRAN program to fit a function by the method of least squares to a polynomial of the form

$$f(x) = A + Bx^2 + Cx^4$$

where f(x) is a known function, and a starting value of x, an ending value of x, and an increment in x are input quantities.

6. Execute the program in Exercise 5 with the cosine function as the given function. Execute for each of the following sets of input parameters:

$x_{initial}$	x_{final}	$x_{increment}$
0.0	1.60	0.10
0.0	1.60	0.01

7. Prepare a flowchart and write a FORTRAN program to fit, by the method of least squares, a set of data points with a function of the form

$$y = Ae^{B(x-C)^2}$$

Allow for as many as 100 data points. *Note*: Linearize the function first.

8. Execute the program in Exercise 7 for the data of Exercise 2.

9. Prepare a flowchart and write a FORTRAN program that will fit, by the method of differential corrections, a set of data points to a function having three unknown coefficients. That is, write the program so that it can be easily modified to fit any function of the form

$$y = F(A,B,C,x)$$

given initial estimates of A, B, and C.

10. Execute the program in Exercise 9 using the data of Exercise 2, the function of Exercise 7, and the answers to Exercise 8 as initial estimates. Note that the final answers will be different from those of Exercise 8, since the methods of Exercises 7 and 9 have different criteria of best fit.

11. By the method of least squares for continuous functions given in Sec. 8.3, determine the coefficients of

$$\cos z = A + Bz + Cz^2 \qquad 0 \leqq z \leqq \pi/2$$

12. Using the methods of Sec. 8.4, find the coefficients of the function

$$\sin^{-1}(x) = AT_1(x) + BT_3(x) + CT_5(x) \qquad -1 \leqq x \leqq 1$$

13. By the methods of Sec. 8.5, find the first nine half-range cosine series Fourier coefficients for the function

$$f(x) = 0 \qquad 0 \leqq x \leqq \pi/8$$

$$f(x) = \frac{4x}{\pi} - \frac{1}{2} \qquad \pi/8 \leqq x \leqq 3\pi/8$$

$$f(x) = 1 \qquad 3\pi/8 \leqq x \leqq 5\pi/8$$

$$f(x) = \frac{7}{2} - \frac{4x}{\pi} \qquad 5\pi/8 \leqq x \leqq 7\pi/8$$

$$f(x) = 0 \qquad 7\pi/8 \leqq x \leqq \pi$$

14. Plot the function given in Exercise 13, and compute and plot the Fourier half-range cosine series determined in Exercise 13.

15. Determine the coefficients for the function given in Exercise 11 by Chebyshev economization of the Taylor's series expansion of the cosine function. Assume that terms involving T_6 and higher-order Chebyshev polynomials may bc disregarded.

9

Smoothing

9.1 ERRORS IN DATA

When observing physical phenomena, it is seldom possible to make a "correct" measurement. *Correct* here may mean: true, theoretical, or average. The difference between the observed value and the correct value is called an *error*, even though it may not necessarily be a mistake. There are several sources for such errors:

1. *Physical Anomalies that Mask the Phenomenon Being Observed* — For example, if we want to determine experimentally the time of high tide at some locality, we can measure the water line at various times. However, weather conditions may be such that wave conditions make our measurements difficult.
2. *Errors Introduced by Instrumentation* — In example 1, measurements can be made more easily if the water is allowed to run into a container through a small opening. This will damp out much of the wave action, but it may introduce a time lag in the measurements.
3. *Inaccuracies of the Measuring Equipment* — If the calibration of the measuring equipment is inaccurate, the readings made will be inaccurate.
4. *Human Errors* — The person making the measurement may make mistakes; or due to his own physical characteristics, he may introduce a systematic bias in the measurements.

If the phenomenon being observed is known to behave according to some general mathematical formula, and it is assumed that the errors are random, the observations can be "fit" to the formula to obtain a probable set of specific parameters that describe the particular experiment under observation. In Chap. 8, we discussed several methods of fitting experimental data.

In many cases, no general mathematical formula is known that describes the experiment. In such a case, it is desirable to eliminate as much of the error as possible so that the modified data would still accurately depict the phenomenon under observation. Such a process is known as *smoothing*. Frequently, the scientist graphs his data, and by a visual inspection draws in a curve that in his judgment "best" represents the data. This is adequate in many cases, but in some, especially those that involve large amounts of data, a more orderly method is required. The following sections will describe such methods.

The smoothing of digital data is analogous to filtering a continuous signal by means of a hardware device. The mathematical similarities of the two processes will be discussed in Sec. 9.3.

9.2 POLYNOMIAL SMOOTHING

If no mathematical expression is known to describe a phenomenon, we can select some arbitrary mathematical form, and by determining appropriate parameter values, construct a function that approximates the behavior of that phenomenon. A very useful mathematical function is the polynomial. The general form is

$$y = a_0 + a_1x + a_2x^2 + \cdots + a_nx^n \tag{9.1}$$

Although, in this discussion, we will confine our studies to functions of a single independent variable, the methods used here can be expanded to accomodate functions of more than one independent variable.

There are three ways that we might use the polynomial, Eq. (9.1), to represent the data

$$y_i = f(x_i) \tag{9.2}$$

These are:

1. If there are N data points, we can determine an N–1 order polynomial that exactly passes through all of the observed points. This procedure would not do any smoothing.
2. If there are N data points, we can select a polynomial of order less than N–1 and determine the coefficients by the method of least squares. This procedure may do too much smoothing.
3. Another method, which is to be discussed here in detail, uses least squares to fit only a portion of the data at one time. By repeating at several sets of data points, it is possible to eliminate small variations while still retaining the large variations.

There are many options available when applying method 3. We will see in Sec. 9.3 how to analyze the effects of the number of data points used and the degree of the polynomial used. Although there are several ways that data points can be grouped into sets and corresponding smoothed data points found, we will confine our discussion to only one method. In this method, we will find a polynomial to fit the first k data points; from this polynomial, we will find a smoothed value corresponding to one of the given data points; the next unused data point will be added to the previous group of data points and the first data point of that group will be discarded; then another polynomial and smoothed value will be found. The last two steps will be repeated until all data points have been used. In addition, if required, special steps must be taken to find smoothed values near the beginning and end of the data. An example will help explain this procedure. Suppose that we elect to fit groups of five data points to second-degree polynomials. Given the data points $(x_1, y_1), (x_2, y_2), (x_3, y_3), \ldots, (x_k, y_k)$, or more briefly $P_1, P_2, P_3, \ldots, P_k$, we would first select P_1, P_2, P_3, P_4, P_5; find a quadratic

$y = a_0^{(1)} + a_0^{(1)}x + a_2^{(1)}x^2$ that† fits these five points; and then find the smoothed values Y corresponding to the first three x's by

$$\begin{aligned} Y_1 &= a_0^{(1)} + a_1^{(1)}x_1 + a_2^{(1)}x_1^2 \\ Y_2 &= a_0^{(1)} + a_1^{(1)}x_2 + a_2^{(1)}x_2^2 \\ Y_3 &= a_0^{(1)} + a_1^{(1)}x_3 + a_2^{(1)}x_3^2 \end{aligned} \tag{9.3}$$

Then, using P_2, P_3, P_4, P_5, P_6, we obtain a new polynomial $y = a_0^{(2)} + a_1^{(2)}x + a_2^{(2)}x^2$, and from this find

$$Y_4 = a_0^{(2)} + a_1^{(2)}x_4 + a_2^{(2)}y_4^2 \tag{9.4}$$

Steps similar to these can be performed to determine $Y_5, Y_6, \ldots, Y_{k-3}$. Finally, the polynomial $y = a_0^{(k-2)} + a_1^{(k-2)}x + a_2^{(k-2)}x^2$ is obtained using the last set of data points $P_{k-4}, P_{k-3}, P_{k-2}, P_{k-1}, P_k$, and the last three smoothed values can be found by

$$\begin{aligned} Y_{k-2} &= a_0^{(k-2)} + a_1^{(k-2)}x_{k-2} + a_2^{(k-2)}x_{k-2}^2 \\ Y_{k-1} &= a_0^{(k-2)} + a_1^{(k-2)}x_{k-1} + a_2^{(k-2)}x_{k-1}^2 \\ Y_k &= a_0^{(k-2)} + a_1^{(k-2)}x_k + a_2^{(k-2)}x_k^2 \end{aligned} \tag{9.5}$$

The foregoing procedure, even for the simple case described, appears to be complicated. However, we will find that the procedure is not as complex as it appears and that, by imposing some restrictions, the procedure can be simplified further. These restrictions are 1) we will only consider data equally spaced in the independent variable x, and 2) we will always fit an *odd* number of data points. Therefore, to fit an nth-degree polynomial to the odd number of points $k = 2m + 1$, by the method of least squares, the equation of condition would be Eq. (9.1) and the normal equations would be††

$$\begin{aligned} \sum_{i=-m}^{m} y_i &= a_0(2m+1) + a_1 \sum_{i=-m}^{m} x_i + a_2 \sum_{i=-m}^{m} x_i^2 + \cdots + a_n \sum_{i=-m}^{m} x_i^n \\ \sum_{i=-m}^{m} x_i y_i &= a_0 \sum_{i=-m}^{m} x_i + a_1 \sum_{i=-m}^{m} x_i^2 + a_2 \sum_{i=-m}^{m} x_i^3 + \cdots + a_n \sum_{i=-m}^{m} x_i^{n+1} \\ \sum_{i=-m}^{m} x_i^2 y_i &= a_0 \sum_{i=-m}^{m} x_i^2 + a_1 \sum_{i=-m}^{m} x_i^3 + a_2 \sum_{i=-m}^{m} x_i^4 + \cdots + a_n \sum_{i=-m}^{m} x_i^{n+2} \\ &\vdots \\ \sum_{i=-m}^{m} x_i^n y_i &= a_0 \sum_{i=-m}^{m} x_i^n + a_1 \sum_{i=-m}^{m} x_i^{n+1} + a_2 \sum_{i=-m}^{m} x_i^{n+2} + \cdots + a_n \sum_{i=-m}^{m} x_i^{2n} \end{aligned} \tag{9.6}$$

†For simplicity in writing the normal equations, it is assumed that the k values of the independent variable are labeled $x_{-m}, x_{-m+1}, \ldots, x_{-1}, x_0, x_1, \ldots, x_{m-1}, x_m$.

††The superscripts in parentheses, (j), indicate the coefficients found when using the jth set of data points. Also, small y is used to denote observed values and large Y is used to denote smoothed values.

These equations can be solved for $a_0, a_1, a_2, \ldots, a_n$. Then, to find a smoothed value corresponding to a value of the independent variable x_i, the equation of condition is evaluated for that x_i:

$$Y_i = a_0 + a_1 x_i + a_2 x_i^2 + \ldots + a_n x_i^n \tag{9.7}$$

The simplifying assumptions made here makes the solution of the normal equations relatively easy. Furthermore, it can be shown that without loss of generality, we can assume that $x_0 = 0$ for each set of 2m+1 data points. Therefore,

$$x_i = ih \tag{9.8}$$

where h is the interval between values of the independent variable.

Now, in the normal equations, we have sums of the form

$$S_j = \sum_{i=-m}^{m} x_i^j$$

and (9.9)

$$T_j = \sum_{i=-m}^{m} x_i^j y_i$$

Since $x_i = ih$, these become

$$S_j = h^j \sum_{i=-m}^{m} i^j \tag{9.10a}$$

and

$$T_j = h^j \sum_{i=-m}^{m} i^j y_i \tag{9.10b}$$

Also, since we are considering only those cases with an odd number of points:

$$S_j = h^j \sum_{i=-m}^{-1} i^j + 0 + h^j \sum_{i=-1}^{m} i^j \tag{9.11a}$$

For j odd,

$$\sum_{i=-m}^{-1} i^j = -\sum_{i=1}^{m} i^j$$

Therefore,

$$S_j = 0 \tag{9.11b}$$

For j even,

$$\sum_{i=-m}^{-1} i^j = \sum_{i=1}^{m} i^j$$

Therefore,

$$S_j = 2h^j \sum_{i=1}^{m} i^j \qquad j \neq 0 \tag{9.11c}$$

Note that for the special case $j = 0$,

$$S_0 = 2m + 1 \tag{9.11d}$$

(For actual values of these sums, consult a mathematical handbook for the sums of the powers of consecutive integers.)

Applying (9.11b), (9.11c), and (9.11d) to Eqs. (9.6),

$$\begin{aligned}
T_0 &= a_0S_0 + 0 + a_2S_2 + 0 + \cdots \\
T_1 &= 0 + a_1S_2 + 0 + a_3S_4 + \cdots \\
T_2 &= a_0S_2 + 0 + a_2S_4 + 0 + \cdots \\
T_3 &= 0 + a_1S_4 + 0 + a_3S_6 + \cdots \\
&\vdots
\end{aligned} \tag{9.12}$$

These equations can be separated into two sets:

$$\begin{aligned}
T_0 &= a_0S_0 + a_2S_2 + \cdots \\
T_2 &= a_0S_2 + a_2S_4 + \cdots \\
&\vdots
\end{aligned} \tag{9.13}$$

and

$$\begin{aligned}
T_1 &= a_1S_2 + a_3S_4 + \cdots \\
T_3 &= a_1S_4 + a_3S_6 + \cdots \\
&\vdots
\end{aligned} \tag{9.14}$$

Once the coefficients $a_0, a_1, a_2, \ldots, a_n$ have been determined from (9.13) and (9.14) they can be used in the equation of condition to find the smoothed values corresponding to various x_i; thus

$$\begin{aligned}
&\vdots \\
Y_{-2} &= a_0 + a_1(-2h) + a_2(-2h)^2 + \cdots \\
Y_{-1} &= a_0 + a_1(-h) + a_2(-h)^2 + \cdots \\
Y_0 &= a_0 \\
Y_1 &= a_0 + a_1(h) + a_2(h)^2 + \cdots \\
Y_2 &= a_0 + a_1(2h) + a_2(2h)^2 + \cdots \\
&\vdots
\end{aligned} \tag{9.15}$$

This still seems complex, but for a specific number of data points and a specific degree polynomial, (9.15) can be expressed as functions of the observed ordinate values only.

9.2.1 Specific Smoothing Formulas

Let us solve and tabulate a few smoothing formulas for various degree polynomials.

9.2.1.1. *Linear Smoothing Formulas*

For a first-order smoothing formula (linear), the equation of condition is

$$y = a_0 + a_1 x \tag{9.16}$$

and the normal equations are

$$\begin{aligned} T_0 &= a_0 S_0 \\ T_1 &= a_1 S_2 \end{aligned} \tag{9.17}$$

The T's and S's can be evaluated by Eqs. (9.10b) and (9.11c) and (9.11d), respectively. In order to obtain further literal expressions, this must be done for specific numbers of data points k. If $k = 3$ $(m = 1)$,

$$\begin{aligned} T_0 &= y_{-1} + y_0 + y_1 \\ S_0 &= 3 \\ T_1 &= h(-y_{-1} + y_1) \\ S_2 &= 2h^2 \end{aligned} \tag{9.18}$$

Therefore

$$\begin{aligned} a_0 &= (y_{-1} + y_0 + y_1)/3 \\ a_1 &= (-y_{-1} + y_1)/(2h) \end{aligned} \tag{9.19}$$

Then

$$\begin{aligned} Y_{-1} &= \frac{1}{6}(5y_{-1} + 2y_0 - y_1) \\ Y_0 &= \frac{1}{3}(y_{-1} + y_0 + y_1) \\ Y_1 &= \frac{1}{6}(-y_{-1} + 2y_0 + 5y_1) \end{aligned} \tag{9.20}$$

If $k = 5(m = 2)$, we find that

$$\begin{aligned} Y_{-2} &= \frac{1}{5}(3y_{-2} + 2y_{-1} + y_0 - y_2) \\ Y_{-1} &= \frac{1}{10}(4y_{-2} + 3y_{-1} + 2y_0 + y_1) \\ Y_0 &= \frac{1}{5}(y_{-2} + y_{-1} + y_0 + y_1 + y_2) \\ Y_1 &= \frac{1}{10}(y_{-1} + 2y_0 + 3y_1 + 4y_2) \\ Y_2 &= \frac{1}{5}(-y_{-2} + y_0 + 2y_1 + 3y_2) \end{aligned} \tag{9.21}$$

If $k = 7 (m = 3)$, then

$$
\begin{aligned}
Y_{-3} &= \frac{1}{28}(13y_{-3} + 10y_{-2} + 7y_{-1} + 4y_0 + y_1 - 2y_2 - 5y_3) \\
Y_{-2} &= \frac{1}{14}(5y_{-3} + 4y_{-2} + 3y_{-1} + 2y_0 + y_1 - y_3) \\
Y_{-1} &= \frac{1}{28}(7y_{-3} + 6y_{-2} + 5y_{-1} + 4y_0 + 3y_1 + 2y_2 + y_3) \\
Y_0 &= \frac{1}{7}(y_{-3} + y_{-2} + y_{-1} + y_0 + y_1 + y_2 + y_3) \\
Y_1 &= \frac{1}{28}(y_{-3} + 2y_{-2} + 3y_{-1} + 4y_0 + 5y_1 + 6y_2 + 7y_3) \\
Y_2 &= \frac{1}{14}(-y_{-3} + y_{-1} + 2y_0 + 3y_1 + 4y_2 + 5y_3) \\
Y_3 &= \frac{1}{28}(-5y_{-3} - 2y_{-2} + y_{-1} + 4y_0 + 7y_1 + 10y_2 + 13y_3)
\end{aligned}
\tag{9.22}
$$

9.2.1.2 *Quadratic Smoothing Formulas*

For a second-order smoothing formula (quadratic), the equation of condition is

$$y = a_0 + a_1x + a_2x^2 \tag{9.23}$$

and the normal equations are

$$
\begin{aligned}
T_0 &= a_0S_0 + a_2S_2 \\
T_2 &= a_0S_2 + a_2S_4
\end{aligned}
\tag{9.24}
$$

and

$$T_1 = a_1S_2 \tag{9.25}$$

Solving Eqs. (9.24), we have

$$
\begin{aligned}
a_0 &= (T_0S_4 - T_2S_2)/(S_0S_4 - S_2^2) \\
a_2 &= (T_2S_0 - T_0S_2)/(S_0S_4 - S_2^2)
\end{aligned}
\tag{9.26}
$$

From Eq. (9.25),

$$a_1 = T_1/S_2 \tag{9.27}$$

If $k = 5$ ($m = 2$), then

$$\begin{aligned}
Y_{-2} &= \frac{1}{35}(31y_{-2} + 9y_{-1} - 3y_0 - 5y_1 + 3y_2) \\
Y_{-1} &= \frac{1}{35}(9y_{-2} + 13y_{-1} + 12y_0 + 6y_1 - 5y_2) \\
Y_0 &= \frac{1}{35}(-3y_{-2} + 12y_{-1} + 17y_0 + 12y_1 - 3y_2) \\
Y_1 &= \frac{1}{35}(-5y_{-2} + 6y_{-1} + 12y_0 + 13y_1 + 9y_2) \\
Y_2 &= \frac{1}{35}(3y_{-2} - 5y_{-1} - 3y_0 + 9y_1 + 31y_2)
\end{aligned} \tag{9.28}$$

If $k = 7$ ($m = 3$), then

$$\begin{aligned}
Y_{-3} &= \frac{1}{42}(32y_{-3} + 15y_{-2} + 3y_{-1} - 4y_0 - 6y_1 - 3y_2 + 5y_3) \\
Y_{-2} &= \frac{1}{28}(10y_{-3} + 8y_{-2} + 6y_{-1} + 4y_0 + 2y_1 - 2y_3) \\
Y_{-1} &= \frac{1}{28}(2y_{-3} + 6y_{-2} + 8y_{-1} + 8y_0 + 6y_1 + 2y_2 - 4y_3) \\
Y_0 &= \frac{1}{21}(-2y_{-3} + 3y_{-2} + 6y_{-1} + 7y_0 + 6y_1 + 3y_2 - 2y_3) \\
Y_1 &= \frac{1}{28}(-4y_{-3} + 2y_{-2} + 6y_{-1} + 8y_0 + 8y_1 + 6y_2 + 2y_3) \\
Y_2 &= \frac{1}{28}(-2y_{-3} + 2y_{-1} + 4y_0 + 6y_1 + 8y_2 + 10y_3) \\
Y_3 &= \frac{1}{42}(5y_{-3} - 3y_{-2} - 6y_{-1} - 4y_0 + 3y_1 + 15y_2 + 32y_3)
\end{aligned} \tag{9.29}$$

9.2.1.3 *Cubic Smoothing Formulas*

For a third-order smoothing formula (cubic), the equation of condition is

$$y = a_0 + a_1x + a_2x^2 + a_3x^3 \tag{9.30}$$

and the normal equations are

$$\begin{aligned}
T_0 &= a_0S_0 + a_2S_2 \\
T_2 &= a_0S_2 + a_2S_4
\end{aligned} \tag{9.31}$$

and

$$\begin{aligned}
T_1 &= a_1S_2 + a_3S_4 \\
T_3 &= a_1S_4 + a_3S_6
\end{aligned} \tag{9.32}$$

Since Eqs. (9.31) are identical to Eqs. (9.24), the solution for a_0 and a_2 are given by Eqs. (9.26). However, from (9.32),

$$\begin{aligned} a_1 &= (T_1S_6 - T_3S_4)/(S_2S_6 - S_4^2) \\ a_3 &= (T_3S_2 - T_1S_4)/(S_2S_6 - S_4^2) \end{aligned} \tag{9.33}$$

If $k = 5(m = 2)$, then

$$\begin{aligned} Y_{-2} &= \frac{1}{210}(207y_{-2} - 58y_{-1} - 18y_0 + 82y_1 - 3y_2) \\ Y_{-1} &= \frac{1}{210}(12y_{-2} + 127y_{-1} + 72y_0 - 13y_1 + 12y_2) \\ Y_0 &= \frac{1}{35}(-3y_{-2} + 12y_{-1} + 17y_0 + 12y_1 - 3y_2) \\ Y_1 &= \frac{1}{210}(12y_{-2} - 13y_{-1} + 72y_0 + 127y_1 + 12y_2) \\ Y_2 &= \frac{1}{210}(-3y_{-2} + 82y_{-1} - 18y_0 - 58y_1 + 207y_2) \end{aligned} \tag{9.34}$$

If $k = 7(m = 3)$, then

$$\begin{aligned} Y_{-3} &= \frac{1}{42}(39y_{-3} + 8y_{-2} - 4y_{-1} - 4y_0 + y_1 + 4y_2 - 2y_3) \\ Y_{-2} &= \frac{1}{42}(8y_{-3} + 19y_{-2} + 16y_{-1} + 6y_0 - 4y_1 - 7y_2 + 4y_3) \\ Y_{-1} &= \frac{1}{42}(-4y_{-3} + 16y_{-2} + 19y_{-1} + 12y_0 + 2y_1 - 4y_2 + y_3) \\ Y_0 &= \frac{1}{21}(-2y_{-3} + 3y_{-2} + 6y_{-1} + 7y_0 + 6y_1 + 3y_2 - 2y_3) \\ Y_1 &= \frac{1}{42}(y_{-3} - 4y_{-2} + 2y_{-1} + 12y_0 + 19y_1 + 16y_2 - 4y_3) \\ Y_2 &= \frac{1}{42}(4y_{-3} - 7y_{-2} - 4y_{-1} + 6y_0 + 16y_1 + 19y_2 + 8y_3) \\ Y_3 &= \frac{1}{42}(-2y_{-3} + 4y_{-2} + y_{-1} - 4y_0 - 4y_1 + 8y_2 + 39y_3) \end{aligned} \tag{9.35}$$

9.2.2 Common Features of Smoothing Formulas

From the previous examples, some common features of polynomial smoothing functions can be deduced. The first and probably most important feature is that, for equally spaced arguments and an odd number of data points, the independent variable does not appear in the smoothing formulas. Furthermore, any of the given formulas can be written in the form

$$Y_p^{(n)} = \sum_{i=-m}^{m} w_{pi}^{(n)} y_i \tag{9.36}$$

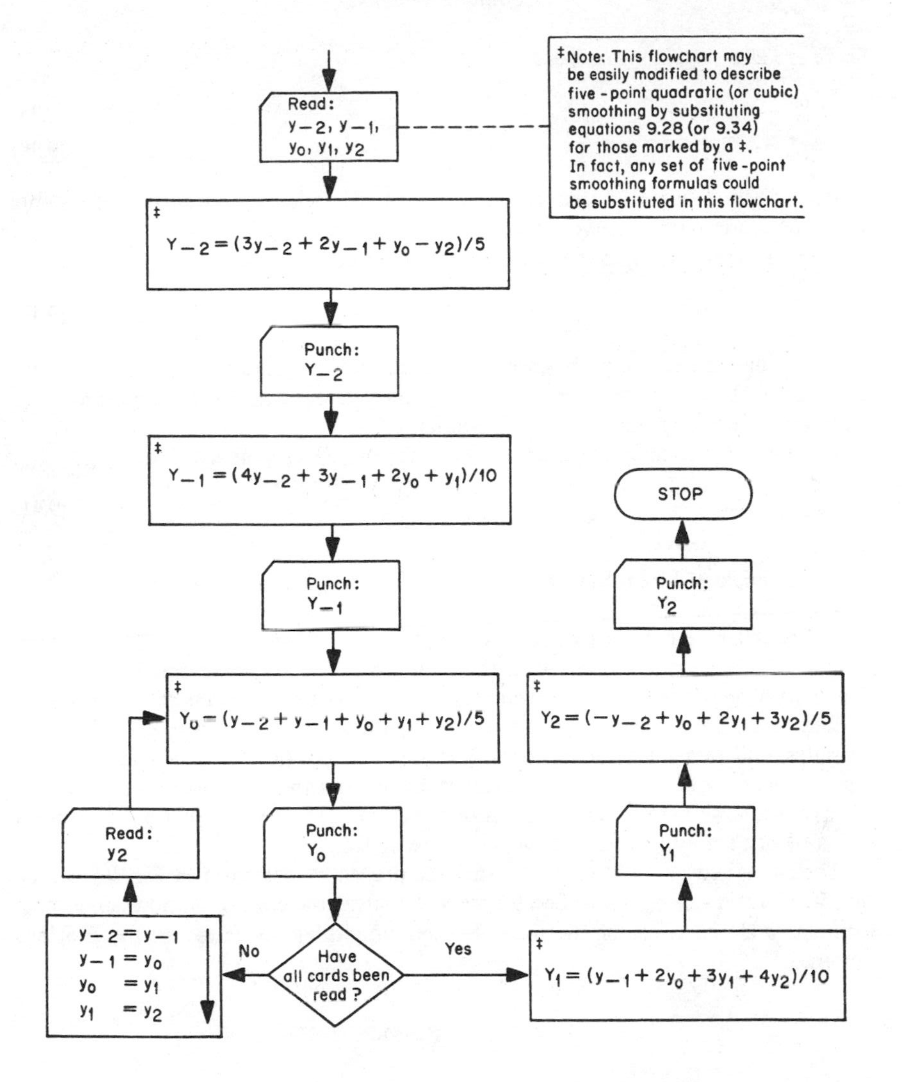

Fig. 9.1. Flowchart for five-point linear smoothing.

where the $w_{pi}^{(n)}$ are a set of weighting constants. The superscript (n) denotes the order of the smoothing formula; m specifies the number of data points $k = 2m + 1$ to be used; and p denotes which smoothed value will be obtained using these weighting constants.

We also have the condition

$$\sum_{i=-m}^{m} w_{pi}^{(n)} = 1 \qquad (9.37)$$

There are also certain symmetries:

$$w_{oi}^{(n)} = w_{o,-i}^{(n)} \tag{9.38}$$

$$w_{pi}^{(n)} = w_{-p,-i}^{(n)} \tag{9.39}$$

Actually (9.38) is a special case of (9.39). But since it relates the weighting constants within one formula, it deserves some special notice.

Another interesting fact to note is that for n even, say $n = 2q$,

$$w_{oi}^{(2q)} = w_{oi}^{(2q+1)} \tag{9.40}$$

This says that the formula obtained for finding the smoothed value at the midpoint of the interval by using an even-degree polynomial is the same as that obtained when using the next higher odd-degree polynomial.

For linear smoothing formulas, we also have the convenient relation

$$w_{oi}^{(1)} = 1/(2m + 1) \tag{9.41}$$

9.2.3 Example of Linear Smoothing

In Sec. 9.2, a verbal description of a smoothing procedure was given. Now that the formulas have been derived, the method can be described more concisely using a flowchart of a specific example. Figure 9.1 is a flowchart of a process for reading a set of data cards, finding the smoothed values using a set of five-point linear smoothing formulas, and punching out the smoothed answers. We have included in the figure a procedure for obtaining smoothed values for the first two and last two data points. If there is a large amount of data, loss of these few data points would not be serious and only one smoothing formula would be needed.

The smoothed data obtained by using the procedure described in Fig. 9.1 can be smoothed again and again. In Sec. 9.3 we will analyze the effect of various smoothing formulas and the effect of the multiple pass smoothing suggested in the previous sentence.

9.3 EFFECTS OF SMOOTHING

In this section, we will derive a method for analyzing the effects of various smoothing formulas. This derivation will involve the use of the Laplace transform. If the reader is unfamiliar with the Laplace transform, he may skip this section and go on to subsequent sections where the results of this analysis are applied.

First, let us assume that the function to be smoothed is a function of time. The function need not be a function of time for the following analysis to be valid, but the technical jargon is more meaningful for functions where time is the independent variable. The purpose in smoothing experimental data is to remove errors. These errors can be regarded as a second set of data that has been added to our "correct" set of data. The error data usually has different characteristics from the "correct" data. (If it does not, we have no way of removing it.) Generally, the errors vary in a somewhat random fashion at high frequencies. However, in the case where errors are introduced by instrumentation and human bias, the errors usually vary at low frequencies. Therefore, it is desirable to use a smoothing function that filters out data that varies at frequencies higher than the frequencies expected to exist in the phenomenon being observed. Such a smoothing function is called a *lowpass filter*. If a smoothing function removes low-frequency data, it is called a *highpass filter*. If the smoothing function filters out both low- and high-frequency data, but allows some data to pass through, it is called a *bandpass filter*. Further discussion of lowpass and bandpass filters is presented in Sec. 9.4. Expressing the smoothing function, Eq. (9.36), as a function of time, we can write

$$\overline{Y}(t) = \sum_{i=-m}^{m} w_i Y(t + i\Delta t) \tag{9.42}$$

It is possible to express the Laplace transform of a translated time function in terms of the transform of the untranslated time function. However, since the function of time to be transformed must be zero for $t < 0$, we must define our original time function $Y(t)$ to be zero for $t < m\Delta t$ so that the transform of $Y(t - m\Delta t)$ can be taken. Making this assumption, the transform of Eq. (9.42) becomes

$$\bar{y}(s) = \sum_{i=-m}^{m} w_i e^{is\Delta t} y(s) \tag{9.43}$$

where $\bar{y}(s)$ and $y(s)$ are the Laplace transforms of $\overline{Y}(t)$ and $Y(t)$, respectively. The purpose in taking the Laplace transform is to obtain functions, with frequency as the independent variable, that correspond to the original functions of time. An analysis of the ratio of these functions $\bar{y}(s)/y(s)$ will give us information on how the smoothing function affects the frequency content of the original data.

We will only consider those smoothing functions that give the smoothed function value at the midpoint of the smoothing interval. In those functions, by Eq. (9.38), $w_i = w_{-i}$. Then we can write (9.43) as

$$\bar{y}(s) = y(s)\left[w_o + \sum_{i=1}^{m} w_i\,(e^{is\Delta t} + e^{-is\Delta t})\right] \tag{9.44}$$

Since the Laplace transform is valid for complex as well as real values of s, we can let $s = j\omega$ ($j^2 = -1$) in Eq. (9.44):

$$\bar{y}(j\omega) = y(j\omega)\left[w_o + \sum_{i=1}^{m} w_i\,(e^{ij\omega\Delta t} + e^{-ij\omega\Delta t})\right] \tag{9.45}$$

Since

$$e^{jx} + e^{-jx} = 2\cos x \tag{9.46}$$

then

$$e^{ij\omega\Delta t} + e^{-ij\omega\Delta t} = 2\cos i\omega\Delta t \tag{9.47}$$

and thus (9.45) becomes

$$\bar{y}(j\omega) = y(j\omega)\left[w_o + 2\sum_{i=1}^{m} w_i \cos i\omega\Delta t\right] \tag{9.48}$$

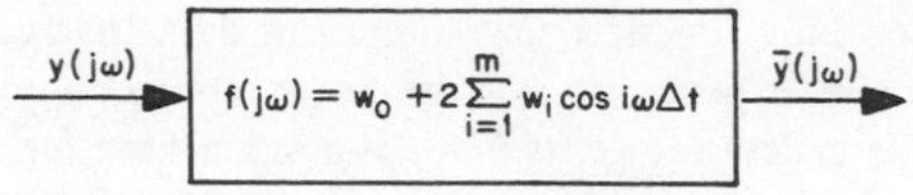

$y(j\omega) \rightarrow \boxed{f(j\omega) = w_o + 2\sum_{i=1}^{m} w_i \cos i\omega\Delta t} \rightarrow \bar{y}(j\omega)$

Fig. 9.2. Block diagram of a filter function.

The term in brackets on the right side of Eq. (9.48) then tells us the effect, as a function of frequency, of the given smoothing formula. This term is comparable to the transform of a filter in an electrical circuit. This concept is shown pictorially in Fig. 9.2 and a plot of an actual filter function is shown in Fig. 9.3.

9.3.1 Effects of Polynomial Smoothing

The function derived in Sec. 9.3

$$f(j\omega) = w_o + 2\sum_{i=1}^{m} w_i \cos i\omega\Delta t \tag{9.49}$$

will be referred to herein as the *filter function*. Even though (9.49) describes the effects of the smoothing, we need to look at some specific cases to better understand these effects.

Let us first consider the effects of applying a *seven-point linear* smoothing formula. In that formula,

$$w_{oi}^{(1)} = 1/7 \qquad (i = -3, -2, -1, 0, 1, 2, 3) \tag{9.50}$$

Applying Eq. (9.50) to (9.49) gives

$$f(j\omega) = (1 + 2\cos\omega\Delta t + 2\cos 2\omega\Delta t + 2\cos 3\omega\Delta t)/7 \tag{9.51}$$

Figure 9.3 plots this function with $\omega\Delta t$ as the abscissa. This function (and the ones to be shown later) can be interpreted as follows. For the frequency(s) where the filter function value is 1, the original data is passed by the smoothing function with no change. For frequencies where the filter function value is less than 1, the original data is attenuated (i.e., diminished) by the smoothing function. For frequencies where the filter function value is greater than 1 (see Figs. 9.6 and 9.8), the original data is amplified by the smoothing function. For those frequencies where the filter function value is negative, the attenuation (or amplification) factor is the absolute value of the filter function. The negative sign only relates to phase shift in the data.

Since the data to be smoothed Y(t) is only available at time intervals of Δt, it would appear that components of the data (or noise) with frequencies greater than $1/\Delta t$ would be lost. Actually, these components are not lost. They will be passed, to some extent, to the smoothed data. Figure 9.3 illustrates just the first half cycle ($0 \leq \omega\Delta t \leq \pi$) of the function given by Eq. (9.51); the second half cycle ($\pi \leq \omega\Delta t \leq 2\pi$) of the function is a mirror reflection of the first half cycle. Successive cycles are duplicates of the first cycle. It is this cyclic nature of the filter function that allows some higher frequencies to be passed by the smoothing formula. It would be desirable to eliminate that part of the data with frequencies greater than $1/\Delta t$, i.e.,

$$\omega > 2\pi/\Delta t \tag{9.52}$$

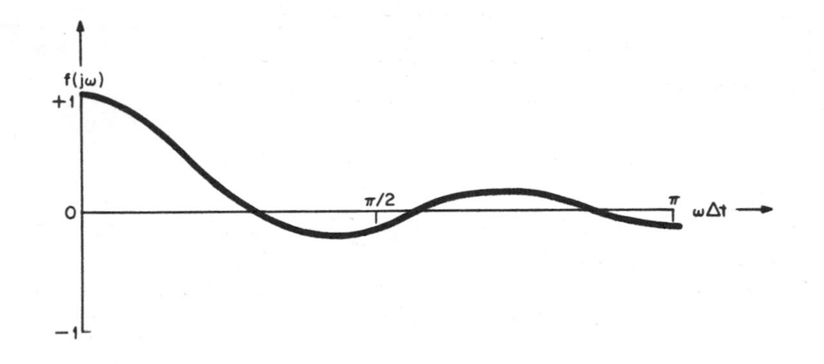

Fig. 9.3. Filter function of seven-point linear smoothing.

In fact, we would like to confine our attention to the frequencies for which

$$\omega < \pi/\Delta t \tag{9.53}$$

or

$$\omega\Delta t < \pi \tag{9.54}$$

We would then only need to consider the first half cycle of the filter function.

It may be known that the data does not contain high frequencies or if present are trivial. If this is not the case, the data must either be filtered by analog methods before the sampling is done or the sampling time (Δt) must be decreased so that the higher frequencies fall in the region defined by Eq. (9.54) and can be attenuated by digital techniques. For the remainder of this discussion, we will assume that frequencies for which $\omega\Delta t$ is greater than π can be safely ignored.

For the smoothing formula being considered here (seven-point linear) we see by Fig. 9.3 that at $\omega\Delta t = 0$ there is no attenuation; as $\omega\Delta t$ increases to about 0.93, the attenuation increases to 100 percent. For the remainder of the plot, the attenuation is never less than 80 percent. Therefore, this smoothing formula is a fairly good lowpass filter as we suspected it would be.

Suppose we next examine the effects of a seven-point second-order smoothing function. By the fourth equation of Eq. (9.29), we have the weighting coefficients $w_o = 7/21$, $w_1 = 6/21$, $w_2 = 3/21$, $w_3 = -2/21$. Applying these to the general form of the filter function, Eq. (9.49), we can write

$$f(j\omega) = (7 + 12\cos\omega\Delta t + 6\cos 2\omega\Delta t - 4\cos 3\omega\Delta t)/21 \tag{9.55}$$

A plot of this is shown in Fig. 9.4. In this example, the attenuation does not reach 100 percent until $\omega = 1.6/\Delta t$. This is also as expected, because a second-order curve will more closely approximate the data than a first-order curve. Therefore, more irregularities (meaning higher frequencies) will be passed by the filter function.

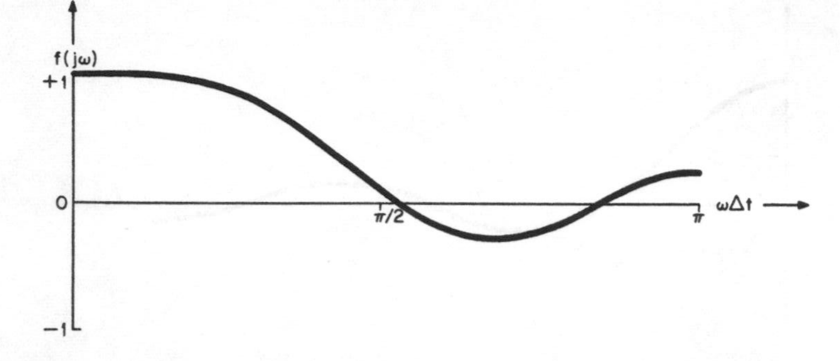

Fig. 9.4. Filter function of seven-point, second-order smoothing.

9.3.2 Effects of Multiple-Pass Smoothing

Multiple-pass smoothing is the process of applying a given smoothing formula more than once. That is, after an initial smoothing of the raw data, the resultant data is smoothed. This process can be repeated a number of times. Two methods can be used to analyze the effects of multiple-pass smoothing. One method is to analytically substitute the smoothing formula into itself repeatedly to obtain a new single-pass smoothing formula. Then this formula can be analyzed as was done previously. For example, on the first smoothing we have

$$\overline{Y}(t) = \sum_{i=-m}^{m} w_i Y(t + i\Delta t) \tag{9.56}$$

and on the second smoothing,

$$\overline{\overline{Y}}(t) = \sum_{i=-m}^{m} w_i \overline{Y}(t + i\Delta t) \tag{9.57}$$

For a specific formula, the value of $\overline{Y}(t)$ obtained in Eq. (9.56) can be substituted into (9.57) to yield $\overline{\overline{Y}}$ as a function of Y. The transform of this function can then be found. A second, simpler method that will yield the same results is to first find the transform of the smoothing formula and then raise it to a power corresponding to the number of smoothings. This gives the required filter function because we can write

$$\overline{y}(s) = f(s)\, y(s) \tag{9.58}$$

and

$$\overline{\overline{y}}(s) = f(s)\, \overline{y}(s) \tag{9.59}$$

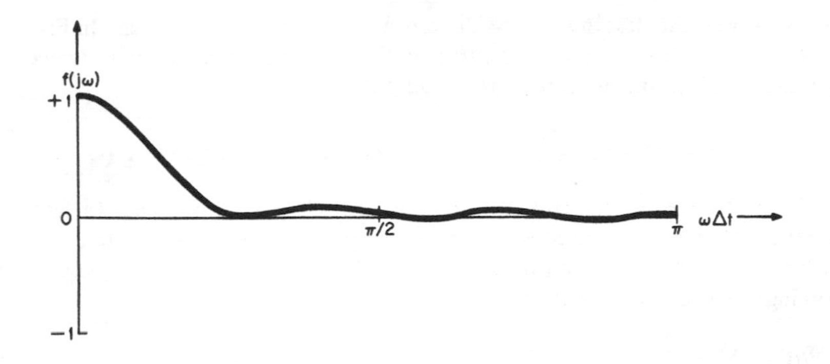

Fig. 9.5. Filter function of two passes of seven-point linear smoothing.

Hence,

$$\bar{\bar{y}}(s) = [f(s)]^2\, y(s) \tag{9.60}$$

Therefore, the filter function of n smoothings is $[f(s)]^n$, where f(s) is the filter function for one smoothing. As an example, consider the seven-point linear smoothing formula whose filter function was given by Eq. (9.51). The filter function for two passes is then

$$\begin{aligned}[f(j\omega)]^2 = (1 &+ 4\cos^2 \omega\Delta t + 4\cos^2 2\omega\Delta t + 4\cos^2 3\omega\Delta t + 4\cos\omega\Delta t \\ &+ 4\cos 2\omega\Delta t + 4\cos 3\omega\Delta t + 8\cos\omega\Delta t\cos 2\omega\Delta t \\ &+ 8\cos\omega\Delta t\cos 3\omega\Delta t + 8\cos 2\omega\Delta t\cos 3\omega\Delta t)/49\end{aligned} \tag{9.61}$$

Using trigonometric identities to eliminate all terms involving squares of cosines and products of cosines, (9.61) can be reduced to

$$\begin{aligned}f^2(j\omega) = \frac{1}{49}(7 &+ 12\cos\omega\Delta t + 10\cos 2\omega\Delta t + 8\cos 3\,\omega\Delta t + 6\cos 4\omega\Delta t \\ &+ 4\cos 5\omega\Delta t + 2\cos 6\,\omega\Delta t)\end{aligned} \tag{9.62}$$

In a similar fashion, the filter function for n passes can be determined.

The filter function given by (9.62) would have resulted if we smoothed one time with the smoothing formula:

$$\begin{aligned}\bar{Y}_0 = (Y_{-6} &+ 2Y_{-5} + 3Y_{-4} + 4Y_{-3} + 5Y_{-2} + 6Y_{-1} + 7Y_0 + 6Y_1 \\ &+ 5Y_2 + 4Y_3 + 3Y_4 + 2Y_5 + Y_6)/49\end{aligned} \tag{9.63}$$

This can be verified by applying the weighting coefficients from this equation to the general form of the filter function, Eq. (9.49). Applying this formula once is probably faster than applying (9.51) twice.

The purpose of multiple-pass smoothing is to remove more irregularities. That this occurs can be verified by plotting the filter function given by Eq. (9.62). This plot is shown in Fig. 9.5; we see that the attenuation is greater at high frequencies than that obtained by using (9.51) once, as shown in Fig. 9.3. It is also interesting to note that Fig. 9.5 can be obtained from Fig. 9.3 by squaring each ordinate. Note that Fig. 9.5

has no negative ordinates; this follows directly from the previous statement. In Figs. 9.3 and 9.4, the negative portions of the curves indicate there is attenuation accompanied by a phase shift into the third or fourth quadrant.

9.4 DERIVATION OF OTHER SMOOTHING FORMULAS

In Sec. 9.3.2 we considered the effects of multiple-pass smoothing. In fact, we demonstrated that a new one-pass smoothing formula, Eq. (9.63), can be derived from the filter function, Eq. (9.62), for a two-pass smoothing process. Since our polynomial smoothing functions all have the form

$$\overline{Y}(t) = \sum_{i=-m}^{m} w_i Y(t + i\Delta t) \qquad (9.56)$$

it seems reasonable that we can construct other smoothing formulas also of this form.† Although the polynomial smoothing formulas already given are useful (primarily because of their simplicity), the filtering obtained may not be what is needed. In order to obtain formulas that filter in the manner we desire, we must select weighting constants that will yield the desired filter functions. Fortunately, the transform of Eq. (9.56) can be expressed as a half-range Fourier cosine series.

$$f(j\omega) = w_0 + 2 \sum_{i=-m}^{m} w_i \cos i\omega t \qquad (9.49)$$

Therefore, since (9.49) is in the Fourier form, Eq. (8.65), the Fourier analysis techniques of Sec. 8.5 can be applied to find the weighting constants. Comparing Eq. (9.49) with (8.65), we note that

$$w_i = a_i/2 \qquad (9.64)$$

and hence the equations for finding the Fourier series coefficients, Eqs. (8.72), can be rewritten as follows to obtain the weighting constants:

$$w_0 = \int_0^{\pi} f(z)dz/\pi$$

$$w_i = \int_0^{\pi} f(z)\cos izdz/\pi \qquad (9.65)$$

where f(z) is the desired filter function and

$$z = \omega\Delta t \qquad (9.66)$$

Actually, the Fourier series is an infinite series that can be constructed to represent any function that does not have any infinite discontinuities in the given interval. In order to use a Fourier series, we must truncate it at a finite number of terms. In most cases, we can obtain a satisfactory approximation in this way.

Now that we have a technique for approximating a filter function, we must decide which filter function we wish to approximate. Probably the most desirable filter is the

†This is a special case of a more general form that will be given at the end of this section.

bandpass filter. This is one that passes data in a certain range of frequencies with no attenuation, but does not pass data at any other frequencies. Such a function can be defined by

$$\begin{aligned} f(z) &= 0. \quad & 0 \leqq z \leqq A \\ f(z) &= 1. \quad & A \leqq z \leqq B \\ f(z) &= 0. \quad & B \leqq z \leqq \pi \end{aligned} \tag{9.67}$$

By Eqs. (9.65) we have

$$\begin{aligned} w_0 &= (B - A)/\pi \\ w_i &= (\sin iB - \sin iA)/(i\pi) \end{aligned} \tag{9.68}$$

Suppose for example, we wish to pass all frequencies up to $(\pi/3)/\Delta t$ radians per second. Then $A = 0$ and $B = \pi/3$. By Eqs. (9.68) we have

$$\begin{aligned} w_0 &= 0.33333333 \\ w_1 &= 0.27566445 \\ w_2 &= 0.13783222 \\ w_3 &= 0 \\ w_4 &= -0.06891611 \\ w_5 &= -0.05921719 \\ w_6 &= 0 \end{aligned} \tag{9.69}$$

Figure 9.6 illustrates this example. The desired filter is the line with a value of +1, from 0 to $\pi/3$, and with a value of 0, from $\pi/3$ to π. Figure 9.6 also shows the plot of thc Fourier series given by (9.49) and with coefficients given by (9.69).

Although the fit is far from perfect, it is better to use the smoothing formula obtained than a polynomial smoothing formula, because we have attained some control of the smoothing process.

In the case of a lowpass filter as shown in Fig. 9.6, since we cannot use an infinite number of terms, the filter function will not be exactly 1 at zero frequency. Therefore,

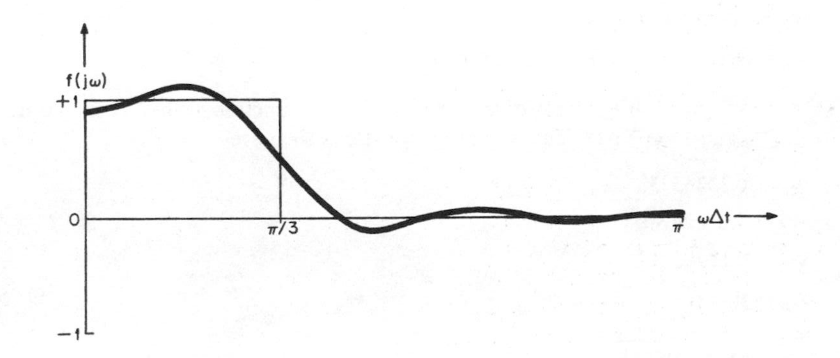

Fig. 9.6. Lowpass filter function with Fourier approximation.

the corresponding smoothing formula will introduce some d-c (direct-current) scaling that can be objectionable. This problem can be alleviated by a normalizing process. The problem arises because the sum of the weights is no longer unity as was the case for polynomial smoothing [Eq. (9.37)]. The weights can be adjusted so that their sum is 1 in the following manner.

Find the actual sum K of the weights by

$$K = \sum_{i=-m}^{m} w_i \tag{9.70}$$

Then, adjust each weight by

$$w_i' = w_i/K \tag{9.71}$$

Although the new filter function, defined by the new weights, does not fit the desired filter function as well as before, the d-c scaling is eliminated.

Other filter functions can be devised. One, which is of some interest, is the trapezoidal filter function. Consider the filter function which is zero up to some frequency A; then rises linearly to 1 at frequency B; remains at 1 until frequency C; descends linearly to zero at frequency D; and then remains at zero to $\pi/\Delta t$. This function defined mathematically is given by

$$\begin{aligned}
f(z) &= 0. && 0 \leqq z \leqq A \\
f(z) &= (z - A)/(B - A) && A \leqq z \leqq B \\
f(z) &= 1. && B \leqq z \leqq C \\
f(z) &= 1. - (z - C)/(D - C) && C \leqq z \leqq D \\
f(z) &= 0. && D \leqq z \leqq \pi
\end{aligned} \tag{9.72}$$

and is illustrated in Fig. 9.7.

Applying Eqs. (9.65) to the preceding function gives

$$\begin{aligned}
w_0 &= (D + C - B - A)/(2\pi) \\
w_i &= [(\cos iB - \cos iA)/(B-A) - (\cos iD - \cos iC)/(D-C)]/(i^2\pi)
\end{aligned} \tag{9.73}$$

For the special case of A = B = 0, these reduce to

$$\begin{aligned}
w_0 &= (D + C)/2\pi \\
w_i &= -(\cos iD - \cos iC)/(D-C)/(i^2\pi)
\end{aligned} \tag{9.74}$$

Figure 9.8 gives an eyample of a trapezoidal filter function with A = B = 0 and $C = 5\pi/18$, and $D = 7\pi/18$. The weighting constants then are

$$\begin{aligned}
w_0 &= 0.33333333 \\
w_1 &= 0.27426708 \\
w_2 &= 0.13505013 \\
w_3 &= 0 \\
w_4 &= -0.06345281 \\
w_5 &= -0.05198217 \\
w_6 &= 0
\end{aligned} \tag{9.75}$$

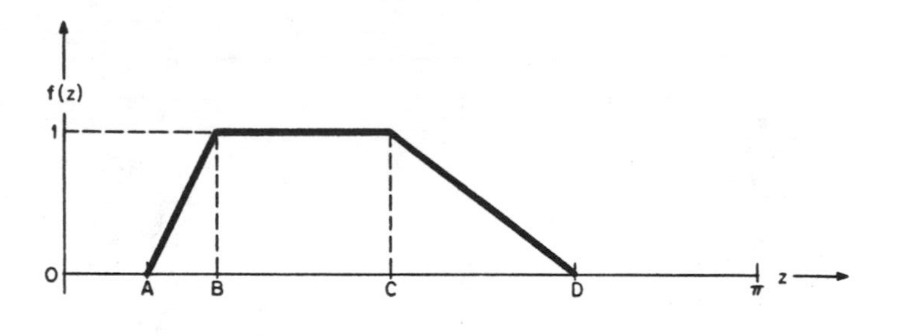

Fig. 9.7. Trapezoidal filter function.

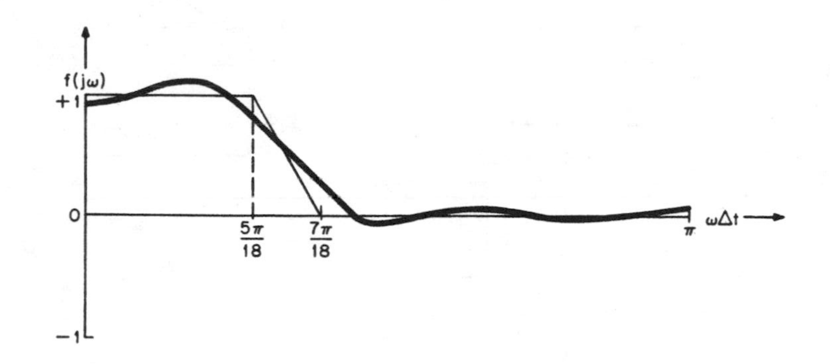

Fig. 9.8. Trapezoidal filter function with Fourier approximation.

Figure 9.8 also shows a plot of the Fourier series with the foregoing coefficients. In this case, for the same number of terms, the approximation is closer to the desired function than in the previous example. However, the approximation of the previous example is almost as good an approximation to the trapezoidal function as the approximation of this example.

It was noted in the beginning of this section that Eq. (9.56) is a special case of a more general form, which is

$$\overline{Y}(t) = \sum_{i=-p}^{q} w_i Y(t + i\Delta t) + \sum_{i=-r}^{-1} u_i \overline{Y}(t + i\Delta t) \quad (9.76)$$

The first sum of (9.76) is similar to the right side of (9.56), but the index i of the sum may not have the same number of positive values as it has of negative values, and the weighting coefficients w_i are not necessarily symmetric as given by (9.38). Furthermore, (9.76) also contains a weighted sum of the r previous *smoothed* values.

The transform of (9.76) is not as easy to fit to a given filter function as the transform of (9.56), but a better fit may usually be obtained with fewer terms.

The flowchart of Fig. 9.9 describes the smoothing of an array Y of numbers using the general smoothing formula of Eq. (9.76). The left side of the flowchart provides for

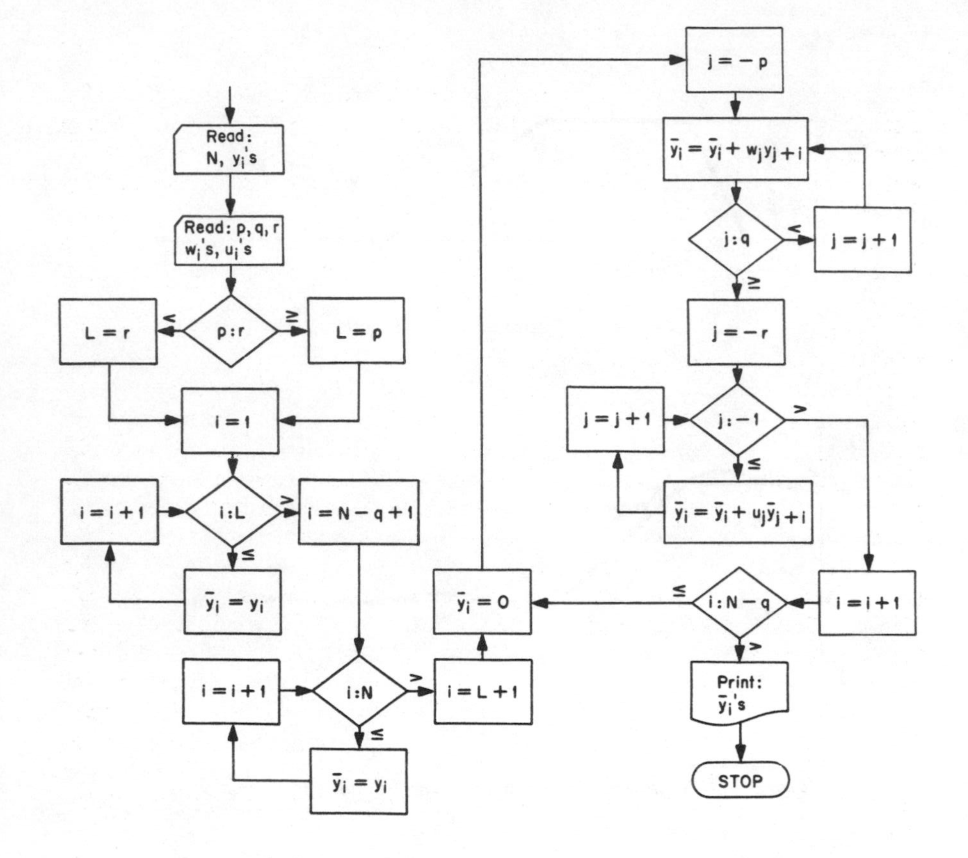

Fig. 9.9. Flowchart for smoothing by use of general smoothing formula.

reading in the array of numbers to be smoothed and the parameters of the smoothing formula. Provision is also included for moving values of the input array to the output array for those elements that cannot be smoothed by the given formula. The right side of the flowchart contains a loop for calculating the individual smoothed values. This loop contains two other loops. The first loop calculates the first sum of (9.76) and the second loop calculates the second sum of (9.76). If a specific smoothing formula does not contain any terms of the second summation of (9.76), r can be set to zero and the flowchart of Fig. 9.9 can still be used. Printout of the smoothed values is also included.

Although the flowchart of Fig. 9.9 can be used for any combination of parameters for Eq. (9.76), a more efficient program will result if the user specializes the flowchart for his particular smoothing formula.

The FORTRAN programmer is advised to note that some of the weighting coefficients w_i and all of the weighting coefficients u_i have nonpositive subscripts. Therefore, the subscripts in the FORTRAN coding must be offset by some positive amount to assure that all subscripts are greater than zero.

9.5 SUMMARY

The necessity for smoothing data was presented in Sec. 9.1. In addition, polynomial smoothing and the effects of such smoothing have also been described. These subjects are of interest in themselves and they also serve as the basis for deriving certain differentiation and integration methods in Chaps. 10 and 11, respectively. They also lay the groundwork for a method of developing smoothing formulas with known frequency domain characteristics, as given in Sec. 9.4. These methods are applicable only for developing smoothing formulas that are symmetric. Therefore, it is necessary to use other techniques to develop smoothing formulas that are asymmetric.

We have seen by Eqs. (9.20), (9.21), (9.22), (9.28), (9.29), (9.34), and (9.35) that the smoothing formulas needed at the start and at the end of the data are not symmetric. Therefore, the methods of Sec. 9.4 cannot be used to determine such formulas. Although other techniques can be developed for finding the coefficients of such formulas, the smoothing of data at each end of our data set is usually not important enough to justify the development of special formulas. If it is necessary to smooth this data, several methods can be used that do not require deriving asymmetric smoothing formulas. For example, the data can be passed to the smoothed set without change; one or more known polynomial smoothing formulas for data near the end of the data set can be employed; or a set of symmetric smoothing formulas of different lengths can be used. To clarify the last example, suppose that an 11-point smoothing formula is to be used to smooth a set of data. Further suppose that the coefficients to be used are those given by Eqs. (9.69) as modified by (9.70) and (9.71). The 11-point smoothing formula cannot be used to smooth the first five or the last five of the data points. However, a nine-point formula can be used to find a smoothed value for the fifth point (from the beginning or the end); a seven-point formula can be used to find a smoothed value for the fourth point; etc. The set of coefficients for each of these formulas can be a subset of those given by Eqs. (9.69) and modified by (9.70) and (9.71).

The general smoothing formula, Eq. (9.76), is also not symmetric except for the special cases where it reduces to (9.56). Therefore, the Fourier techniques of Sec. 9.5 cannot be applied to determine the weighting coefficients. If data is to be smoothed in real time (i.e., smoothed as it is received), no future data is available, and the general form with $q = 0$ must be used. A method for finding the coefficients involves designing a continuous filter having the desired frequency domain characteristics and then transforming it into a digital filter (smoothing function).

EXERCISES

1. Prepare a flowchart and write a FORTRAN program that will perform a nine-point smoothing on an array of numbers. Assume that the array of numbers and its size N is available in the computer memory. Provide for the coefficients of the smoothing formula to be input from cards. Output the smoothed values to another array. Print the array of smoothed values. Assume also that the smoothing formula gives a value only for the midpoint of the nine points. Since special formulas for the first four points and the last four points are not available, carry the raw values of these points to the array of smoothed values.

2. Derive the coefficients for a second-order, nine-point smoothing formula that gives a smoothed value at the midpoint of the nine points.

3. Execute the program in Exercise 1 using the coefficients obtained in Exercise 2 and the data in Table 9.1.

TABLE 9.1

i	y_i									
1–10	287	311	335	354	367	377	376	375	372	368
11–20	364	357	352	343	337	341	366	399	428	443
21–30	446	443	440	437	441	441	450	459	473	491
31–40	518	549	590	640	685	724	749	766	774	780
41–50	782	772	764	750	730	705	668	647	638	646
51–60	677	717	795	865	939	977	1000	992	980	966
61–70	948	931	918	905	899	895	897	899	906	921
71–80	942	963	1001	1034	1063	1091	1111	1123	1128	1128
81–90	1123	1116	1106	1097	1087	1078	1073	1073	1078	1086
91–100	1093	1105	1119	1130	1143	1155	1170	1196	1207	1206
101–110	1213	1231	1248	1262	1271	1269	1271	1293	1317	1350
111–120	1378	1393	1402	1402	1402	1404	1406	1412	1438	1476
121–130	1491	1493	1485	1486	1490	1503	1524	1543	1563	1576
131–140	1586	1590	1591	1589	1590	1588	1587	1588	1590	1586
141–150	1585	1586	1591	1617	1650	1681	1696	1714	1715	1714

4. Derive the coefficients for a fourth-order, nine-point smoothing formula that gives a smoothed value at the midpoint of the nine points.

5. Execute the program in Exercise 1 using the coefficients obtained in Exercise 4 and the data of Exercise 3.

6. Modify the flowchart and program in Exercise 1 to permit multipass smoothing. That is, allow an additional input M, the number of smoothing passes; move the output array to the input array at the end of each pass except the last pass; print the results of the final smoothing only.

7. Repeat Exercise 3 using the program in Exercise 6 with M = 2.

8. Repeat Exercise 5 using the program in Exercise 6 with M = 2.

9. Calculate and plot the filter function for the formula derived in Exercise 2 for both one- and two-pass smoothing.

10. Calculate and plot the filter function for the formula derived in Exercise 4 for both one- and two-pass smoothing.

11. Derive a 25-point smoothing formula whose filter function is a lowpass filter with cutoff at $\omega\Delta t = \pi/6$. (Normalize to prevent d-c scaling.)

12. Prepare a flowchart and write a FORTRAN program to apply the formula derived in Exercise 11. (*Note*: This program can be similar to that of Exercise 1.)

13. Execute the program in Exercise 12 using the data in Exercise 3.

14. Derive the filter function for the smoothing formula

$$\overline{Y_0} = wY_0 + (1 - w)\overline{Y}_{-1}$$

Compute and plot the absolute value of the filter function over the interval $[0, \pi]$ for $w = 0.3$ and $w = 0.7$.

15. Prepare a flowchart and write a FORTRAN program that will smooth an array of data by the formula of Exercise 14.

16. Execute the program in Exercise 15 using the data in Exercise 3. Use $w = 0.3$.

17. Repeat Exercise 16 but with $w = 0.7$.

10

Numerical Differentiation

10.1 INTRODUCTION

In many engineering and science applications, it is necessary to find the derivative of a function. If the function can be expressed in mathematical form, a derivative can usually be found by analytic methods. If the function is given in graphical or tabular form only, then graphical or numerical methods must be employed to obtain a derivative. The purpose of this chapter is to discuss some of the available methods for obtaining derivatives. Chap. 11 will discuss numerical integration. However, this is a convenient place to compare the relative ease and accuracy of differentiation and integration by analytic and numerical methods. Table 10.1 makes this comparison.

10. 1 COMPARISON OF ANALYTIC AND NUMERICAL DIFFERENTIATION AND INTEGRATION

Type\Operation	*Differentiation*	*Integration*
Analytic	Usually possible and relatively easy	Not always possible; when possible may be difficult
Numerical	Difficult to do accurately	Almost any accuracy desired may be attained

As an example of a problem where numerical differentiation is required, consider the following. Suppose we wish to determine the drag function (deceleration due to air resistance versus velocity) for a certain projectile. This can be determined by test firing the projectile and observing position as a function of time. From these observations, the first and second derivatives of position with respect to time must be found numerically to obtain velocity and acceleration, respectively.

The basic definition of differentiation is given by

$$f'(x) = \lim_{h \to 0} \frac{f(x + h) - f(x)}{h} \tag{10.1}$$

This definition can be used to obtain the formulas for the derivatives of algebraic expressions. We will see later that (10.1) can also be used in various forms to obtain numerical derivatives.

When the function f(x) is graphed, the derivative of that function, for some given value of x, can be interpreted as the slope of the line which is tangent to the function at the given value of x. This is illustrated in Fig. 10.1.

A useful tool for numerical differentiation is the *mean value theorem*. This theorem states that if a function is continuous and differentiable between two points on the function, then the slope of the line joining these two points is equal to the derivative of the function at at least one other point between the two given points. Expressed mathematically, this is

$$f'(\xi) = \frac{f(x_2) - f(x_1)}{x_2 - x_1} \tag{10.2}$$

where $x_1 \leqq \xi \leqq x_2$. This is illustrated graphically in Fig. 10.2.

10.2 GRAPHICAL DIFFERENTIATION

The relationship between the derivative of a function and the tangent to a curve suggests a method for graphically finding a derivative. That is, as in Fig. 10.1, draw the graph of the function and then, at the point in question, construct a tangent line. The slope of the tangent line can then be determined by reading the values of x and y at two points (x_1, y_1 and x_2, y_2) on the tangent line and using these in

$$f'(x_A) = \frac{y_2 - y_1}{x_2 - x_1} \tag{10.3}$$

The accuracy of graphical differentiation at its best is limited by the accuracy with which we can read the graph. A more serious disadvantage is the inherent inaccuracy of determining the tangent line.

Various mechanical devices are available for determining an approximate tangent line. The simplest (and perhaps most accurate) such device is a polished bar. The use of this device is illustrated in Fig. 10.3. The bar is placed on the graph of f(x) at the point $[x_A, f(x_A)]$ where the derivative is desired. The bar is rotated about this point until the reflection of the curve and the curve itself line up so that they appear to be a single smooth continuous curve through the reflective interface. The face of the bar is then normal to the curve. A line can be drawn along this edge; then by measuring two points on this line (x_1, y_1 and x_2, y_2)†, the slope of the normal is found by

$$S_N = \frac{y_2 - y_1}{x_2 - x_1} \tag{10.4}$$

and the slope of the tangent by

$$S_T = -1./S_N \tag{10.5}$$

Therefore the required derivative is given by

$$f'(x_A) = \frac{x_1 - x_2}{y_2 - y_1} \tag{10.6}$$

†Both x and y must be measured in the same units. After determining a dimensionless slope, the appropriate scale factors must be applied to get the actual derivative.

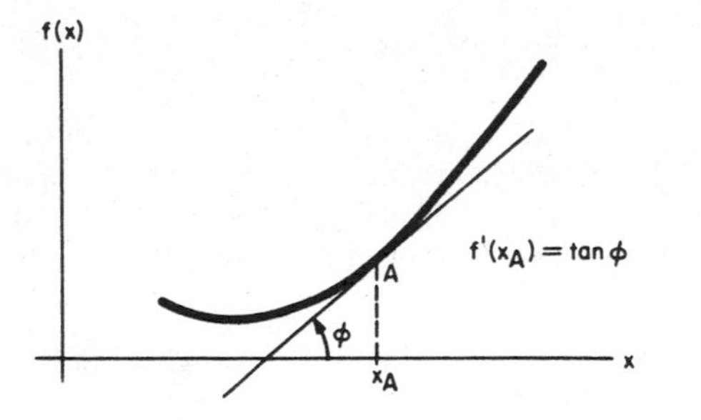

Fig. 10.1. Geometrical interpretation of a derivative.

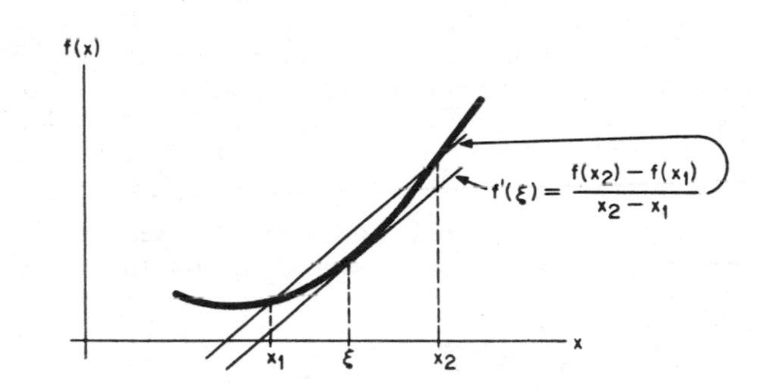

Fig. 10.2. Illustration of mean value theorem.

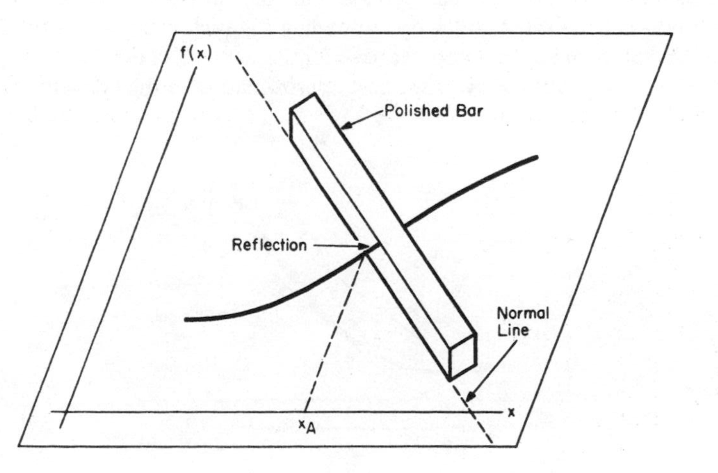

Fig. 10.3. Graphical determination of a normal.

This method of graphically determining the derivative is generally better than others because

1. It is fairly easy to determine if the curve and its reflection do not join smoothly.
2. Rotation of the bar through the angle θ causes a rotation of the image through 2θ. This means that an error of alignment $\Delta\theta$ in the curve and its image results in an error of only $\Delta\theta/2$ in the slope of the normal line.

10.3 NUMERICAL DIFFERENTIATION BY SECANT LINE APPROXIMATION

The basic definition of differentiation, Eq. (10.1), suggests a numerical method of approximating a derivative. Substituting a particular value of x, say x_A, into (10.1), we have

$$f'(x_A) \doteq [f(x_A + h) - f(x_A)]/h \tag{10.7}$$

The slope so defined is the same as that of the line AB in Fig. 10.4.

Whether we approach the point from the right, as in (10.1), or the left, should not make any difference in finding the derivative. Therefore, we can make another approximation:

$$f'(x_A) \doteq [f(x_A) - f(x_A - h)]/h \tag{10.8}$$

This derivative has the same slope as the line CA in Fig. 10.4. Neither of the foregoing two approximations are very good as can be seen by Fig. 10.4. However, we can also approximate Eq. (10.1) by

$$f'(x_A) \doteq [f(x_A + h) - f(x_A - h)]/(2h) \tag{10.9}$$

This gives the same slope as that of line CB. This appears to be a better approximation, and it is. By the mean value theorem, we know that the slope of line CB is the same as that of the function at some point ξ for which $x_C \leqq \xi \leqq x_B$. Since x_A is also in this same range, it is not unreasonable to approximate $f'(x_A)$ with the slope of the line CB. A similar analysis can be made using line CA or AB, but since x_A is an end point of these, it is usually less likely that the corresponding ξ is as close to x_A as in the previous analysis. We will see in Sec. 10.4 another basis for the fact that, of the three cases considered, the slope of the line CB gives the best approximation to the derivative.

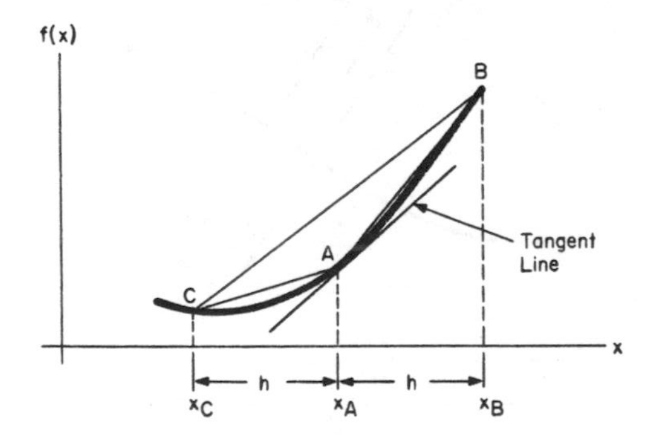

Fig. 10.4. Comparison of secant line approximation for a derivative.

10.4 NUMERICAL DIFFERENTIATION BY USE OF INTERPOLATING POLYNOMIALS

In Chap. 4, we discussed various methods of interpolation. These methods were primarily methods in which a polynomial of degree N was passed exactly through N+1 table values. In this manner, a mathematical expression (a polynomial) was obtained to represent some tabular data. This function could then be used to find values of the function for which there was no table entry. It follows then that these same polynomials can be differentiated to obtain the derivatives of tabular data. Each of the following subsections will describe the various differentiation methods that correspond to the interpolation methods of Chap. 4.

10.4.1 Gregory–Newton Methods

In Sec. 4.6, the Gregory–Newton forward interpolation method was presented. It is based on the interpolating polynomial

$$y = A_0 + A_1(a-x_k) + A_2(a-x_k)(a-x_{k+1}) + A_3(a-x_k)(a-x_{k+1})(a-x_{k+2}) \quad (10.10)$$
$$+ \cdots + A_N(a-x_k)(a-x_{k+1})(\cdots)(a-x_{k+N-1}) \quad (4.24)$$

It was shown in Sec. 4.6 that this form, Eq. (10.10), greatly facilitates the solution for the A's. Although the A's can be found for any set of tabular data, the formula is greatly simplified by assuming the data has equally spaced arguments. Then (10.10) can be reduced to

$$y = f_k + \frac{\Delta f_k}{1!}u + \frac{\Delta^2 f_k}{2!}u(u-1) + \cdots + \frac{\Delta^N f_k}{N!}u(u-1)(\cdots)(u-N+1) \quad (10.11)\ (4.32)$$

where

$$u = (a-x_k)/h \quad (10.12)$$

h is the interval between consecutive x's and the $\Delta^i f_k$ are the Newton forward differences as defined in Sec. 4.5.

Differentiating this function with respect to *a*, we have

$$y'(a) \doteq \frac{1}{h}\left\{\Delta f_k + \frac{\Delta^2 f_k}{2!}[(u-1) + (u)] + \frac{\Delta^3 f_k}{3!}[(u-1)(u-2) + u(u-2) + u(u-1)] + \cdots + \frac{\Delta^N f_k}{N!}[(u-1)(u-2)(\cdots)(u-N+1) + u(u-2)(\cdots)(u-N+1) + \cdots]\right\} \quad (10.13)$$

This formula is interesting and useful in itself if we desire derivatives at other than tabulated points. However, when differentiating numerically, we are usually only interested in the derivatives at tabulated points.

If we wish to find a derivative at a tabulated point, except near the end of the table, we can select k so that it corresponds to the point of interest. Therefore, $u = 0$ and (10.13) simplifies to

$$y'(x_k) = \frac{1}{h}\left[\Delta f_k + \frac{\Delta^2 f_k}{2!}(-1) + \frac{\Delta^3 f_k}{3!}(-1)(-2) + \cdots + \frac{\Delta^N f_k}{N!}(-1)(-2)(\cdots)(-N+1)\right] \quad (10.14)$$

which further simplifies to

$$y'(x_k) = \frac{1}{h}\left[\Delta f_k - \Delta^2 f_k/2 + \Delta 3 f_k/3 - \cdots (-1)^{N+1} \Delta^N f_k/N\right] \tag{10.15}$$

Suppose, for example, we wish to find the derivative of θ with respect to x using the data of Table 4.3. For the table entry x = 300 ft and using only the first three terms of (10.15), we have

$$\theta'(300) \doteq \frac{1}{40}[-0.443673 - (0.092460)/2 + (-0.026110)/3)]$$

$$\doteq -0.0124652 \text{ deg/ft} \tag{10.16}$$

This compares favorably with the exact derivative

$$\theta'(300) = -0.0125504 \text{ deg/ft} \tag{10.17}$$

which was obtained by analytic means. However, (10.13) can yield better results. The Gregory–Newton interpolation formula passes exactly through the points used in obtaining the formula. This does *not* imply that the differentiation formula based on the Gregory–Newton interpolation formula will give exact values of the derivatives at these same points. This can be illustrated as follows. We can write

$$f(a) = y(a) + \varepsilon(a) \tag{10.18}$$

where f(a) is the exact function, y(a) is the interpolation polynomial, and $\varepsilon(a)$ is the error f(a) − y(a). The error function for this example is shown in Fig. 10.5.

We can differentiate Eq. (10.18) to give

$$f'(a) = y'(a) + \varepsilon'(a) \tag{10.19}$$

We see then that the true derivative f′(a) differs from the polynomial derivative y′(a) [Eq. (10.13)] by the derivative of the error function $\varepsilon(a)$. By inspection of Fig. 10.5, we can see that at u = 0, 1, 2, 3 (corresponding to points 4, 5, 6, 7 of Table 4.3), the derivative is not zero. In fact, it is only zero in the neighborhood of u = 0.5, 1.5, 2.5. The error in the derivative appears to be less for u = 1 and u = 2 than for u = 0. This is the usual case. When making a polynomial approximation to some tabular data, the excursions from the true values are least near the center of the fitted data and worst near the ends (and beyond). In Sec. 4.8, methods were suggested for selecting k, such that the error in the interpolation procedure would be minimized.

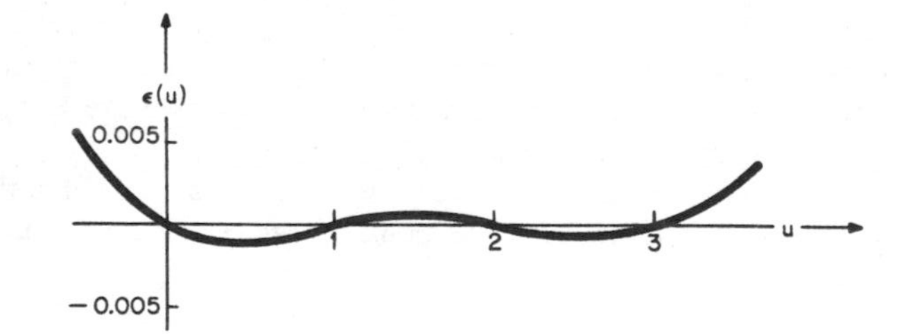

Fig. 10.5. Example error curve for a third-order differentiation formula.

The same methods should be applied when finding derivatives by use of interpolating polynomials. Applying the methods of Sec. 4.8 to the preceding example, we obtain either $k = 3$ and $u = 1$ or $k = 2$ and $u = 2$. Choosing $k = 3$ and $u = 1$ for an example, Eq. (10.13) reduces to

$$y'(x_{k+1}) \doteq \frac{1}{h}(\Delta f_k + \Delta^2 f_k/2 - \Delta^3 f_k/6) \tag{10.20}$$

Substituting values from Table 4.3,

$$\theta'(300) \doteq -0.0125957 \tag{10.21}$$

We see then that this is a closer approximation to the value given by Eq. (10.17) than the value given by (10.16).

A differentiation formula can also be derived from the Gregory–Newton backward interpolation formula. Differentiating Eq. (4.41) with respect to a,

$$y'(a) = \frac{1}{h}\left\{ \nabla f_k + \frac{\nabla^2 f_k}{2!}(u+1+u) + \frac{\nabla^3 f_k}{3!}[(u+1)(u+2) + u(u+2) + u(u+1)] + \cdots + \frac{\nabla^N f_k}{N!}[(u+1)(\cdots)(u+N-1) + u(\cdots)(u+N-1) + \cdots] \right\} \tag{10.22}$$

Selecting $k = 6$ and $u = -2$, (10.22) then becomes

$$y'(x_{k-2}) \doteq \frac{1}{h}(\nabla f_k - 3\nabla^2 f_k/2 + \nabla^3 f_k/3) \tag{10.23}$$

Then,

$$\theta'(300) \doteq -0.0125957 \tag{10.24}$$

which is the same result as given by Eq. (10.21). This is expected because both examples use the same data (points 3, 4, 5, 6) and hence the polynomial must be identical if reduced to standard form.

10.4.2 Central Difference Methods

Several central difference interpolation formulas were given in Chap. 4. The central difference formulas are most accurate near the center of the tabulated data used, i.e., for $-1 \leqq u \leqq +1$. Hence, the formulas resulting from differentiating the interpolation polynomials are also most accurate in this central range. Therefore, it is more meaningful to specialize these formulas to the case $u = 0$ than it was for the Gregory–Newton differentiation formulas.

Two formulas by Gauss [Eqs. (4.44) and (4.46)] and one by Stirling [Eq. (4.48)] will be used here to derive differentiation formulas. Differentiating and substituting $u = 0$ gives

Gauss forward:

$$y'(x_k) = \frac{1}{h}(\delta_f f_k - \delta_f^2 f_k/2! - \delta_f^3 f_k/3! + 2\delta_f^4 f_k/4! + \cdots) \tag{10.25}$$

Gauss backward:

$$y'(x_k) = \frac{1}{h}(\delta_b f_k + \delta_b^2 f_k/2! - \delta_b^3 f_k/3! - 2\delta_b^4 f_k/4! + \cdots) \qquad (10.26)$$

Stirling:

$$y'(x_k) = \frac{1}{h}(\delta f_k - \delta^3 f_k/3! + 4\delta^5 f_k/5! - \cdots) \qquad (10.27)$$

Using only through third differences and applying these to the example in Sec. 10.4, we have respectively,

$$\theta'(300) \doteq -0.0125957 \qquad (10.28)$$

$$\theta'(300) \doteq -0.0124734 \qquad (10.29)$$

$$\theta'(300) \doteq -0.0125346 \qquad (10.30)$$

Note that the value found by the Gauss forward method [Eq. (10.28)] is exactly the same as the values found by Gregory–Newton forward and backward methods [Eqs. (10.21) and (10.24)]. This is again because the resulting polynomial passes exactly through entries 3, 4, 5, and 6 of Table 4.3. The value obtained by the Gauss backward method is different, since it is derived from entries 2, 3, 4, and 5 of Table 4.3. Stirling's formula gives the most accurate results of all even though only odd differences are involved [Eq. (10.27)]. But this result is based on five entries of Table 4.3 (2, 3, 4, 5, 6) rather than four entries as in the other cases.

Any of the foregoing differentiation formulas can be rewritten in terms of ordinate values only. However, we will only do this for Eq. (10.27), derived from Stirling's interpolation formula, since it yields a symmetric form.

The derivative of a second-degree Stirling polynomial is

$$y'(x_k) = (f_{k+1} - f_{k-1})/(2h) \qquad (10.31)$$

Although the second-degree interpolation polynomial is derived from three data points $(k-1, k, k+1)$, only two points $(k-1, k+1)$ are needed to find the derivative. Equation (10.31) is essentially the same as (10.9). In Sec. 10.3, it was asserted that the slope of line CB in Fig. 10.4 was a better approximation of the derivative at point A than either the slope of line CA or the slope of line AB. We can now see that the slope of line CB is the same as the slope (at point A) of a parabola passing through points C, A, and B.

The derivative of a fourth-degree Stirling polynomial can be written as

$$y'(x_k) = (-f_{k+2} + 8f_{k+1} - 8f_{k-1} + f_{k-2})/(12h) \qquad (10.32)$$

Again, the data point k is not needed to find the derivative.

10.5 NUMERICAL DIFFERENTIATION BY MEANS OF SMOOTHING POLYNOMIALS

In Sec. 10.4 we demonstrated that an approximation to a set of table values will have an error in the derivative even when there is no error in the function value. In addition, experimental data will contain errors in the data. Therefore, any derivative calculated from this data will also be in error. In general, differentiation is an *error-*

amplifying process.[†] It is imperative that we take precautions to avoid introducing any more error than is absolutely necessary.

In Chap. 9 the smoothing of experimental data was discussed. We stated there that it is better to use smoothing formulas determined from frequency considerations than to use polynomial smoothing formulas. However, a large portion of Chap. 9 was devoted to polynomial smoothing formulas. One reason for that was that such formulas can be a basis for differentiation (and integration) formulas.

The polynomial smoothing formula

$$y = a_0 + a_1x + a_2x^2 + \cdots + a_nx^n \tag{10.33}$$
(9.1)

when differentiated gives

$$y'(x) = a_1 + 2a_2x + 3a_3x^2 + \cdots + na_nx^{n-1} \tag{10.34}$$

Making the same simplifying assumptions as in Sec. 9.2, the a_i's can be determined by the methods of that section. Substituting the derived values of the a_i's and the abscissa values of the data points into Eq. (10.34), the derivative at any data point can be found. The usual procedure in finding such derivative values is to select an odd number of data points and find the derivative at the midpoint of this data; next, discard the first point of the set and add a new point at the end of the set; and then find a derivative for the midpoint of the new set. Since such a procedure is concerned only with the derivative at the midpoint of a set of data ($x_i = 0$), it is only necessary to determine a_1 since

$$Y'_0 = a_1 \tag{10.35}$$

10.5.1 Specific Smoothing Differentiation Formulas

We will now apply the results already obtained in Sec. 9.2.1 to obtain specific smoothing formulas.

10.5.1.1 *Derivatives From First-Order Smoothing Formulas*

In the case of first-order smoothing formulas, the derivative found for any abscissa will be the same as that found for the midpoint of the set of data points, since the slope of a straight line is constant. From Sec. 9.2.1.1, we can say that

$$a_1 = T_1/S_2 \tag{10.36}$$

Then, for a three-point linear smoothing,

$$a_1 = (-y_{-1} + y_1)/(2h) \tag{10.37}$$
(9.19)

and hence

$$Y'_0 = (-y_{-1} + y_1)/(2h) \tag{10.38}$$

This again is essentially the same formula as (10.9) or (10.31). In other words, the slope of the "best" straight line through three data points is the same as the slope of the line joining the two end points.

---

†See Sec. 10.7.

For a five-point linear smoothing (m = 2),

$$a_1 = (-2y_{-2} - y_{-1} + y_1 + 2y_2)/(10h) \tag{10.39}$$

and hence

$$Y'_0 = (-2y_{-2} - y_{-1} + y_1 + 2y_2)/(10h) \tag{10.40}$$

For a seven-point linear smoothing (m = 3),

$$Y'_0 = a_1 = (-3y_{-3} - 2y_{-2} - y_{-1} + y_1 + 2y_2 + 3y_3)/(28h) \tag{10.41}$$

10.5.1.2 *Derivatives From Second-Order Smoothing Formulas*

From Eq. (9.27) we still have

$$a_1 = T_1/S_2 \tag{10.36}$$

Therefore, the differentiation formulas (only for the midpoints of the sets of data points used) will be the same as derived in Sec. 10.5.1.1.

10.5.1.3 *Derivatives From Third-Order Smoothing Formulas*

From Eqs. (9.33) we have

$$a_1 = (T_1S_6 - T_3S_4)/(S_2S_6 - S_4^2) \tag{10.42}$$

and hence for a five-point (m = 2) cubic smoothing formula,

$$Y'_0 = a_1 = (y_{-2} - 8y_{-1} + 8y_1 - y_2)/(12h) \tag{10.43}$$

For a seven-point (m = 3) cubic smoothing formula

$$Y'_0 = a_1 = (22y_{-3} - 67y_{-2} - 58y_{-1} + 58y_1 + 67y_2 - 22y_3)/(252h) \tag{10.44}$$

Note that Eq. (10.43) is the same as (10.32). This results in part because a differentiation formula derived from an odd-order smoothing polynomial will be the same as one derived from the next higher even-order smoothing polynomial. (See Sec. 10.5.2.) In this case, Eq. (10.43) was derived from a third-order smoothing polynomial. The same formula would result if derived from a fourth-order smoothing polynomial. *But* five points are satisfied exactly by a fourth-order polynomial; therefore a differentiation formula based on any fourth-order polynomial (such as Stirling's interpolation formula) will result in the same differentiation formula.

10.5.2 Common Features of Smoothing Differentiation Formulas

As in the case of the smoothing formulas themselves, the differentiation formulas have certain common features. The resulting differentiation formulas (for the mid-abscissa point) are of the form

$$Y'_0 = \sum_{i=-m}^{m} w_i y_i$$

$$w_i = -w_{-i} \tag{10.45}$$

$$w_0 = 0$$

which implies the condition

$$\sum_{i=-m}^{m} w_i = 0 \tag{10.46}$$

and we also have

$$\sum_{i=-m}^{m} i w_i = 1 \tag{10.47}$$

Using a superscript, as in Sec. 9.2.2, to denote the order of the smoothing polynomial used, for n even (n = 2q), then,

$$w_i^{(2q)} = w_i^{(2q-1)} \tag{10.48}$$

This says that the differentiating formula obtained from an even-order polynomial is the same as the one obtained from the next lower odd-degree polynomial.

10.6 HIGHER DERIVATIVES

If first derivatives can be found numerically, second and higher derivatives can also be found numerically. However, since numerical differentiation tends to *amplify errors*,‡ it is difficult to determine higher derivatives accurately. Higher derivatives can be found numerically by two different methods.

10.6.1 Higher Derivatives by Repeated Numerical Differentiation

Any of the previously discussed methods can be used repetitively to obtain higher derivatives. That is, the original tabular data can be differentiated numerically for each tabular entry. These results then form a new table of data, which can be differentiated again numerically. Even very simple methods (for example, first order) can be applied repeatedly. The accuracy may not be very good, especially near the ends of the table, but the process would be easy to use.

10.6.2 Higher Derivatives by Extension of First Derivative Methods

Any of the previously discussed polynomial differentiation methods can be extended to obtain second and higher order derivatives. Since these methods are based on polynomials, the polynomial can be differentiated repeatedly to obtain the desired derivative. If an Nth derivative is required, at least an Nth-order polynomial must be used.

There are numerous differentiation methods that could be considered, but the reader can easily develop his own higher derivative methods based on previous first derivative methods. We will develop one here as an example, which is based on Stirling's central difference formula. That formula is given by

$$y = f_k + \frac{\delta f_k}{1!}u + \frac{\delta^2 f_k}{2!}u^2 + \frac{\delta^3 f_k}{3!}u(u^2-1) + \frac{\delta^4 f_k}{4!}u^2(u^2-1) + \frac{\delta^5 f_k}{5!}u(u^2-1)(u^2-4) + \cdots \qquad (10.49)\ (4.48)$$

‡See Sec. 10.7.

where $u = (a-x_k)/h$ and the δ's are the Stirling central differences as defined in Chap. 4. Differentiating (10.49) with respect to *a*,

$$y'(a) = \frac{1}{h}\left[\frac{\delta f_k}{1!} + \frac{2\delta^2 f_k}{2!}u + \frac{\delta^3 f_k}{3!}(3u^2-1) + \frac{\delta^4 f_k}{4!}(4u^3-2u) + \frac{\delta^5 f_k}{5!}(5u^4-15u^2+4) + \cdots\right] \tag{10.50}$$

Evaluating this at $u = 0$ gives Eq. (10.27). Repeated differentiation gives

$$y''(a) = \left[\frac{1}{h^2}\quad \delta^2 f_k + \delta^3 f_k u + \frac{\delta^4 f_k}{4!}(12u^2-2) + \frac{\delta^5 f_k}{5!}(20u^3-30u) + \cdots\right] \tag{10.51}$$

$$y'''(a) = \left[\frac{1}{h^3}\quad \delta^3 f_k + \delta^4 f_k u + \frac{\delta^5 f_k}{5!}(60u^2-30) + \cdots\right] \tag{10.52}$$

$$\vdots$$

Evaluating these for $u = 0$ gives

$$y''(x_k) = \frac{1}{h^2}(\delta^2 f_k - \delta^4 f_k/12 + \cdots) \tag{10.53}$$

$$y'''(x_k) = \frac{1}{h^3}(\delta^3 f_k - \delta^5 f_k/4 \ + \cdots) \tag{10.54}$$

$$\vdots$$

It can be seen from the preceding examples that differentiation formulas (for $u = 0$) obtained from Stirling's interpolation formula only involve odd differences for odd-order derivatives and even differences for even-order derivatives. Furthermore, if the derivatives are approximated by a single term, we have

$$y^n(x_k) \doteq \delta^n f_k/h^n \tag{10.55}$$

10.7 INHERENT ERROR OF DIFFERENTIATION FORMULAS

It was stated in Sec. 10.5 that differentiation is an *error-amplifying* process. That is, if errors exist in the original data, numerical differentiation of this data will usually result in larger relative errors in the derivatives than those that existed in the data. The error in a derivative found numerically is made up of two parts: that which results from errors in the data and that which results from the numerical method used. Since the errors in the data are usually not known, the resulting errors in the derivatives will not be known. Although such errors are usually the greater part of the total error, they cannot be conveniently analyzed. However, the error resulting from the numerical method can be analyzed conveniently using Taylor's series. Although not the major source of error, such errors, for some selected numerical differentiation methods, will be analyzed in the following subsections.

10.7.1 Error in the Gregory–Newton Method

Consider Eq. (10.15), which is a differentiation formula derived from the Gregory–Newton forward interpolation formula and evaluated for $u = 0$. Retaining only the first three terms from Eq. (10.15) and expressing in terms of ordinates results in

$$y'_k = (2f_{k+3} - 9f_{k+2} + 18f_{k+1} - 11f_k)/(6h) \tag{10.56}$$

Expanding each term in a Taylor's series gives

$$f_{k+3} = f_k + 3hf'_k + 9h^2f''_k/2 + 27h^3f'''_k/6 + 81h^4f^{iv}(\xi_3)/24 \qquad x_k \leqq \xi_3 \leqq x_{k+3} \tag{10.57}$$

$$f_{k+2} = f_k + 2hf'_k + 4h^2f''_k/2 + 8h^3f'''_k/6 + 16h^4f^{iv}(\xi_2)/24 \qquad x_k \leqq \xi_2 \leqq x_{k+2} \tag{10.58}$$

$$f_{k+1} = f_k + hf'_k + h^2f''_k/2 + h^3f'''_k/6 + h^4f^{iv}(\xi_1)/24 \qquad x_k \leqq \xi_1 \leqq x_{k+1} \tag{10.59}$$

Substituting the values obtained from the preceding equations into (10.56) results in

$$y'_k = f'_k + h^3[9f^{iv}(\xi_3) - 8f^{iv}(\xi_2) + f^{iv}(\xi_1)]/8 \tag{10.60}$$

The Taylor's series expansion of y' about the point x_k is

$$y' = f'_k + uf''_k + u^2f'''_k/2 + u^3f^{iv}(\xi_T)/6 \tag{10.61}$$

where $u = (a\text{-}x_k)/h$ and $x_k \leqq \xi_T \leqq a$. But for $a = x_k$, (10.61) reduces to $y' = f'_k$ and the error is given by

$$\varepsilon = y' - y'_k = -h^3[9f^{iv}(\xi_3) - 8f^{iv}(\xi_2) + f^{iv}(\xi_1)]/8 \tag{10.62}$$

This can be reduced to the more compact form

$$\varepsilon = -h^3f^{iv}(\xi)/4 \tag{10.63}$$

where $x_k \leqq \xi \leqq x_{k+3}$.

In Sec. 10.4.1, it was demonstrated that for a four-point differentiation formula, the error in the derivative is less when k is selected so that $u = 1$ than when k is selected so that $u = 0$. The differentiation formula for $u = 1$ is given by Eq. (10.20). Expressing this formula in terms of ordinates, we can write

$$y'_{k+1} = (-f_{k+3} + 6f_{k+2} - 3f_{k+1} - 2f_k)/(6h) \tag{10.64}$$

This implies

$$y'_k = (-f_{k+2} + 6f_{k+1} - 3f_k - 2f_{k-1})/(6h) \tag{10.65}$$

By Taylor's series,

$$f_{k-1} = f_k - hf'_k + h^2f''_k/2 - h^3f'''_k/6 + h^4f^{iv}(\xi_{-1})/24 \tag{10.66}$$

where $x_{k-1} \leqq \xi_{-1} \leqq x_k$.

Substituting Eqs. (10.58), (10.59), and (10.66) into (10.65) yields

$$y'_k = f'_k - h^3[8f^{iv}(\xi_2) - 3f^{iv}(\xi_1) + f^{iv}(\xi_{-1})]/72 \tag{10.67}$$

Therefore the error is

$$\varepsilon = y' - y'_k = h^3[8f^{iv}(\xi_2) - 3f^{iv}(\xi_1) + f^{iv}(\xi_{-1})]/72 \qquad (10.68)$$

This also can be re-expressed in more compact form, namely

$$\varepsilon = h^3 f^{iv}(\xi)/12 \qquad (10.69)$$

where $x_{k-1} \leqq \xi \leqq x_{k+2}$.
Assuming that the fourth derivative does not vary markedly in the interval x_{k-1} to x_{k+3}, then the absolute value of the error given by (10.69) is only one-third that given by (10.63). This then verifies the contention made earlier that the error that results from using (10.20) is less than that which results from using (10.15).

10.7.2 Error in Central Difference Methods

Equations (10.25), (10.26), and (10.27) give three differentiation formulas based on central difference interpolation methods. An error analysis can be performed for each of these in the manner given in Sec. 10.7.1. For consistency in comparing with other methods, we will truncate each of the foregoing formulas after the term containing the third difference and express each in terms of ordinates.

From the Gauss forward formula

$$y'_k = (-f_{k+2} + 6f_{k+1} - 3f_k - 2f_{k-1})/(6h) \qquad (10.70)$$

which is identical to Eq. (10.65).
From the Gauss backward formula,

$$y'_k = (2f_{k+1} + 3f_k - 6f_{k-1} + f_{k-2})/(6h) \qquad (10.71)$$

From Stirling's formula,

$$y'_k = (-f_{k+2} + 8f_{k+1} - 8f_{k-1} + f_{k-2})/(12h) \qquad (10.72)$$

Since (10.70) is identical to (10.65), the error will be the same as that expressed by (10.69).
An error analysis of (10.71) requires the Taylor's series expansion for f_{k-2}.

$$f_{k-2} = f_k - 2hf'_k + 4h^2 f''_k/2 - 8h^3 f'''_k/6 + 16h^4 f^{iv}(\xi_{-2})/24 \qquad (10.73)$$

where $x_{k-2} \leqq \xi_{-2} \leqq x_k$.
Substituting Eqs. (10.59), (10.66), and (10.73) into (10.71) yields

$$y'_k = f'_k + h^3[f^{iv}(\xi_1) - 3f^{iv}(\xi_{-1}) + 8f^{iv}(\xi_{-2})]/72 \qquad (10.74)$$

Simplifying as before, the error is

$$\varepsilon = -h^3 f^{iv}(\xi)/12 \qquad (10.75)$$

where $x_{k-2} \leqq \xi \leqq x_{k+1}$.
Since Stirling's differentiation formula [Eqs. (10.72)] is based on five points, it is necessary to carry one more term in Eqs. (10.58), (10.59), (10.66), and (10.73). If this is done and the results applied to (10.72), the equation becomes

$$y'_k = f'_k + h^4[-4f^{v}(\xi_2) + f^{v}(\xi_1) + f^{v}(\xi_{-1}) - 4f^{v}(\xi_{-2})]/180 \qquad (10.76)$$

From this, the simplified form of the error expression is

$$\varepsilon = -h^4 f^{v}(\xi)/30 \tag{10.77}$$

where $x_{k-2} \leqq \xi \leqq x_{k+2}$.

10.7.3 Error in Smoothing Differentiation Methods

As in the previous subsections, we will derive the error expressions for some selected examples only.

Consider Eq. (10.40); each y_{-2}, y_{-1}, y_1, and y_2 can be expanded in a Taylor's series about x_0, y_0. It is only necessary to carry terms through the third derivative in these expansions. Omitting the details, Eq. (10.40) can then be written as

$$Y_0' = y_0' + h^2[16y'''(\xi_{-2}) + y'''(\xi_{-1}) + y'''(\xi_1) + 16y'''(\xi_2)]/60 \tag{10.78}$$

Then the error is

$$\varepsilon = -17h^2 y'''(\xi)/30 \tag{10.79}$$

where $-2h \leqq \xi \leqq 2h$.

In a similar fashion, we can analyze the error in Eq. (10.43). However, it was already noted that (10.43) is the same as (10.32), which is the same as (10.72). The error is the same as that given by (10.77).

10.7.4 Error in Second-Order Numerical Differentiation Methods

In Sec. 10.6 two methods of finding higher-order derivatives were suggested. In this subsection, examples of the inherent error of the two methods will be presented for second-order differentiation based on the Stirling interpolation formula.

The first of the methods given, Sec. 10.6.1, describes the use of repeated numerical differentiation to obtain higher-order derivatives. Such a method can be based on a formula such as Eq. (10.31). This equation can be used to find the derivatives at three consecutive points to form a table for a second first-order differentiation. Although this would be done with numerical values, for the sake of this analysis it will be done literally. Hence,

$$y_{k-1}' = (f_k - f_{k-2})/(2h) \tag{10.80a}$$

$$y_k' = (f_{k+1} - f_{k-1})/(2h) \tag{10.80b}$$

$$y_{k+1}' = (f_{k+2} - f_k)/(2h) \tag{10.80c}$$

By Eq. (10.31) we can also write

$$y_k'' = (y_{k+1}' - y_{k-1}')/(2h) \tag{10.81}$$

Substituting (10.80a) and (10.80c) into (10.81) yields

$$y_k'' = (f_{k+2} - 2f_k + f_{k-2})/(4h^2) \tag{10.82}$$

The inherent error of this equation can be determined by use of its Taylor's series expansion through the fourth-order term. Such an analysis will yield the error expression

$$\varepsilon = -h^2 f^{iv}(\xi)/3 \tag{10.83}$$

The second method of obtaining higher-order derivatives was discussed in Sec. 10.6.2. In this method, higher-order derivative formulas are found from lower-order formulas by analytic means. Then the derived formula(s) can be used to obtain the required derivative(s). As a comparable example to the analysis of the previous paragraph, consider Eq. (10.53) truncated after one term and expressed in terms of ordinates.

$$y_k'' = (f_{k+1} - 2f_k + f_{k-1})/h^2 \tag{10.84}$$

By the use of a Taylor's series expansion of this equation, the following error expression can be derived:

$$\varepsilon = -h^2 f^{iv}(\xi)/12 \tag{10.85}$$

Of these two examples, the second has the smaller error. This comparison does not prove, in general, that the second method is better than the first method. Also, in selecting a numerical differentiation method, factors other than inherent error may be relevant.

10.8 EFFECTS OF RANDOM ERRORS

In Sec. 10.7, the inherent (or truncation) errors for various differentiation methods were derived. The reader is advised that these errors may only represent a small portion of the actual error. Errors in the data can create much larger errors in the derivatives than will be incurred by truncating the differentiation formula. It is for this reason that smoothing polynomials have been suggested as a basis for differentiation formulas. These formulas can have larger inherent errors than the formulas based on interpolating polynomials, but it is hoped that the smoothing may remove some of the random error in the data. As a comparative example, Eq. (10.31) is a differentiation formula based on a second-order interpolating polynomial, and Eq. (10.40) is a differentiation formula based on a five point second-order smoothing polynomial. The inherent error terms are as follows:

For Eq. (10.31),

$$\varepsilon = h^2 f'''(\xi)/6 \tag{10.86}$$

and for Eq. (10.40),

$$\varepsilon = -17h^3 f'''(\xi)/30 \tag{10.87}$$

The above comparison would indicate that the use of (10.31) would incur less error than the use of (10.40). This is true if there is no error in the data points. It is not possible to predict, in general, the effect of random errors in the data. However, by employing simplifying assumptions, some suggestion as to the relative magnitude of such effects can be determined.

Suppose that each ordinate can have an error of $\pm\varepsilon$. A worst case analysis will give the maximum error that can occur. Substituting $y_i \pm \varepsilon$ for each y in (10.31) gives

$$y_k' + E = [y_{k+1} \pm \varepsilon - (y_{k-1} \pm \varepsilon)]/(2h) \tag{10.88}$$

where E is the total error in the derivative. Selecting signs of the ε's so as to give the largest positive total error gives

$$y_k' + E = (y_{k+1} + \varepsilon - y_{k-1} + \varepsilon)/(2h) \tag{10.89}$$

If the exact derivative is given by Eq. (10.31), then by subtracting (10.31) from (10.89) leaves

$$E = \varepsilon/h \tag{10.90}$$

Similar analysis of Eq. (10.40) gives

$$y'_k + E = [-2(y_{k-2} \pm \varepsilon) - (y_{k-1} \pm \varepsilon) + (y_{k+1} \pm \varepsilon) + 2(y_{k+2} \pm \varepsilon)]/(10h) \tag{10.91}$$

Again selecting signs to maximize the total error,

$$y'_k + E = (-2y_{k-2} + 2\varepsilon - y_{k-1} + \varepsilon + y_{k+1} + \varepsilon + 2y_{k+2} + 2\varepsilon)/(10h) \tag{10.92}$$

and

$$E = 6\varepsilon/(10h) \tag{10.93}$$

For the example given and the assumptions made, the use of (10.40) will result in a maximum total random error which is only 0.6 of the maximum total random error that would be incurred by the use of (10.31). Which of these equations gives the more accurate result will depend on the actual values of the truncation error and the random error. Neither one of these errors can be determined exactly, but usually some estimate of their maximum size can be made.

10.9 DERIVATIVES BY SEMI-ANALYTIC METHODS

It may be possible to determine numerical derivatives more accurately by methods other than the ones described in this chapter. In the methods given, the data (or a subset of the data) is approximated by a polynomial, which is then differentiated analytically and evaluated at a given point. The data (or a subset of the data) can be approximated with a function other than a polynomial; this function can then be differentiated analytically and evaluated at a given point. If possible, the approximating function should be of the same form (or nearly the same form) as the function that describes the physical phenomenon that the data represents.

For example, suppose that an automotive engineer wishes to experimentally determine the velocities and accelerations of a part in an automotive suspension system. One method of doing this is to instrument the part in such a manner that its position can be recorded periodically. The resulting tabular data can be differentiated by the methods of this chapter to obtain velocities and accelerations. However, if the engineer can develop a mathematical model describing the position of the part versus time, the observed data can be fit to this function by one of the methods of Chap. 8 and the resulting function differentiated and evaluated to give the velocities and accelerations desired. It may not be meaningful or practical to fit all of the data to a simplified model. However, it may be sufficiently accurate to fit small groups of data to a simplified model and use each set of parameters thus found to determine the derivatives for the corresponding data point sets.

10.10 SUMMARY

Various differentiation methods have been presented in this chapter. Two of the methods given were graphical, while the others were numerical methods based on polynomials. It has been stressed repeatedly that numerical differentiation tends to amplify errors. Therefore, if at all possible, derivatives should be found by analytic means.

The numerical differentiation formulas given in this chapter can be summarized in the following manner. If the formulas are expressed in terms of ordinates, they all have the general form

$$f'_0 = \left[\sum_{i=j}^{k} w_i f_i \right] / (Wh) \tag{10.94}$$

Table 10.2 gives the parameters for each of the formulas presented in this chapter in terms of the parameters of Eq. (10.94). Note that

$$\sum_{i=j}^{k} w_i = 0 \tag{10.95}$$

and

$$\sum_{i=j}^{k} i w_i / W = 1 \tag{10.96}$$

even when $j \neq k$. [See Eqs. (10.46) and (10.47),]

TABLE 10.2 DIFFERENTIATION FORMULA PARAMETERS

		w_i								
j	*k**i*	−3	−2	−1	0	1	2	3	*W*	*Equations*
0	1				−1	1			1	(10.7)
−1	0			−1	1				1	(10.8)
−1	1			−1	0	1			2	(10.9), (10.31), (10.37)
0	3				−11	18	−9	2	6	(10.56)
−1	2			−2	−3	6	−1		6	(10.65)
−2	1		1	−6	3	2			6	(10.71)
−2	2		1	−8	0	8	−1		12	(10.32), (10.43), (10.72)
−2	2		−2	−1	0	1	2		10	(10.40)
−3	3	−3	−2	−1	0	1	2	3	28	(10.41)
−3	3	22	−67	−58	0	58	67	−22	252	(10.44)

Other differentiation formulas can be developed by the techniques of this chapter and added to Table 10.2. In fact, as long as we satisfy the conditions of Eqs. (10.95) and (10.96), we can construct differentiation formulas without going through any lengthy analysis. However, we would not know *a priori* anything about the nature of such formulas. But once the formula has been constructed, the techniques of Sec. 10.7 can be employed to determine the inherent error. Also, the technique of Sec. 10.8 can be employed to obtain some measure of the effect of random errors in the data on the result of the numerical differentiation.

Since the differentiation methods based on polynomials can be represented by the general formula Eq. (10.94), it is possible to prepare a general flowchart for differentiation by any of these methods. Figure 10.6 is such a general flowchart. Any set of

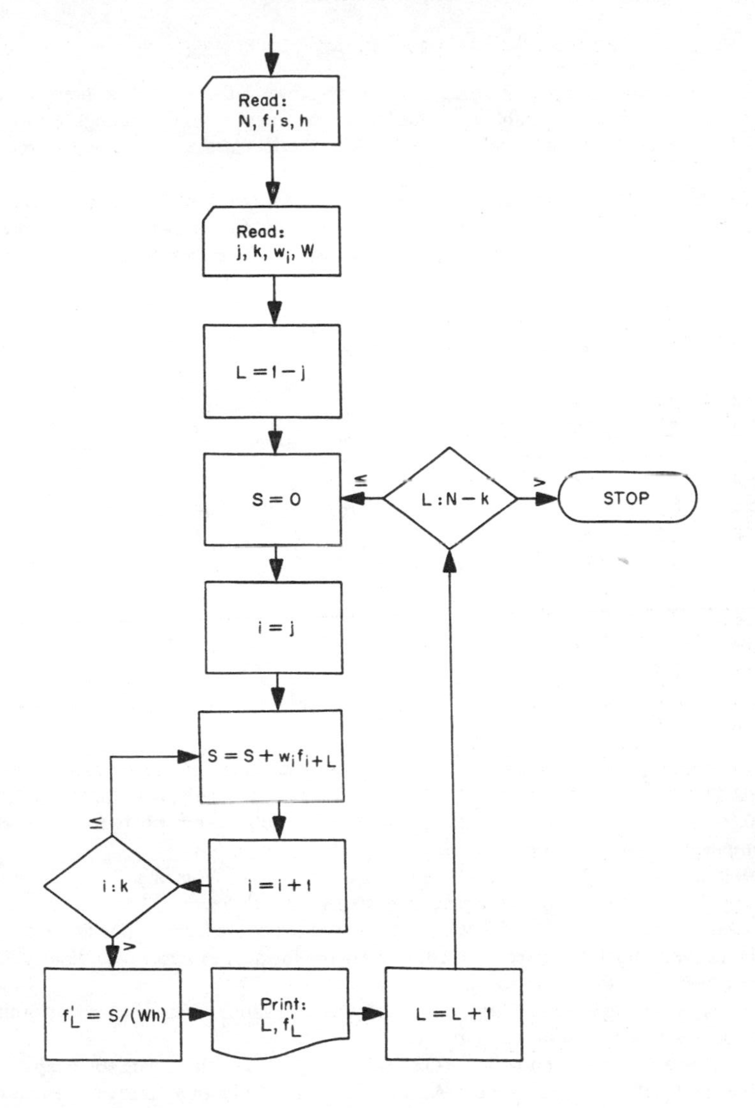

Fig. 10.6. Flowchart for differentiation by general differentiation formula.

differentiation parameters, such as those shown in Table 10.2, can be used with the flowchart to differentiate an array of function values.

A more efficient program will result if this flowchart is specialized for the particular formula that is to be used. Note that some of the weighting coefficients have nonpositive subscripts. A FORTRAN program for this flowchart must offset all subscripts by an amount such that all subscripts are positive.

EXERCISES

1. Prepare a table of sines with arguments ranging from -0.05 to 2.05 at intervals of 0.05. Numerically differentiate this table for values of the argument from 0.0 through 2.0 by use of Eq. (10.9). Also tabulate the error between the exact value of the derivative and the value computed.

2. By the use of Taylor's series, derive the expression for the truncation error for Eq. (10.9). Compare the errors obtained in Exercise 1 with the error predicted by the derived expression.

3. A quantity of air is observed in adiabatic expansion. The measurements of pressure P as the volume V is increased are given by

V	P	$-P'$
2.0	13.4	9.92
2.4	10.6	6.37
2.8	8.7	4.37
3.2	7.2	3.16
3.6	6.1	2.37
4.0	5.1	1.84
4.4	4.5	1.45
4.8	4.0	1.18
5.2	3.4	0.97
5.6	3.1	0.81
6.0	2.6	0.68
6.4	2.3	0.59
6.8	2.0	0.50
7.2	1.7	0.44
7.6	1.3	0.39

Values of the derivative are also tabulated. These were obtained by differentiating a formula that was fit to all of the data using least squares. Calculate the derivatives for this data using Eq. (10.56) for values of V from 2.0 through 6.4. Also compute the error between the given and the computed values of the derivative.

4. Repeat Exercise 3 using Eq. (10.65) instead of (10.56). Note that it will be possible to obtain derivatives for values of V ranging from 2.4 through 6.8.

5. Derive a fourth term for Eq. (10.13) and evaluate the entire formula (the first three terms plus the derived fourth term) for $u = 2.0$. Express the formula in terms of ordinates. Derive the truncation error for this formula.

6. Using the first two terms of Eq. (10.27), write a differentiation formula in terms of ordinates. Derive the truncation error for this formula.

7. Prepare a flowchart and write a FORTRAN program to differentiate some tabular data by a five-point differentiation formula. Also provide for reading in a table of given values of the derivative. Include computation of the error between the given and computed values, and print out derivatives and errors.

8. Execute the program in Exercise 7 using the formula derived in Exercise 5 and the data from Exercise 3. Calculate the values of the derivative and the error for values of V from 2.8 through 6.8.

9. Repeat Exercise 8 using Eq. (10.40) instead of the formula from Exercise 5.

10. Repeat Exercise 8 using Eq. (10.43) instead of the formula from Exercise 5.

11. Prepare a flowchart and write a FORTRAN program to differentiate some tabular data by a seven-point differentiation formula.

12. Execute the program in Exercice 11 using Eq. (10.41) and the data from Exercise 3, Chap. 9.

13. Execute the program in Exercise 11 using Eq. (10.44) and the data from Exercise 3, Chapter 9.

14. By the use of Taylor's series, derive the truncation error for Eq. (10.41).
15. By the use of Taylor's series, derive the truncation error for Eq. (10.44).
16. The pressure P for the adiabatic expansion of a diatomic gas is given by

$$P = CV^{-1.4}$$

Therefore, the derivative with respect to V is given by

$$P' = -1.4P/V$$

Even though air does not behave exactly in the same manner as a pure diatomic gas, use the above formula to calculate the derivative for the data of Exercise 3. Compare these results with the given values of the derivative.

11

Numerical Integration

11.1 INTRODUCTION

The fundamental theorem of integral calculus states that the value of the definite integral of a function between certain limits can be expressed as the value of some other function evaluated at the upper limit minus the value of that same function evaluated at the lower limit. Algebraically, this can be stated as

$$\int_a^b f(x)dx = F(b) - F(a) \tag{11.1}$$

where $f(x)$ is the derivative $F'(x)$ of $F(x)$; or $F(x)$ is known as the antiderivative of $f(x)$. Although the derivative of almost any mathematically expressible function can be found, the converse of this (finding the antiderivative) is not always possible. In some cases, the antiderivative of a function may exist but be unknown to the user. Therefore, it may not be possible to obtain the value of a given integral by mathematical analysis.

In this chapter, methods for finding the values of integrals by graphical and numerical means are presented. In Sec. 11.2 two methods of graphical integration are mentioned. However, the subject of this chapter is Numerical Integration and hence most of the chapter is devoted to that subject. Each of the methods given in this chapter is based on integration of a polynomial approximation of the function to be integrated. The methods differ in the type and degree of polynomial approximation used. In Sec. 11.3 a class of methods is given that is based on interpolating polynomials; Sec. 11.4 discusses a class of methods based on smoothing polynomials; and in Sec. 11.5, another class of methods, Gaussian quadrature, which is also based on polynomial approximation, is presented. In addition, error analyses are given for some of the methods.

A useful concept to employ in the discussion of integration is to interpret the definite integral of a function as an area. If the units of the function and its independent variable are both length, then the integral of the function does represent an area. However, no matter what units are actually involved, each one can be represented with a length and the result plotted. The area under the curve between the two limits then represents the integral of the function. This is illustrated in Fig. 11.1.

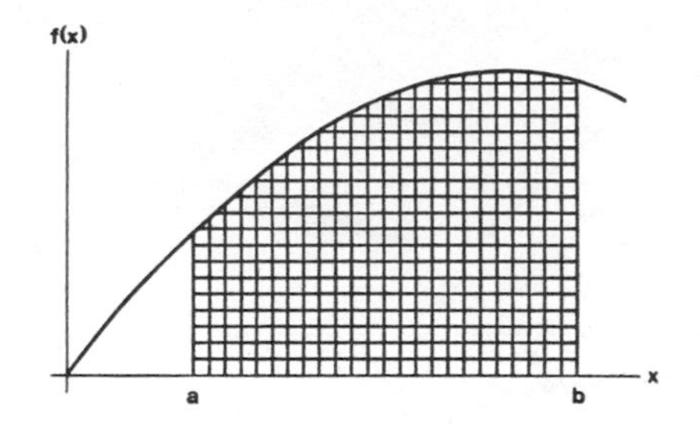

Fig. 11.1. Graphical representation of integral of a function.

The integral of a function of two independent variables can be interpreted as a volume. For functions of more than two independent variables, it is not possible to make a visual representation of the integral of the function. To integrate functions of more than one independent variable requires multiple integration. Multiple numerical integration will not be discussed explicitly in this chapter, but the numerical methods for single integration can be extended for multiple integration.

11.2 GRAPHICAL INTEGRATION

As with the case of differentiation, integration can also be carried out graphically.

In one method, the function is plotted on graph paper and the area determined by counting squares. From the example in Fig. 11.1, we see that there are some parts of squares along the curve which complicate the counting. If these fragments are ignored, the result will be too small. However, an estimate can be made for the size of each fragment and totaled to give an estimate for the area of all fragments. This area in terms of numbers of squares must then be converted to actual units. As an example, suppose the curve of Fig. 11.1 represents the instantaneous velocity of a rocket as a function of time. If each division in the vertical direction represents 50 feet/second in velocity and each division in the horizontal direction represents 0.5 second, then each square represents 25 feet. The area enclosed by a, b, the curve, and the x-axis is approximately 378 squares, hence, the distance traversed between time a and time b is 9450 feet.

A mechanical device also exists for finding areas of plane figures.This device is called a *planimeter*. Since it is designed to be used to find areas, it can be used to find the value of a definite integral. The operator of the device need only trace the outline of the area whose size is required. A value corresponding to the area is read directly from a counter. This value must then be scaled appropriately to account for the scaling of the graph and the physical setup of the device.

11.3 NUMERICAL INTEGRATION USING INTERPOLATING POLYNOMIALS

In this section, several numerical integration methods that are based on interpolating polynomials will be presented. These methods are known generally as the Newton–Cotes formulas. However, we will see that some of these have their own specific names.

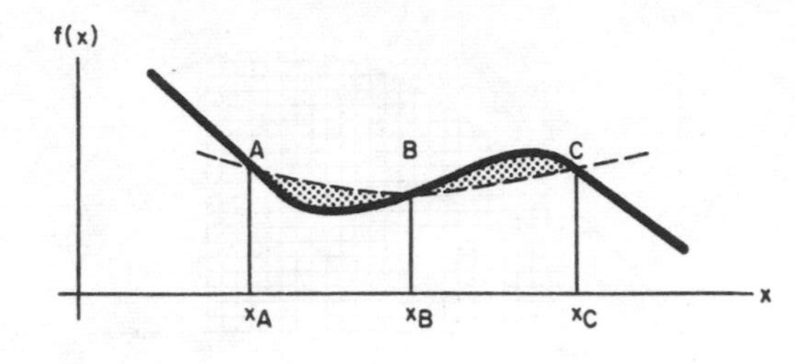

Fig. 11.2. Error incurred by itegrating a second-degree approximation to a function instead of integrating the function itself.

In the case of integration (as opposed to differentiation), we not only need to resort to numerical methods for integrating tabular functions, but we may also need to use numerical integration when the function to be integrated can be expressed mathematically. This is, as stated earlier, because the antiderivative of the function does not exist, is not known, or is very complicated. What we can do is to approximate the function with a form that we can integrate analytically. Again, the polynomial is a convenient form to use. Unlike differentiation, integration is a smoothing process. This is because errors usually tend to cancel each other out. As an extreme example, consider Fig. 11.2. The solid curve represents a function that is to be integrated from x_A to x_C. The dashed curve is a parabola that passes exactly through points A, B, and C on the original function. Even though the parabola is a rather poor approximation to the function, the area under the parabola, between the limits x_A and x_C, is a fairly good approximation of the area under the given function between the same limits. The shaded area between A and B is the error in area between the exact and approximate areas between the limits x_A and x_B. The shaded area between B and C is likewise the error in area between the limits x_B and x_C. These two errors have opposite signs and have about equal magnitude so they almost cancel each other out.

Not all cases work out this nicely. An error can still accumulate due to the nature of the function and the integration method used. However, if the data contains random errors, the effects of these errors will still tend to cancel out.

To find an integral over a large interval, it is best to divide this large interval into several smaller intervals and approximate the function in each of the smaller intervals. In Fig. 11.2, if we had passed a parabola through points A, B, and some intermediate point, we could have approximated the function and its integral between x_A and x_B more accurately. The same could have been done between x_B and x_C. Totaling the two areas thus obtained will result in a much smaller error. If the function is only tabulated at a finite number of points, then we are limited to this set of points for our integration scheme.

11.3.1 Trapezoidal Rule

A constant is the simplest polynomial to use in approximating a function. We will ignore this possibility in the discussion of integration because we will find that more sophisticated approximations will yield considerably better results for very little added

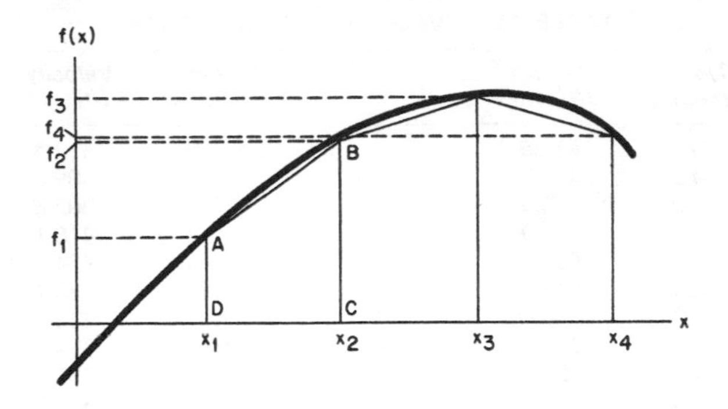

Fig. 11.3. Illustration of trapezoidal rule.

effort. The next polynomial to consider is the first-order polynomial, which is a straight line. By the Gregory–Newton interpolation formula, we have

$$y = f_k + \Delta f_k u \tag{11.2}$$

Substituting $\Delta f_k = f_{k+1} - f_k$ and $u = (a\text{-}x_k)/h$,

$$y = f_k + (f_{k+1} - f_k)(a\text{-}x_k)/h \tag{11.3}$$

If this approximation is to be used to find the integral over the limits of x_k to x_{k+1}, then

$$I_k = \int_{x_k}^{x_{k+1}} y\,da = [f_k - (f_{k+1} - f_k)x_k/h] \int_{x_k}^{x_{k+1}} da + [(f_{k+1} - f_k)/h] \int_{x_k}^{x_{k+1}} a\,da \tag{11.4}$$

Upon integrating,

$$I_k = [f_k - (f_{k+1} - f_k)x_k/h](x_{k+1} - x_k) + [(f_{k+1} - f_k)/h](x_{k+1}^2 - x_k^2)/2 \tag{11.5}$$

Since $h = x_{k+1} - x_k$,

$$I_k = f_k(x_{k+1} - x_k) - (f_{k+1} - f_k)x_k + (f_{k+1} - f_k)(x_{k+1} + x_k)/2 \tag{11.6}$$

This reduces to

$$I_k = (f_{k+1} + f_k)(x_{k+1} - x_k)/2 \tag{11.7}$$

This method is called the *trapezoidal rule* because we could have used trapezoids to arrive at the same result by geometry. Consider Fig. 11.3. Suppose we wish to find the area under the curve from x_1 to x_2. If we choose to approximate this area by the quadrilateral ABCD, we have, since the quadrilateral is a trapezoid,

$$\text{Area ABCD} = 1/2(\overline{DA} + \overline{CB})\,\overline{DC} \tag{11.8}$$

This can be written as

$$\text{Area ABCD} = 1/2(f_1 + f_2)(x_2 - x_1) \tag{11.9}$$

which is just a special case of Eq. (11.7) for $k = 1$.

TABLE 11.1 VALUES FOR FIG. 11.1

Time (sec)	*Velocity (ft/sec)*	*Time (sec)*	*Velocity (ft/sec)*
a + 0.0	439.6	a + 6.5	874.4
+ 0.5	477.5	+ 7.0	889.6
+ 1.0	515.5	+ 7.5	902.8
+ 1.5	563.9	+ 8.0	916.1
+ 2.0	603.8	+ 8.5	924.7
+ 2.5	647.5	+ 9.0	930.4
+ 3.0	686.4	+ 9.5	932.3
+ 3.5	718.7	+10.0	934.2
+ 4.0	754.7	+10.5	931.3
+ 4.5	785.1	+11.0	927.5
+ 5.0	810.8	+11.5	918.0
+ 5.5	836.4	+12.0	908.5
+ 6.0	855.4		

No restriction has been imposed on the interval size $h = x_{k+1} - x_k$, and therefore the trapezoidal rule is very convenient to use if data points are spaced at unequal intervals.

However, if the data points are spaced at equal intervals, then the area over several strips can be written in the simplified form

$$I = \sum_{k=i}^{j-1} I_k = \frac{h}{2}(f_i + 2f_{i+1} + 2f_{i+2} + \cdots + 2f_{j-1} + f_j) \tag{11.10}$$

From Fig. 11.3, we can see that for this example, the area obtained by this method will be too small. However, if the curvature of the function is opposite to that of Fig. 11.3, then the trapezoidal rule will yield a result that is too large. If some portions of the function have positive curvature and others have negative curvature, the errors will partially cancel out.

Figure 11.1 was constructed from the experimentally determined data given in Table 11.1. This data can be used to find the integral of the function, from *a* to *b*, by the trapezoidal rule. Using the values of Table 11.1 in Eq. (11.10) yields the result I = 9506. By counting squares (Sec. 11.2), we obtained the result I = 9450. Assuming the result obtained by the trapezoidal method is more accurate, it appears that our earlier estimate of the number of squares was too low by about two squares. This is only about a 1/2 percent error in the total number of squares.

In Sec. 11.3.4 we will consider the accuracy of the trapezoidal method. However, it is convenient to present here the results of using fewer data points. The results for various Δt's are given in Table 11.2.

TABLE 11.2 RESULTS OF APPLYING THE TRAPEZOIDAL RULE TO DATA OF TABLE 11.1

Δt	I
0.5	9506
1.0	9499
2.0	9477
3.0	9439
6.0	9177

11.3.2 Simpson's Rule

Although the trapezoidal rule is often adequate, increased accuracy can be obtained by using integration rules based on higher-degree polynomials. The next polynomial to consider is the second-degree, or quadratic, polynomial. Again we can write the approximating polynomial in the Gregory–Newton forward form, although any second-degree form will yield the same result. We will assume equal intervals of the abscissa, but a similar more complicated rule can be derived for unequal intervals.

$$y = f_k + \Delta f_k u + \Delta^2 f_k\, u(u-1)/2 \qquad (11.11)\ (4.32)$$

If we wish to integrate from the first point used to the last point used, we would integrate from $a = x_k$ to $a = x_{k+2}$. This is equivalent to integrating from $u = 0$ to $u = 2$. Then, since $da = h du$,

$$I_k = \int_{x_k}^{x_{k+2}} y\,da = h\int_0^2 y\,du = h\left[f_k\int_0^2 du + \Delta f_k\int_0^2 u\,du + \Delta^2 f_k\int_0^2 (u^2-u)\,du/2\right] \qquad (11.12)$$

Then,

$$I_k = h[2f_k + 2\Delta f_k + \Delta^2 f_k(8/3 - 2)/2] \qquad (11.13)$$

Writing (11.13) in terms of ordinate values [see Eqs. (4.20) and (4.21)],

$$I_k = h[2f_k + 2f_{k+1} - 2f_k + f_{k+2}/3 - 2f_{k+1}/3 + f_k/3] \qquad (11.14)$$

which reduces to

$$I_k = (f_k + 4f_{k+1} + f_{k+2})h/3 \qquad (11.15)$$

Equation (11.15) is called *Simpson's rule.*

Simpson's rule is considerably better than the trapezoidal rule. Since an interpolating polynomial is forced to pass exactly through the table points, the sign of the error will be opposite on two consecutive intervals. This means that the sign of the error of the integral will also be opposite on two consecutive intervals. This is illustrated in Fig. 11.2. Therefore, since Simpson's rule uses two consecutive intervals, the errors tend to cancel out.

Simpson's rule, like the trapezoidal rule, can be used repetitively to obtain the integral over numerous intervals. However, there must be an even number of intervals (odd number of points). In general, we can write

$$I = \sum_{n=0}^{(j-i-2)/2} I_{2n+i} = \frac{h}{3}(f_i + 4f_{i+1} + 2f_{i+2} + 4f_{i+3} + 2f_{i+4} + \cdots + 2f_{j-2} + 4f_{j-1} + f_j) \qquad (11.16)$$

where j-i must be even. Figure 11.4 is a flowchart for evaluating Eq. (11.16). This flowchart reduces all multiplications (except the final one) to additions. The use of a variable connector (computed GO TO in FORTRAN) determines whether a term is to be added once or twice.

In the event the number of intervals is not even, Simpson's rule can be used for an even number of the available intervals and some other method (for example, the trapezoidal rule) can be used for the remaining interval(s).

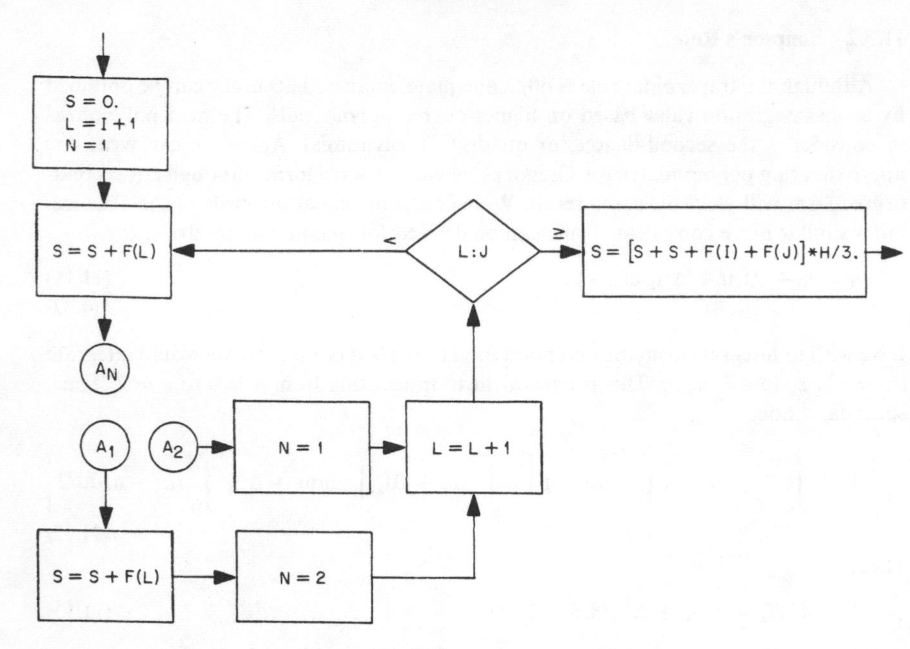

Fig. 11.4. Flowchart for evaluating Simpson's rule.

As in Sec. 11.3.1, the data from Table 11.1 can be used with (11.16). The results of applying that formula to the data of Table 11.1 for various Δt's is given in Table 11.3. Also repeated in this table are the results from using the trapezoidal rule. Even a cursory inspection reveals that Simpson's rule yields much better results than the trapezoidal rule. In Sec. 11.3.4, the inherent error for both of these methods will be given.

The fact that the results by Simpson's rule are nearly the same for $\Delta t = 0.5$, 1.0, and 2.0 indicates that the "correct" answer is in the neighborhood of the numbers given, and the small variations are due to rounding errors and the fact that different sets of data (and their associated errors) were used in each case.

TABLE 11.3 COMPARISON OF RESULTS USING BOTH THE TRAPEZOIDAL RULE AND SIMPSON'S RULE ON THE DATA OF TABLE 11.1

Δt	I	
	Trapezoidal	*Simpson's*
0.5	9506	9508
1.0	9499	9506
2.0	9477	9509
3.0	9439	9526
6.0	9177	9539

11.3.3 Integration by Use of Higher-Order Interpolating Polynomials

We can devise other numerical integration schemes based on higher-order interpolating polynomials. Only one (third order) will be treated here in detail. Again, using the Gregory–Newton forward interpolating polynomial, but now including third differences, we have

$$y = f_k + \Delta f_k u + \Delta^2 f_k u(u-1)/2 + \Delta^3 f_k u(u-1)(u-2)/6 \tag{11.17}$$

The indefinite integral of this is

$$\int y da = h[f_k u + \Delta f_k u^2/2 + \Delta^2 f_k(u^3/3\text{-}u^2/2)/2 + \Delta^3 f_k(u^4/4\text{-}u^3 + u^2)/6] \tag{11.18}$$

To integrate over all points used in the interpolation, we would evaluate (11.18) from $a = x_k$ to $a = x_{k+3}$, or $u = 0$ to $u = 3$. Hence,

$$I_k = \int_{x_k}^{x_{k+3}} y da = h(3f_k + 9\Delta f_k/2 + 9\Delta^2 f_k/4 + 3\Delta^3 f_k/8) \tag{11.19}$$

Again substituting ordinate expressions for differences and simplifying,

$$I_k = 3h(f_k + 3f_{k+1} + 3f_{k+2} + f_{k+3})/8 \tag{11.20}$$

This formula is sometimes called *Simpson's three-eights rule* or *Newton's three-eights rule*. Equation (11.20) is an interesting and useful equation in itself. However, its derivation here has been primarily to illustrate another point. In each of the methods discussed thus far, the integrations have been over limits that included all points used by the interpolating formula. However, this is not necessary. Suppose that we use the third-order interpolating polynomial, Eq. (11.17), to find the integral between the limits x_k and x_{k+2}. In terms of u, the limits would be 0 and 2. Applying these limits to Eq. (11.18) gives

$$\int_{x_k}^{x_{k+2}} y da = h(2f_k + 2\Delta f_k + \Delta^2 f_k/3 + (0)\Delta^3 f_k) \tag{11.21}$$

Note that the term involving the third difference vanishes. Since the remaining terms are identical to those in Eq. (11.13), Eq. (11.21) reduces to Simpson's rule:

$$\int_{x_k}^{x_{k+2}} y da = h(f_k + 4f_k + f_{k+2})/3 \tag{11.22}$$

The significance of this fact is that Simpson's rule, although based on a second-degree polynomial, gives exact results if the function being integrated is actually a third – (or lower) – degree polynomial.

Other integration expressions based on interpolating polynomials are presented in Table 11.4.

TABLE 11.4 NUMERICAL INTEGRATION FORMULAS BASED ON INTERPOLATING POLYNOMIALS OF ORDERS ONE THROUGH FIVE

Order	Rule
1	$h(f_k + f_{k+1})/2$ (Trapezoidal)
2	$h(f_k + 4f_{k+1} + f_{k+2})/3$ (Simpson's)
3	$3h(f_k + 3f_{k+1} + 3f_{k+2} + f_{k+3})/8$ (Three-eights)
4	$2h(7f_k + 32f_{k+1} + 12f_{k+2} + 32f_{k+3} + 7f_{k+4})/45$
5	$5h(19f_k + 75f_{k+1} + 50f_{k+2} + 50f_{k+3} + 75f_{k+4} + 19f_{k+5})/288$

11.3.4 Inherent Error of Integration Methods Based on Interpolating Polynomials

The *inherent error* of a computational method is the error of the method itself. It does not include errors such as random errors in the data, rounding errors, or mistakes. The inherent error can usually be found from the truncation error of the Taylor's series expansion of the formula(s) of the method. Therefore, the inherent errors are also known as *truncation errors*. To analyze the inherent errors in integration methods, we employ the fundamental theorem of integral calculus (Sec. 11.1). By this theorem,

$$\int_{x_k}^{x_{k+1}} f(x)dx = F(x_{k+1}) - F(x_k) \tag{11.23}$$

or since $x_{k+1} = x_k + h$,

$$I_k = \int_{x_k}^{x_k+h} f(x)dx = F(x_k + h) - F(x_k) \tag{11.24}$$

This integral, when approximated by the trapezoidal rule, is

$$I_k = h[f(x_k + h) + f(x_k)]/2 + \varepsilon \tag{11.25}$$

The quantity ε, which is the error, is inserted so that (11.25) can be written as an equality. From Eqs. (11.24) and (11.25), we can determine the following expression for the error function:

$$\varepsilon = F(x_k + h) - F(x_k) - h[f(x_k + h) + f(x_k)]/2 \tag{11.26}$$

By Taylor's series,

$$F(x_k + h) = F(x_k) + hF'(x_k) + h^2F''(x_k)/2 + h^3F'''(\xi_1)/6 \tag{11.27}$$

where $x_k \leqq \xi_1 \leqq x_k+h$. Since by the fundamental theorem $f(x) = F'(x)$, and if we assume that the function and all its derivatives are continuous, then $f'(x) = F''(x)$, $f''(x) = F'''(x)$, etc. Then, from (11.27) we have

$$F(x_k+h) - F(x_k) = h[f(x_k) + hf'(x_k)/2 + h^2f''(\xi_1)/6] \tag{11.28}$$

Also, by Taylor's series we can obtain

$$h[f(x_k+h) + f(x_k)]/2 = h[f(x_k) + hf'(x_k)/2 + h^2f''(\xi_2)/4] \tag{11.29}$$

where $x_k \leqq \xi_2 \leqq x_k+h$. Substituting Eqs. (11.28) and (11.29) into (11.26) yields

$$\varepsilon = h^3f''(\xi_1)/6 - h^3f''(\xi_2)/4 \tag{11.30}$$

It can be shown that a ξ exists such that (11.30) can be rewritten as

$$\varepsilon = -h^3 f''(\xi)/12 \tag{11.31}$$

Therefore, the inherent error in the trapezoidal rule is dependent on the second derivative of the function being integrated, evaluated someplace on the interval in question ($x_k \leqq \xi \leqq x_k + h$), and is also dependent on the third power of the interval size.

Although we generally will not know the value of the second derivative required to evaluate Eq. (11.31), we can at least see that decreasing the interval size will decrease the error. To determine the error incurred over say n intervals, we must sum up the errors on each interval.

$$E = \sum_{i=1}^{n} \varepsilon_i = -(h^3/12) \sum_{i=1}^{n} f''(\xi_i) \tag{11.32}$$

Defining $\bar{f}''$ by

$$\bar{f}'' = (1/n) \sum_{i=1}^{n} f''(\xi_i) \tag{11.33}$$

Eq. (11.32) then becomes

$$E = -nh^3 \bar{f}''/12 \tag{11.34}$$

and since

$$x_{i+n} = x_i + nh \tag{11.35}$$

Eq. (11.34) becomes

$$E = -(x_{i+n} - x_i)h^2 \bar{f}''/12 \tag{11.36}$$

Therefore, the total error over an interval (x_i, x_{i+n}) is proportional to the square of the interval size. This implies that dividing the interval size by 2 will divide the total error by 4, if the average second derivative does not vary markedly with interval size. As an example, consider the results given in Table 11.2. Assuming that 9506 is the correct result, then the total errors are E(1) = 7, which is less than one fourth of E(2) = 29, and E(3) = 67, which is less than one fourth of E(6) = 329.

In a similar fashion, the error for Simpson's (or any other) rule can also be found. The error expression for Simpson's rule is given by

$$\varepsilon = -h^5 f^{iv}(\xi)/90 \tag{11.37}$$

In this case, the error given by (11.37) is the error over two intervals, each of width h. An expression for the total error over an interval (x_i, x_{i+n}) can be derived as before, but remembering that each integration involves two intervals, there will be only h/2 error terms to be summed. Hence,

$$E = -(x_{i+n} - x_i)h^4 \bar{f}^{iv}/180 \tag{11.38}$$

This implies that dividing the interval size by 2 will cut the total error by a factor of 16. Table 11.3 does not corroborate this implication. The reason for this is that the errors are already so small that rounding error is significant. Even though the truncation error may diminish rapidly as the interval size is decreased, at some point rounding errors will begin to increase with decreasing interval size.

The foregoing error analysis considers only the inherent error of the methods and assumes perfect data. If in fact the data is not exact, other errors will also be incurred. A brief discussion of the effect of random error is provided in Sec. 11.5.

11.4 NUMERICAL INTEGRATION USING SMOOTHING FORMULAS

In Sec. 11.3, we considered integration by use of interpolating polynomials. Even if the data contains errors, those methods tend to smooth the results. In this section, we will consider integration methods based on the smoothing polynomials of Chap. 9. These methods will provide even more smoothing than the previous methods. However, we will see later that the methods based on smoothing polynomials have larger *inherent* errors than methods based on interpolating polynomials. Therefore, the methods of this section are recommended only for integrating tabular data that may contain errors.

In Chap. 9, the smoothing methods derived were based on finding a polynomial, by the method of least squares, that would "best" fit 2m+1 consecutive data points. That polynomial is given by

$$Y(x_i) = a_0 + a_1x_i + a_2x_i^2 + \cdots + a_nx_i^n \tag{11.39}$$

To find the integral of this polynomial over all 2m+1 points, it is necessary to integrate from x_{-m} to x_m. Since we can define $x_i = ih$, then the integral is

$$I = \int_{x_{-m}}^{x_m} Y(x_i)dx = h\int_{-m}^{m} (a_0 + a_1ih + a_2i^2h^2 + \cdots + a_ni^nh^n)di \tag{11.40}$$

which becomes

$$I = 2hm(a_0 + a_2h^2m^2/3 + a_4h^4m^4/5 + \cdots) \tag{11.41}$$

Integrals over other limits can also be taken, but in this discussion, as in Sec. 11.3, we will only consider the integral over all data points used to find the polynomial.

Since Eq. (11.41) contains only those coefficients with even subscripts, it is only necessary to find those coefficients. In Chap. 9, we found that, under certain simplifying assumptions, the normal equations could be separated into two sets [Eqs. (9.13) and (9.14)]. One of these sets [(9.13)] involves only those coefficients with even subscripts while the other [(9.14)] involves only coefficients with odd subscripts. Therefore, to construct particular integration formulas, it is only necessary to solve one set of equations, namely (9.13).

We also note from Eqs. (9.13) and (9.14) that the set (9.13) remains the same for finding the coefficients of a polynomial of order 2k+1 as for finding the coefficients of a polynomial of order 2k. Therefore, it is only necessary to consider using polynomials of even order.

11.4.1 Integration Formulas from Zero-Order Smoothing Polynomials

Using the symbol $I_m^{(n)}$ to indicate the integral obtained by use of an nth-order polynomial integrated over the 2m+1 points from −m to +m, we can write from Eq. (11.41) for the special case n = 0

$$I_m^{(0)} = 2hma_0 \tag{11.42}$$

Since $a_0 = T_0/S_0$ [Eq. (9.17)], we can simplify (11.42) to

$$I_m^{(0)} = 2hm \sum_{i=-m}^{m} y_i/(2m+1) \tag{11.43}$$

In particular we have

$$I_1^{(0)} = 2h(y_{-1} + y_0 + y_1)/3 \tag{11.44a}$$

$$I_2^{(0)} = 4h(y_{-2} + y_{-1} + y_0 + y_1 + y_2)/5 \tag{11.44b}$$

$$I_3^{(0)} = 6h(y_{-3} + y_{-2} + y_{-1} + y_0 + y_1 + y_2 + y_3)/7 \tag{11.44c}$$

Although these formulas were derived by use of a zero-order polynomial, the same results would be obtained from the use of a first order polynomial. Therefore, expressions for $I_1^{(1)}$, $I_2^{(1)}$, and $I_3^{(1)}$ would be identical, respectively, to (11.44).

In Chap. 9, the formulas derived were simplified by only considering sets of data containing an odd number of points. This was primarily so that each set would have a midpoint whose abscissa would coincide with that of a member of the set. This is not necessary for integration and therefore similar formulas can be derived for sets containing an even number of data points. However, in this discussion, we will use the same restrictions as used in Chap. 9 so that we can capitalize on the analysis already performed.

11.4.2 Integration Formulas Based on Second-Degree Smoothing Polynomials

If our data is to be fit by a second-degree polynomial, then Eq. (11.41) becomes

$$I_m^{(2)} = 2hm(a_0 + a_2h^2m^2/3) \tag{11.45}$$

We have by Eqs. (9.26):

$$\begin{aligned} a_0 &= (T_0S_4 - T_2S_2)/(S_0S_4 - S_2^2) \\ a_2 &= (T_2S_0 - T_0S_2)/(S_0S_4 - S_2^2) \end{aligned} \qquad \begin{matrix}(11.46)\\(9.26)\end{matrix}$$

For m=1[Eqs. (9.11c) and (9.11d)],

$$\begin{aligned} S_0 &= 3 \\ S_2 &= 2h^2 \\ S_4 &= 2h^4 \\ \therefore S_0S_4 - S_2^2 &= 2h^4 \end{aligned} \tag{11.47}$$

Also since [see Eq. (9.10b)]

$$\begin{aligned} T_0 &= y_{-1} + y_0 + y_1 \\ T_2 &= h^2(y_{-1} + y_1) \end{aligned} \tag{11.48}$$

then

$$\begin{aligned} a_0 &= y_0 \\ a_2 &= (y_{-1} - 2y_0 + y_1)/(2h^2) \end{aligned} \tag{11.49}$$

Substituting Eqs. (11.49) into (11.45) gives

$$I_1^{(2)} = h(y_{-1} + 4y_0 + y_1)/3 \tag{11.50}$$

which is Simpson's rule since, when m = 1, there are only three points, and a second-degree equation will satisfy these exactly. We can also derive

$$I_2^{(2)} = 4h(11y_{-2} + 26y_{-1} + 31y_0 + 26y_1 + 11y_2)/105 \tag{11.51a}$$

$$I_3^{(2)} = h(7y_{-3} + 12y_{-2} + 15y_{-1} + 16y_0 + 15y_1 + 12y_2 + 7y_3)/14 \tag{11.51b}$$

Note also that $I_1^{(3)}$ is the same as $I_1^{(2)}$ [Eq. (11.50)], and $I_2^{(3)}$ and $I_3^{(3)}$ are the same as $I_2^{(2)}$ and $I_3^{(2)}$, respectively [Eqs. (11.51)].

11.4.3 Inherent Error of Integration Methods Based on Smoothing Polynomials

The purpose in using the formulas derived in Sec. 11.4 is to attempt to smooth out errors caused by random errors in the data. Since we do not know *a priori* the nature of such errors, it is only convenient to analyze the inherent error of the methods, assuming perfect data. Since, in general, we cannot expect the fitted polynomial to pass exactly through the given data points, neither can we expect the inherent error to be any better than that found by the use of interpolating polynomials.

As in Sec. 11.3.4, we can use a Taylor's series expansion to determine the inherent error. Suppose, for example, we derive the inherent error incurred by using a three-point zero-order integration method [Eq. (11.44a)]. By the fundamental theorem of integral calculus, we can write

$$I = \int_{x_k}^{x_k+2h} f(x)dx = F(x_k + 2h) - F(x_k) \tag{11.52}$$

Rewriting Eq. (11.44a) in the same symbolism as used in Sec. 11.3.4,

$$I_1^{(0)} = 2h[f(x_k) + f(x_k + h) + f(x_k + 2h)]/3 \tag{11.53}$$

By the use of Taylor's series, Eq. (11.52) is expressed as

$$I = 2hf(x_k) + 2h^2f'(x_k) + 4h^3f''(\xi_1)/3 \tag{11.54}$$

where $x_k \leqq \xi_1 \leqq x_k + 2h$. Similarly we can expand the last two terms of Eq. (11.53) by Taylor's series:

$$f(x_k + h) = f(x_k) + hf'(x_k) + h^2f''(\xi_2)/2 \tag{11.55}$$

$$f(x_k + 2h) = f(x_k) + 2hf'(x_k) + 2h^2f''(\xi_3)$$

where $x_k \leqq \xi_2 \leqq x_k + h$ and $x_k \leqq \xi_3 \leqq x_k + 2h$. Applying Eqs. (11.55) to (11.53) gives

$$I_1^{(0)} = 2hf(x_k) + 2h^2f'(x_k) + h^3[f''(\xi_2)/3 + 4f''(\xi_3)/3] \tag{11.56}$$

Since the error is given by

$$\varepsilon = I - I_1^{(0)} \tag{11.57}$$

then

$$\varepsilon = h^3[4f''(\xi_1) - f''(\xi_2) - 4f''(\xi_3)]/3 \tag{11.58}$$

It can be shown that a ξ exists such that (11.58) can be written as

$$\varepsilon = -h^3f''(\xi)/3 \tag{11.59}$$

This is very similar to Eq. (11.31) which is the error term for the trapezoidal rule. Although (11.59) indicates the error of (11.53) is four times that of the trapezoidal rule, it must be noted that this error is over two intervals. An analysis of the error over a range (x_i, x_{i+n}) yields the following expression for the total error:

$$E = -(x_{i+n} - x_i)h^2\bar{f}''/6 \qquad (11.60)$$

(*Note:* The derivation of this equation and Eqs. (11.62) and (11.64) are similar to the derivation of (11.36) in Sec. 11.3.4.)
In a similar fashion, we can find the error terms for Eqs. (11.44b) and (11.44c):

$$\begin{aligned} \varepsilon &= -4h^3f''(\xi)/3 \\ \varepsilon &= -3h^3f''(\xi) \end{aligned} \qquad (11.61)$$

These represent the error incurred in integrating over four and six intervals, respectively. Therefore, the corresponding total errors over a range (x_i, x_{i+n}) are

$$\begin{aligned} E &= -(x_{i+n} - x_i)h^2\bar{f}''/3 \\ E &= -(x_{i+n} - x_i)h^2\bar{f}''/2 \end{aligned} \qquad (11.62)$$

Similar analyses can be carried out for the second-order methods. The results are:

$$\begin{aligned} &\text{The error in } I_1^{(2)} \text{ is} \quad \varepsilon = -h^5f^{iv}(\xi)/90 \\ &\text{The error in } I_2^{(2)} \text{ is} \quad \varepsilon = -34h^5f^{iv}(\xi)/315 \\ &\text{The error in } I_3^{(2)} \text{ is} \quad \varepsilon = -39h^5f^{iv}(\xi)/70 \end{aligned} \qquad (11.63)$$

These represent the error over two, four, and six intervals, respectively. The total error expressions over a given interval (x_i, x_{i+n}) will be, respectively,

$$\begin{aligned} E &= -(x_{i+n} - x_i)h^4\bar{f}^{iv}/180 \\ E &= -17(x_{i+n} - x_i)h^4\bar{f}^{iv}/630 \\ E &= -13(x_{i+n} - x_i)h^4\bar{f}^{iv}/140 \end{aligned} \qquad (11.64)$$

The error expressions that have been given are interesting and useful in themselves. In addition, they can be used to draw general conclusions about the error expressions for integration methods based on polynomials:

1. The methods based on zero- and first-order polynomials have errors that are proportional to the second derivative of the function and the third power of the interval size.
2. The methods based on second- and third-order polynomials have errors that are proportional to the fourth derivative of the function and the fifth power of the interval size.
3. These observations lead to the general conclusion that integration methods based on polynomials of orders $2n$ and $2n+1$ will have inherent errors that are proportional to the $(2n+2)$-th derivative of the function being integrated and to the $(2n+3)$-th power of the interval size.

These conclusions are only relevant for the inherent error of the basic formula for a given integration method. In actual practice, the basic formula of an integration method is usually applied repeatedly to consecutive sets of data points in order to obtain an approximate integral over a large interval. The total error, therefore, will be the sum of the errors incurred at each step of the process. Since the number of steps required is proportional to the interval size, we find that the total error is proportional to the $(2n+2)$-th power of the interval size for methods based on polynomials of degree $2n$ and $2n+1$.

11.5 EXAMPLE EFFECTS OF RANDOM ERROR IN NUMERICAL INTEGRATION

It was stated in Sec. 11.4 that the use of integration methods based on smoothing polynomials will provide more smoothing than methods based on interpolating polynomials even though the latter methods have smaller inherent errors. The effects of random errors are difficult to analyze in general. However, a simple numerical example will serve to illustrate the effects of random error in some selected numerical integration methods. Suppose that a function y is to be integrated over two intervals, that for each interval $h = 1$, and that the true value of y at each point is zero with each point containing an error $\pm\varepsilon$. Therefore, the true answer is zero and the answer given by the numerical method will be the total error caused by random errors in the ordinates. Since there are three ordinates and each can take on three values ($+\varepsilon$, 0, and $-\varepsilon$), there are 27 cases to investigate, which due to symmetry can be reduced to 14 cases. This is because, for any data set given, the signs can be reversed and the result will have the same magnitude as before but with the opposite sign. These cases are listed in Table 11.5 for the trapezoidal rule, the three-point linear smoothing rule, and Simpson's rule.

As shown in Table 11.5, of the 14 cases given, the three-point smoothing method has the least error in seven of the cases, Simpson's rule has the least error in three of

TABLE 11.5 COMPARISON OF EFFECTS OF RANDOM ERRORS FOR THREE INTEGRATION METHODS

			Total Error Divided by ε		
y_1	y_2	y_3	*Trapezoidal Rule* $(y_1 + 2y_2 + y_3)/2$	*3-Point Smoothing* $2(y_1 + y_2 + y_3)/3$	*Simpson's Rule* $(y_1 + 4y_2 + y_3)/3$
ε	ε	ε	2	2	2
ε	ε	0	3/2	4/3[†]	5/3
ε	ε	$-\varepsilon$	1	2/3[†]	4/3
ε	0	ε	1	4/3	2/3[†]
ε	0	0	1/2	2/3	1/3[†]
ε	0	$-\varepsilon$	0	0	0
ε	$-\varepsilon$	ε	0[†]	2/3	−2/3
ε	$-\varepsilon$	0	−1/2	0[†]	−1
ε	$-\varepsilon$	$-\varepsilon$	−1	−2/3[†]	−4/3
0	ε	ε	3/2	4/3[†]	5/3
0	ε	0	1	2/3[†]	4/3
0	ε	$-\varepsilon$	1/2	0[†]	1
0	0	ε	1/2	2/3	1/3[†]
0	0	0	0	0	0

[†] Error with the smallest absolute value in the set.

the cases, and the trapezoidal rule has the least error in only one case. The other three cases are ties. Expanding this study to the full 27 cases possible, all of the figures in the table are doubled except the tie cases of which there are only five. This simple example does not prove in general that a smoothing integration method is better than one based on an interpolating polynomial. But it does show that cases exist where such a method can yield better results than a method having a smaller inherent error.

Table 11.5 also helps to illustrate what happens in the worst case. (See the first line of the table.) In integration, the worst case of the combination of random errors occurs when the error in each ordinate has the same sign as the error for each other ordinate and all errors are of the maximum size. The error in the final result will be the difference between the answer obtained using the data with errors as described in the foregoing and the answer obtained using error-free data. Mathematically this can be shown to be

$$E = \pm\varepsilon(x_{i+n} - x_i) \qquad (11.65)$$

where the integration is over the interval (x_i, x_{i+n}). This result is independent of the integration method used. Graphically, this can be visualized as adding (or subtracting) a band of area of width ε to (or from) the area being determined. Therefore, if we have some idea of the maximum value of ε, Eq. (11.65) sets the maximum limits for the total error.

11.6 GAUSSIAN QUADRATURE

Gaussian quadrature is based on the formula

$$\int_{Z_1}^{Z_2} f(z)dz = \sum_{i=0}^{n} a_i f(z_i) \qquad (Z_1 < z_i < Z_2) \qquad (11.66)$$

This formula has $n+1$ parameters a_i and $n+1$ parameters z_i. Therefore, there are $2n+2$ parameters. These parameters can be found in such a manner that Eq. (11.66) will be exact for $f(z)$, a polynomial of any degree that is $2n+1$ or less. In order to find the $2n+2$ parameters, it is necessary to set up $2n+2$ equations and solve for the parameters. This can be done by selecting $2n+2$ *different* polynomials each of degree $2n+1$, or less, and substituting them one at a time into Eq. (11.66). The parameters found by solving these equations will also make (11.66) exact for $f(z)$ equal to any other polynomial of degree $2n+1$ or less. Since any set of $2n+2$ different polynomials can be used, it is simplest to use the polynomials z^j for $j = 0, 1, 2, \ldots, 2n, 2n+1$. The equations thus obtained will not be linear in the parameters and therefore will be difficult to solve.

To simplify this procedure, we can transform our argument z so that our integral takes on a standard form. This transformation is

$$z = (Z_2 - Z_1)x/2 + (Z_2 + Z_1)/2 \qquad (11.67)$$

Application of this transformation to (11.66) gives the standard form

$$\int_{Z_1}^{Z_2} f(z)dz = \frac{(Z_2 - Z_1)}{2}\int_{-1}^{1} F(x)dx = \frac{(Z_2 - Z_1)}{2}\sum_{i=0}^{n} A_i F(x_i) \qquad (11.68)$$

In this form it is slightly easier to solve for the parameters. As an example, suppose that $n = 1$. Then, upon substituting $F(x) = 1$, $F(x) = x$, $F(x) = x^2$, and $F(x) = x^3$ into Eq. (11.68), our $2n + 2$ (= 4) equations become

$$\begin{aligned} A_0 + A_1 &= 2 \\ A_0x_0 + A_1x_1 &= 0 \\ A_0x_0^2 + A_1x_1^2 &= 2/3 \\ A_0x_0^3 + A_1x_1^3 &= 0 \end{aligned} \tag{11.69}$$

Even for this simple case, it is difficult to solve for the parameters. Techniques other than solving simultaneous equations exist for determining these parameters. These will not be presented in this text because, by using the standard form, Eq. (11,68), there is only one set of parameters for each value of n and these are already available in tabular form.

Table 11.6 gives the parameters for the first four values of n. If other sets of parameters are needed, more extensive tables are available in other texts and reference books.

TABLE 11.6 PARAMETERS FOR GAUSSIAN QUADRATURE

n	i	A_i	x_i
1	0	1.00000000	−0.57735027
	1	1.00000000	0.57735027
2	0	0.55555556	−0.77459667
	1	0.88888889	0.00000000
	2	0.55555556	0.77459667
3	0	0.34785485	−0.86113631
	1	0.65214515	−0.33998104
	2	0.65214515	0.33998104
	3	0.34785485	0.86113631
4	0	0.23692689	−0.90617985
	1	0.47862867	−0.53846931
	2	0.56888889	0.00000000
	3	0.47862867	0.53846931
	4	0.23692689	0.90617985

Gaussian quadrature is usually used to integrate known functions that cannot be integrated analytically. Gaussian quadrature can be used on experimental data, but it requires either interpolating the data to obtain function values at each of the x_i's or conducting the experiment so that readings are made such that the values of the independent variables correspond to the x_i's.

11.6.1 Example of Gaussian Quadrature

Suppose that it is required to find the value of an elliptic integral of the second kind with modulus $k = 0.5$. Then we must evaluate

$$E(0.5) = \int_0^{\pi/2} \sqrt{1. - 0.25 \sin^2 \phi \, d\phi} \tag{11.70}$$

The value of this integral to eight significant figures is

$$E(0.5) = 1.4674622 \tag{11.71}$$

To find the value by Gaussian quadrature, the independent variable must first be transformed using Eq. (11.67) to yield

$$\phi = (\pi/4)x + \pi/4 \tag{11.72}$$

Substituting (11.72) into (11.70) and simplifying gives

$$E(0.5) = (\pi/4) \int_{-1}^{1} \sqrt{0.875 - 0.125 \sin(\pi/2)x\, dx} \tag{11.73}$$

This integral can then be approximated by (11.66). Choosing n = 2 and the associated parameters from Table 11.6, we have

$$\begin{aligned} F(x_0) &= F(-0.77459667) = 0.99611570 \\ F(x_1) &= F(0.00000000) \quad = 0.93541435 \\ F(x_2) &= F(0.77459667) \quad = 0.87049039 \end{aligned} \tag{11.74}$$

Then the approximate value is given by

$$E(0.5) = (\pi/4)[5F(x_0) + 8F(x_1) + 5F(x_2)]/9 = 1.4675030 \tag{11.75}$$

This agrees closely with the value given by Eq. (11.71).

11.7 SUMMARY

The subject of this chapter has been numerical integration. This is a process by which a number of values of the *integrand* (function that is to be integrated) are combined together to approximate the integral of the function. When the integrand is a function of a single variable, numerical integration is sometimes called *mechanical quadrature.* Although we have not covered methods for integrating functions of more than one variable, the methods given can be extended to provide for multiple integration. For example, suppose we wish to integrate the function f(x,y) by Simpson's rule over the limits x_1 to x_5 and y_1 to y_5. First, we can integrate over either x or y. Choosing x, we can form the new functions that are functions of y only:

$$\begin{aligned} F(y_1) &= \Delta x[f(x_1,y_1) + 4f(x_2,y_1) + 2f(x_3,y_1) + 4f(x_4,y_1) + f(x_5,y_1)]/3 \\ F(y_2) &= \Delta x[f(x_1,y_2) + 4f(x_2,y_2) + 2f(x_3,y_2) + 4f(x_4,y_2) + f(x_5,y_2)]/3 \\ F(y_3) &= \Delta x[f(x_1,y_3) + 4f(x_2,y_3) + 2f(x_3,y_3) + 4f(x_4,y_3) + f(x_5,y_3)]/3 \\ F(y_4) &= \Delta x[f(x_1,y_4) + 4f(x_2,y_4) + 2f(x_3,y_4) + 4f(x_4,y_4) + f(x_5,y_4)]/3 \\ F(y_5) &= \Delta x[f(x_1,y_5) + 4f(x_2,y_5) + 2f(x_3,y_5) + 4f(x_4,y_5) + f(x_5,y_5)]/3 \end{aligned} \tag{11.76}$$

Then finally,

$$I = \Delta y[F(y_1) + 4F(y_2) + 2F(y_3) + 4F(y_4) + F(y_5)]/3 \tag{11.77}$$

In the example given, there are the same number of values (five) of x's as there are of y's. This is not necessary, but was chosen for the sake of brevity. The nature of the function may require using more ordinate values in one dimension than in another. The foregoing technique can be extended for functions of three or more variables.

It has been stated in Sec. 11.3 that integration is a smoothing process and discussion and illustration of this was provided. In Sec. 11.5, a brief comparison of the effects of random errors was given.

In general, the integration methods given in this chapter can be written in the form

$$I(k) = h \sum_{i=0}^{k} w_i f_i / W \tag{11.78}$$

In this form, I(k) is defined as the integral over k intervals, each of width h. Even though some of the formulas given earlier were not given in this form, we can always redefine subscripts such that the first function value is always called f_0. Then each formula can be written in the form of Eq. (11.78). Table 11.7 summarizes all the formulas of this chapter (except Gaussian quadrature) and gives the order of the inherent error over a total interval of integration.

TABLE 11.7 PARAMETERS OF VARIOUS NUMERICAL INTEGRATION METHODS

Order of Polynomial	Type*	w_i 0	1	2	3	4	5	6	W	Order of Error
1	I	1	1						2	$O(h^2)$
1	S	2	2	2					3	$O(h^2)$
1	S	4	4	4	4	4			5	$O(h^2)$
1	S	6	6	6	6	6	6	6	7	$O(h^2)$
2	I	1	4	1					3	$O(h^4)$
2	S	44	104	124	104	44			105	$O(h^4)$
2	S	7	12	15	16	15	12	7	14	$O(h^4)$
3	I	3	9	9	3				8	$O(h^4)$
4	I	14	64	24	64	14			45	$O(h^6)$
5	I	95	375	250	250	375	95		288	$O(h^6)$

* Type: I — Based on interpolating polynomial.
S — Based on smoothing polynomial.

When performing a numerical integration, the basic formula used is usually applied a number of times, say N. The N results are summed to give the integral over the whole interval. When selecting parameters for a numerical integration, N, k, and h must be selected such that their product is equal to the difference between the upper and lower limits of integration. Fig. 11.5 is a flowchart for numerical integration over N intervals between the limits x_L and x_U by any numerical integration formula of the form of Eq. (11.78). This flowchart is primarily for use with a known function. It can also be used with tabulated functions, if acceptable accuracy can be obtained by interpolation. If a tabulated function is to be numerically integrated using only the tabulated values (i.e., without interpolation), this flowchart can be simplified by eliminating the computation of x and substituting f_{i+j} for f(x). Instead of reading in the limits and calculating h, the tabulated values of f and the value of h must be read into memory. There must be exactly Nk + 1 tabulated values, and they must correspond to values of the independent variable spaced at intervals of width h between the appropriate limits.

For a specific integration formula, Fig. 11.5 can usually be simplified because of the simplicity and symmetry of the weighting coefficients. Such a simplified flowchart is shown for Simpson's rule in Fig. 11.4.

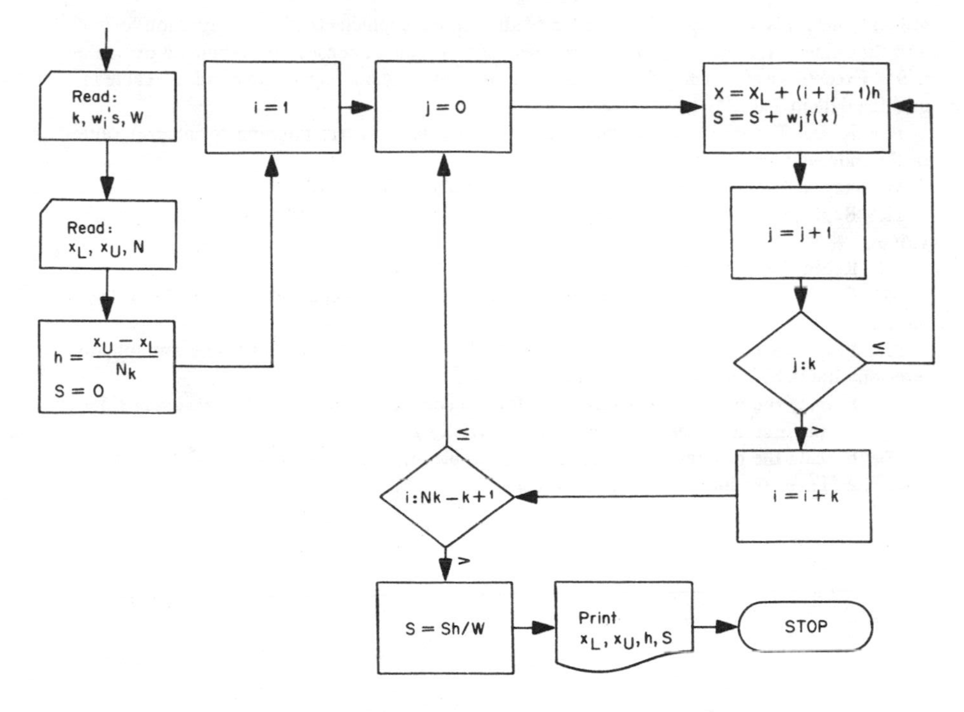

Fig. 11.5. General flowchart for numerical integration.

EXERCISES

1. Prepare a flowchart and write a FORTRAN program that will integrate a function by the trapezoidal rule. Enter the function as an arithmetic statement function so that it may be easily changed, and read the integration limits and number of intervals to be used from a data card.

2. Execute the program in Exercise 1 using the function $F(x) = \sin x$ and the limits $x = 0$ to $x = \pi$ using each of the following integration intervals: 10, 100, and 1000.

3. Execute the program in Exercise 1 using the function

$$F(x) = (1. -0.25 \sin^2 x)^{-1/2}$$

(elliptic integral of the first kind) and the limits $x = 0$ to $x = \pi/2$ using each of the following integration intervals: 2, 5, and 10.

4. Repeat Exercise 3, but integrate over the limits $x = 0$ to $x = \pi/4$ using each of the following integration intervals: 10, 100, and 1000. Why does it take many more integration intervals to get an answer of comparable accuracy to those obtained for Exercise 3?

5. Repeat Exercise 1 substituting Simpson's rule for the trapezoidal rule.

6. Execute the program in Exercise 5 using the function $F(x) = \sin x$ and the limits $x = \pi/2$ to $x = \pi$ using each of the following integration intervals: 10, 100, and 1000.

7. Execute the program in Exercise 5 using the function of Exercise 3 and the limits and number of integration intervals of Exercise 4.

8. Prepare a flowchart and write a FORTRAN program that will integrate a tabulated function by any five-point integration formula. The tabulated function must have $4n + 1$

values, where n is an integer. Provide for reading in the coefficients of the integration formula and the value of the increment in the independent variable of consecutive tabulated ordinates.

9. Execute the program in Exercise 8 using the fourth-order integration formula in Table 11.4 and the data in Table 11.1.

10. Repeat Exercise 9 using only those ordinate values corresponding to integral values of the independent variable from Table 11.1.

11. Repeat Exercise 9 using the five-point formula of Eq. (11.44b).

12. Repeat Exercise 11 using only the data for which the independent variable has integral values.

13. Repeat Exercise 9 using the five-point formula of Eq. (11.51a).

14. Repeat Exercise 13 using only the data for which the independent variable has integral values.

15. Prepare a flowchart and write a FORTRAN program that will integrate any function over any limits by Gaussian quadrature with $n = 2$.

16. Execute the program in Exercise 15 for the complete (limits $x = 0$ to $x = \pi/2$) elliptic integral of the first kind (see the formula in Exercise 3.)

17. Execute the program in Exercise 15 for an incomplete elliptic integral of the first kind (Exercise 3) over the limits $x = 0$ to $x = \pi/4$.

12

Numerical Solution of Differential Equations

12.1 INTRODUCTION

The solution of differential equations is very important in science and engineering. Many physical phenomena can be described only by a differential equation or a set of differential equations. Quite often, such an equation or set of equations cannot be solved by analytic means. Therefore, the alternative is to obtain a numerical solution. The purpose of this chapter is to present a sampling of various techniques for the numerical solution of ordinary differential equations. A general discussion and the definition of necessary terms is given later in this section.

Section 12.2 presents two methods that are each based on the use of an approximation to the analytic solution for the purpose of obtaining numerical results. Later, in Sec. 12.8, another method that is based on analytic integration, but which requires step-by-step numerical evaluation, is presented. Sections 12.3 and 12.4 give Euler's method and the modified Euler's method, respectively, for the step-by-step numerical solution of differential equations. Other predictor–corrector methods are given in Sec. 12.5, and in Sec. 12.6 the Runge–Kutta methods are presented. The inherent error and the stability of the step-by-step methods of Secs. 12.3 through 12.6 are the subjects of Sec. 12.7. The solution of higher-order equations is discussed in Sec. 12.9.

12.1.1 Differential Equations

First, we must define what is meant by the term *differential equation.* A differential equation is an equation that involves not only independent and dependent variables, but also one or more derivatives of the dependent variables taken with respect to one or more of the independent variables. Such equations may occur in sets. In order to have a solvable set, it is necessary to have as many equations as there are dependent variables.

If the dependent variables are functions of more than one independent variable, the equations will be *partial* differential equations. This is because (except in trivial cases) the derivatives involved must be partial derivatives. Differential equations involving only one independent variable are known as *ordinary* differential equations. In this text, we will confine our discussion to ordinary differential equations.

The general form of a kth-order differential equation is

$$G(x, y, y', y'', \ldots, y^{(k-1)}, y^{(k)}) = 0 \tag{12.1}$$

where the superscript k means the kth derivative with respect to the independent variable, which in this case is x. Therefore, the *order* of a differential equation is the same as the order of the highest derivative appearing in that differential equation.

An example of a second-order differential equation is

$$my'' + y'R(y') - mg = 0 \tag{12.2a}$$

or

$$mv' + v\,R(v) - mg = 0 \tag{12.2b}$$

Both of these equations describe the motion of a falling particle acted on by the forces of gravity and the resistance of the fluid through which the particle is falling. In Eq. (12.2), y is the vertical position coordinate; y′ (or velocity v) and y″ (or acceleration v′) are, respectively, the first and second derivatives with respect to time of the variable y; m is the mass of the particle; R(y′) is a resistance function of the fluid; and g is the acceleration due to gravity. Eq. (12.2a) is a second-order equation in the position coordinate. Since the position coordinate y does not appear explicitly in (12.2a), this equation can be rewritten as a first-order equation in the vertical velocity v. This is shown as Eq. (12.2b). Both of these equations describe fairly accurately the motion of a particle falling a short distance through a homogeneous fluid. For this case, Eq. (12.2b) is preferable. Since it is a first-order equation in v, it can be solved to obtain velocity as a function of time, and velocity can then be integrated to obtain y. If the conditions specified here are not satisfied, (12.2) may require further generalization by expressing R and g as functions of y. Then the equation would have to be used as given in (12.2a). (It can still be reduced to a first-order equation; but another variable, y, will be present.)

Throughout most of this chapter, the discussion of methods for solving differential equations will be confined to the special case of a single first-order differential equation. That is, the general form will be given by

$$G(x, y, y') = 0 \tag{12.3}$$

These methods can be easily expanded to provide for solving sets of equations, each equation being similar to (12.3) and containing more than one dependent variable and its derivative. In some of the methods, illustrations are given for solving sets of equations simultaneously. Also, these methods can be expanded to accomodate second- or higher-order differential equations, or even sets of such equations. This is accomplished by replacing each kth-order equation by k first-order equations. More detailed discussion of how this is done is described in Sec. 12.9.

The general first-order differential equation given by Eq. (12.3) can be rewritten as

$$y' = F(x,y) \tag{12.4}$$

If the function F(x,y) can be expressed in the form

$$F(x,y) = f_1(x)/f_2(y) \tag{12.5}$$

then, since we can substitute dy/dx for y', (12.4) can be written as

$$f_2(y)dy = f_1(x)dx \tag{12.6}$$

If antiderivatives of f_1 and f_2 are known, an analytic solution of (12.6) can be obtained. Even if antiderivatives are not known, the methods of Chap. 11 can be used to integrate (12.6). In order to obtain a specific solution for an equation of the form of (12.4), one variable pair (x_0, y_0) must be specified. Because of this fact, such problems are called *initial value problems*. In general, many different solutions can result depending on which initial values (x_0, y_0) are given.

In many cases, the variables are not separable as in Eq. (12.6). For example, (12.2) is such a case. Matters are further complicated by the fact that it may be necessary to use a table rather than a mathematical expression for the function R. Therefore, numerical methods must be devised that allow us to start with the initial value for a function, and proceed one step at a time to solve for other values of the function which satisfy the differential equation.

12.2 SEMI-NUMERICAL METHODS

In this section, we will discuss two, not strictly numerical, methods for solving differential equations. However, they do provide means for obtaining numerical solutions.

12.2.1 Taylor's Series Method

We can expand any function having continuous derivatives in a Taylor's series. Thus we can write

$$y(x) = y(x_0) + (x\text{-}x_0)y'(x_0) + (x\text{-}x_0)^2y''(x_0)/2! + (x\text{-}x_0)^3y'''(x_0)/3! + \cdots \qquad (12.7)\ (5.6)$$

The problem then in determining the function $y(x)$ is in finding the values of the various derivatives evaluated at x_0. If the differential equation to be solved is given by Eq. (12.4), then to find higher derivatives it is necessary to differentiate the function $F(x,y)$ (and its derivatives) implicitly with respect to x and evaluate each at x_0. As an example, suppose we wish to solve numerically the differential equation

$$y'(x) = xe^y \tag{12.8}$$

Let the initial values be $y = 0$ when $x = 0$. This equation has been chosen because it can be solved by analytic means yielding the result

$$y(x) = -\ln(1 - x^2/2) \tag{12.9}$$

To apply the Taylor's series method, we must repeatedly differentiate (12.8). The first few derivatives so obtained are

$$\begin{aligned} y''(x) &= e^y + y'^2 \\ y'''(x) &= e^y y' + 2y'y'' \qquad (12.10) \\ y^{iv}(x) &= e^y y'^2 + e^y y'' + 2y''^2 + 2y'y''' \end{aligned}$$

Evaluating (12.8) and (12.10) at $x = 0$,

$$\begin{aligned} y'(0) &= 0 \\ y''(0) &= 1 \\ y'''(0) &= 0 \\ y^{iv}(0) &= 3 \end{aligned} \tag{12.11}$$

We can see by Eqs. (12.10) that the derivatives are getting more and more complex with each differentiation. Using the values obtained in (12.11) and $x_0 = 0$, and substituting into (12.7),

$$y(x) = 0 + (x-0)(0) + (x-0)^2(1)/2! + (x-0)(0)/3! + (x-0)^4(3)/4! + \cdots \tag{12.12}$$

which simplifies to

$$y(x) = x^2/2 + x^4/8 + \cdots \tag{12.13}$$

If (12.9) is expanded in a series, (12.13) will result. This verifies that at least the first two terms of the equation are correct.

From this example, we see that such a method is not very amenable to computer solution. Although the resulting series can be easily evaluated by computer, the problem solver must do a considerable amount of analytic work for each problem just to obtain the series approximation. The approximation obtained may be useful only in a small neighborhood of x_0. In fact, the series may not converge. We will not explore the conditions for convergence here, but only note that even if the series does converge over some interval, we may have to restrict our values of x to some smaller interval in order that a reasonable number of terms in the series will suffice to obtain the required accuracy. In the case illustrated, Eq. (12.13) will yield fewer than eight decimal places of accuracy, if the absolute value of x exceeds 0.07. However, we will see in Sec. 12.5 that some methods require more than the initial values to get the method started. A Taylor's series will be very useful in those cases to obtain starting values of sufficient accuracy.

12.2.2 Picard's Method

From the original differential equation, Eq. (12.4), and its initial values, we can write the following expression for the solution:

$$y(x) = y(x_0) + \int_{x_0}^{x} F(x,y)dx \tag{12.14}$$

Since y appears under the integral sign, the integration cannot be performed. However, (12.14) can be used as an iterative form to obtain mathematical (rather than numerical) approximations to the solution. This is done by substituting an approximation for y, which is a function of x only, into (12.14) and integrating. Using superscripts to identify successive mathematical approximations for y, (12.14) is written as

$$y^{(n+1)}(x) = y(x_0) + \int_{x_0}^{x} F(x,y^{(n)}(x))dx \tag{12.15}$$

It may not be possible to carry out the indicated integration. However, if it is possible, then one approximate function $y^{(n)}(x)$ will lead to another approximate function $y^{(n+1)}(x)$. This process may converge to the solution, or to a form that is equivalent to the solution, or to a form that will give reasonable numerical approximations for the solution. Since Picard's method does not have much value for computer solution, we will not explore the conditions for convergence. However, it is of interest to apply this method to an example such as the one in Sec. 12.2.1. This example illustrates some of the difficulties encountered in using this method. Given

$$F(x,y) = xe^y \tag{12.16}$$

and the initial values $y = 0$ when $x = 0$. We can make the initial approximation $y^{(0)}(x) = 0$. Then, from (12.15) and (12.16),

$$\begin{aligned} y^{(1)}(x) &= 0 + \int_0^x x\,dx \\ &= x^2/2 \end{aligned} \tag{12.17}$$

Substituting this result into (12.16) and the result of that substitution into (12.15) gives

$$\begin{aligned} y^{(2)}(x) &= 0 + \int_0^x xe^{(x^2/2)}\,dx \\ &= e^{(x^2/2)} - 1 \end{aligned} \tag{12.18}$$

Further integrations will be even more difficult or even impossible. However, if Eq. (12.18) is expanded in a series, the first two terms agree with (12.13). Because of the algebra involved and because we may obtain forms that have no antiderivative, we can see that Picard's method is not of much use for computer solution. However, the approximation of (12.18) may be sufficiently accurate for use in a small neighborhood of $x = 0$ and hence can be used to obtain starting values for methods to be discussed later. This approximation is slightly more accurate than the two terms given in Eq. (12.13).

Taylor's series and Picard's semi-numerical methods have the added disadvantage that the problem analyst must determine the range of the independent variable over which the function is sufficiently accurate to be used. It may be necessary to develop a set of approximate functions each of which is to be used over a different range of the independent variable. In Sec. 12.8, another method which is based on analytic integration and may be applicable for certain special problems will be discussed.

12.3 EULER'S METHOD

Euler's method is sometimes called a one-step or step-by-step method. That is, values of the desired function are calculated one at a time and the values obtained are used in the calculation of later values. We can write

$$y(x_{k+1}) = y(x_k) + \int_{x_k}^{x_{k+1}} y'(x)\,dx \tag{12.19}$$

It is important to remember that this equation is exact. Therefore, if we can predict the behavior of $y'(x)$ in the interval (x_k, x_{k+1}), then we can obtain the exact value of $y(x_{k+1})$. Further values of y can then be calculated by reapplying Eq. (12.19). However, since we are usually interested in equations where y' is a function of y, the foregoing procedure is not possible.

By the mean value theorem (see Sec. 10.1), we can write

$$y'(\xi) = \frac{y(x_{k+1}) - y(x_k)}{x_{k+1} - x_k} \tag{12.20}$$

where $x_k \leqq \xi \leqq x_{k+1}$. Solving (12.20) for $y(x_{k+1})$, we have

$$y(x_{k+1}) = y(x_k) + y'(\xi)(x_{k+1} - x_k) \tag{12.21}$$

Comparing (12.21) with (12.19), we see that if we can find the proper value of y' to use, the integral can be represented exactly in the following manner:

$$y'(\xi)(x_{k+1} - x_k) = \int_{x_k}^{x_{k+1}} y'(x)\,dx \tag{12.22}$$

That is, the integral of a function over some interval is exactly equal to the interval length times the function evaluated at some point in the interval.

Euler's method, and other methods to be discussed in subsequent sections, are each based on a different technique for obtaining an approximate average derivative to use in place of the term $y'(\xi)$ in Eq. (12.21). The approximation used by Euler's method is the value of the derivative evaluated at the starting point of the interval. Thus, letting $h = x_{k+1} - x_k$, (12.21) is approximated by

$$y(x_{k+1}) = y(x_k) + hy'(x_k) \tag{12.23}$$

Or, letting $y(x_i) = y_i$ and substituting for y',

$$y_{k+1} = y_k + hF(x_k, y_k) \tag{12.24}$$

If we use Euler's method to solve the example of Sec. 12.2.2

$$y'(x) = xe^y \tag{12.8}$$

then we can write Eq. (12.24) as

$$y_{k+1} = y_k + hx_k e^{y_k} \tag{12.25}$$

The results obtained by using various values for h are given in Table 12.1. Although values of y were calculated for x at intervals of h, only the results at intervals of 0.2 in x have been printed in the table. Additional details of how this table was computed are given later in this section.

A graphical presentation of a portion of these results is given in Fig. 12.1. This figure illustrates how the approximate solution tends to deviate from the exact solution. In this example, the approximation is always less than the exact solution. If the function had opposite curvature, the approximation would always be too great. In the example, since the initial slope is zero, the first computed value is the same as the initial value, namely zero. At the first computed point ($x = 0.1$), the slope that is computed for use in the next interval is less than what the average slope should be for the next interval. Furthermore, the first computed point is not on the desired function.

TABLE 12.1 EULER'S METHOD SOLUTIONS OF $y = xe^y$ FOR VARIOUS INTEGRATION INTERVALS

x	Exact	h = 0.1	h = 0.01	h = 0.001	h = 0.0001	h = 0.0001[†]
0.0	0.00000000	0.00000000	0.00000000	0.00000000	0.00000000	0.00000000
0.2	0.02020271	0.01000000	0.01917010	0.02009929	0.02019226	0.02019226
0.4	0.08338161	0.06112085	0.08109345	0.08315195	0.08335767	0.08335788
0.6	0.19845094	0.15910210	0.19432281	0.19803545	0.19840522	0.19840634
0.8	0.38566248	0.31750308	0.37828316	0.38491657	0.38557447	0.38557805
1.0	0.69314718	0.56539226	0.67848340	0.69165320	0.69296152	0.69297055

[†] Includes provision for reducing the rounding error.

In addition to the obvious difficulties of this method, there are also hidden difficulties associated with the particular computer being used. For example, the quantity $h = 0.0001$ cannot be represented exactly in a binary computer. After 10,000 steps, a significant error can accumulate in the independent variable. Likewise, if the interval size is small, the amount added to the ordinate at each step can be small in comparison with the ordinate itself. Therefore, the number of significant places added at each step can be considerably less than the number of places being carried in the ordinate. In the example, for $x = 1.$ and $h = 0.0001$, the amount to be added to y to obtain a value corresponding to $x = 1.0001$ should be $\Delta y = 0.0002$. Because of the numerical approximation, Δy will not be exactly 0.0002, and even if it were, it could not be represented exactly in a binary computer. Since $y(1) = 0.6929+$, three significant decimal places would be truncated from Δy in making the addition.

Certain computing techniques can be employed to reduce these errors. One method uses extended precision arithmetic. Another method uses the approximate y's, as in the foregoing, throughout some large interval Nh, and also accumulates the Δy's so that, at the end of each large interval, the value of y from the beginning of the interval

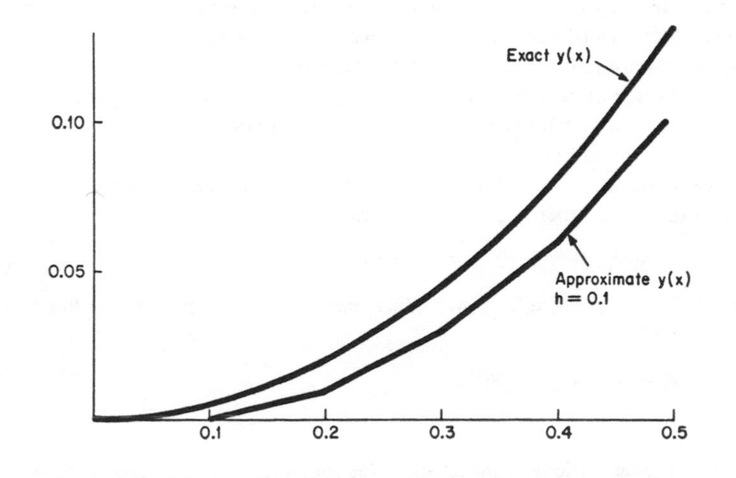

Fig. 12.1. Graph of an Euler's method solution and exact solution of $y' = xe^y$.

is updated by the total accumulated amount to obtain a new, more accurate value. Restating this in more mathematical terms, calculate the y_k's as specified[†] by Eq. (12.24) from $k = 0$ to $k = N$; also accumulate a sum S of the increments $hF(x_k,y_k)$. Then, although a y_N has been found, calculate a new one by adding S to y_0. With this new y_N, calculate another N steps, accumulating a separate sum of the increments. Then a new value of y_{2N} is calculated by adding the new S to y_N. This process can be repeated to the end of the integration. The preceding technique deals with only one aspect of rounding error. For completeness, the entire mathematical process should be reviewed with the goal of finding alternative computational procedures to minimize rounding error.

Figure 12.2(A) is a simple flowchart of Euler's method, while the flowchart in Fig. 12.2(B) includes the provision described in the preceding for reducing rounding error. It is assumed in both cases that

1. the initial values of X and Y,
2. the interval size h,
3. the number of intervals between printouts N, and
4. the final abscissa X_F

have been specified prior to entering this part of the program.

The flowchart in Fig. 12.2(A) was used to calculate the columns of Table 12.1 labeled $h = 0.1$, $h = 0.01$, $h = 0.001$, and $h = 0.0001$. The final column of Table 12.1 was calculated by the use of the flowchart in Fig. 12.2(B). By comparing the last two columns of Table 12.1, we can see that the provision for reducing the rounding error given in Fig. 12.2(B) did reduce the error by a small amount (5 percent). Additional analysis of the truncation errors for this example is given in Sec. 12.7.

12.4 MODIFIED EULER'S METHOD

In Sec. 12.3, we used y'_k to approximate $y'(\xi)$. It is obvious from the example that this is not a very good approximation. Another possibility is to use the derivative at the other end point, y'_{k+1}. This is probably no better an approximation. Furthermore, if y' is dependent on y (the usual case), then we must know y_{k+1} before we can calculate y'_{k+1}. Another reasonable possibility is to use the average of the two values of the derivative obtained at each end point of an interval. This is not only reasonable, but we can show that it is a fairly good approximation under certain conditions. This is demonstrated in the following.

Expanding the function itself in a Taylor's series about one end point of the interval and evaluating at the other end point, we have

$$y_{k+1} = y_k + y'_k h + y''_k h^2/2 + y'''(\xi_1)h^3/6 \qquad (12.26)$$

where $x_k \leqq \xi_1 \leqq x_{k+1}$. Similarly, we can expand the derivative of the function in a Taylor's series also.

$$y'_{k+1} = y'_k + y''_k h + y'''(\xi_2)h^2/2 \qquad (12.27)$$

† Here and in later sections of this chapter, the simplified notation $y_k = y(x_k)$, and similar notation for derivatives, will be employed to increase the readability of equations.

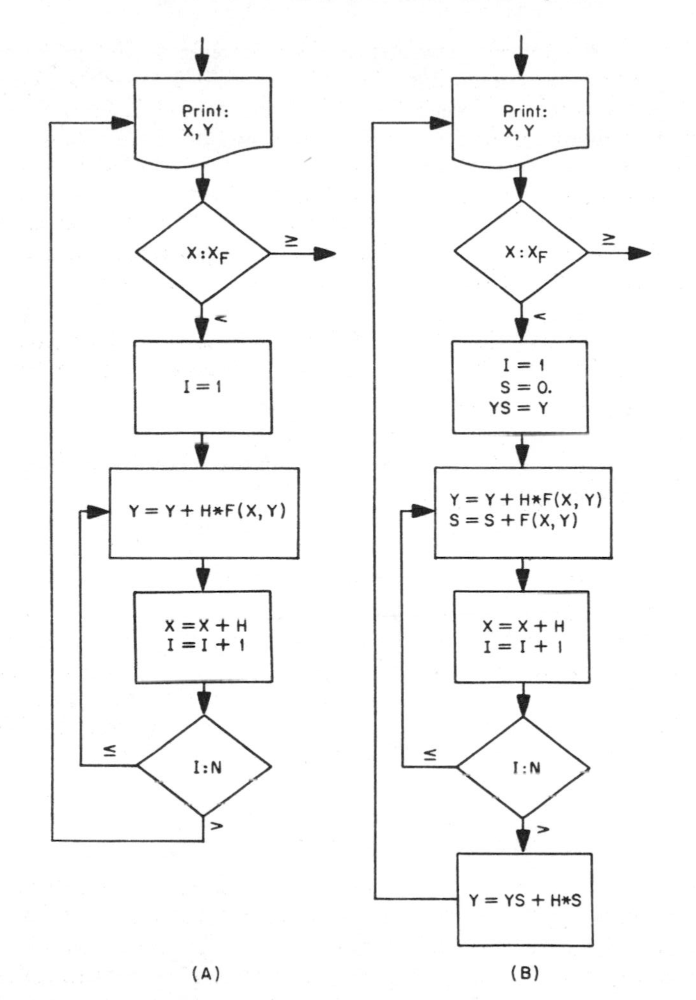

Fig. 12.2. Flowcharts of Euler's method. Both of these flowcharts provide for printout only at every N integration interval. This technique, which is simple to illustrate here, will not be included in subsequent flowcharts since it would tend to divert attention from the method being illustrated.

where $x_k \leqq \xi_2 \leqq x_{k+1}$. Then, using Eq. (12.26) in the mean value theorem, Eq. (12.20),

$$y'(\xi) = y'_k + y''_k h/2 + y'''(\xi_1)h^2/6 \tag{12.28}$$

Finding the average of the two derivatives using (12.27) gives

$$(y'_k + y'_{k+1})/2 = y'_k + y''_k h/2 + y'''(\xi_2)h^2/4 \tag{12.29}$$

Comparing Eqs. (12.28) and (12.29), we see that the first two terms of each series agree. Therefore, the average

$$\bar{y}'_k = (y'_k + y'_{k+1})/2 \tag{12.30}$$

is a better derivative value to use at each step than y'_k. However, we still do not know the value of y'_{k+1}. This value is obtainable by an iterative method. We will not attempt to analyze the conditions for convergence, but only describe the process. Using superscripts to denote iterations, we can describe the process as follows.

Find a starting approximation by Euler's method.

$$y_{k+1}^{(0)} = y_k + hF(x_k, y_k) \tag{12.31}$$

[This is essentially Eq. (12.24).] Then find an approximation to the average derivative by

$$\bar{y}_k'^{(n)} = [F(x_k, y_k) + F(x_{k+1}, y_{k+1}^{(n)})]/2 \tag{12.32}$$

and a new value for the function by

$$y_{k+1}^{(n+1)} = y_k + h\,\bar{y}_k'^{(n)} \tag{12.33}$$

Equations (12.32) and (12.33) are applied repeatedly until two successive values of y_{k+1} agree within some allowable error.

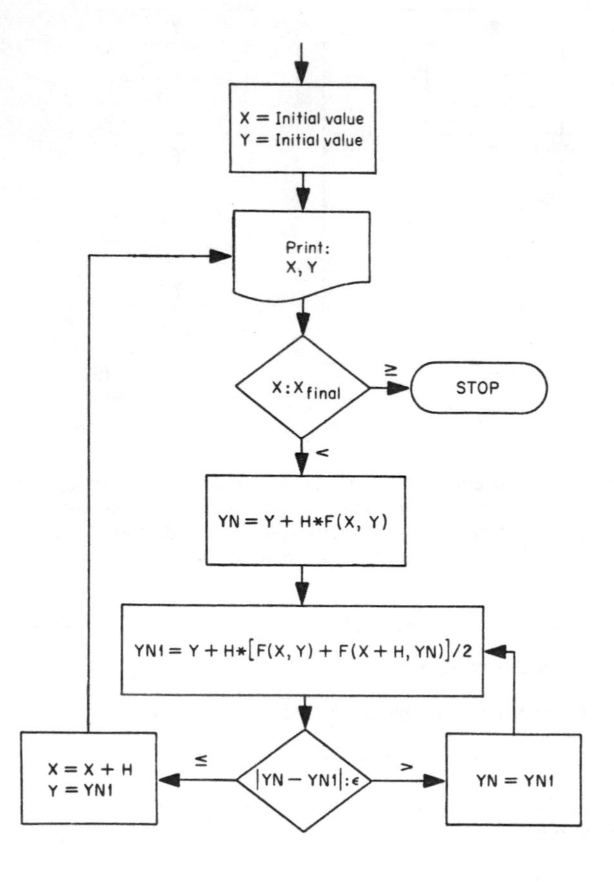

Fig. 12.3. Simplified flowchart of modified Euler's method.

TABLE 12.2 MODIFIED EULER'S METHOD SOLUTIONS OF $y' = xe^y$ FOR VARIOUS INTEGRATION INTERVALS

x	*Exact*	*h = 0.1*	*h = 0.01*	*h = 0.001*
0.0	0.00000000	0.00000000	0.00000000	0.00000000
0.2	0.02020271	0.02025499	0.02020323	0.02020271
0.4	0.08338161	0.08362147	0.08338400	0.08338162
0.6	0.19845094	0.19914326	0.19845784	0.19845095
0.8	0.38566248	0.38750573	0.38568080	0.38566248
1.0	0.69214718	0.69872915	0.69320225	0.69314723

Applying this method to the previous example [Eq. (12.8)], we obtain the results shown in Table 12.2. We can see that, at least for this example, the modified Euler's method is a considerable improvement over the ordinary Euler's method. The error incurred by these two methods will be analyzed in Sec. 12.7.

A simplified flowchart for the modified Euler's method is given in Fig. 12.3. This method is also subject to the same computational problems discussed in Sec. 12.3. However, these errors are usually less pronounced because larger integration intervals are used. The flowchart of Fig. 12.3 does not include any provision for minimizing these computational difficulties or for stopping the iteration if it is not converging.

12.5 PREDICTOR–CORRECTOR METHODS

Predictor–corrector methods for solving differential equations are methods in which two formulas are used. One formula, the *predictor*, is used to predict a future value of the function. and a second formula, the *corrector*, is used in an iterative fashion to calculate better approximations of that future value. Both formulas usually depend on values of the function already obtained for prior points. A general form for each of these formulas is given by Eqs. (12.34) and (12.35). A more general form will be presented in connection with error analysis in Sec. 12.7.

Predictor:

$$y_{k+1}^{(0)} = y_k + hf_1(y'_k, y'_{k-1}, y'_{k-2}, \ldots) \qquad (12.34)$$

Corrector:

$$y_{k+1}^{(n+1)} = y_k + hf_2(y'^{(n)}_{k+1}, y'_k, y'_{k-1}, \ldots) \qquad (12.35)$$

The functions f_1 and f_2 can be interpreted as various approximations for $y'(\xi)$ as defined in Sec. 12.3.

The primary difficulty of predictor–corrector methods is that since they require the use of several prior data values, several values of the function must be known to get the solution started. Usually exact values are not available, and the first few points must be calculated by other integration methods.

The modified Euler's method is essentially a predictor–corrector method. Equation (12.31) is the predictor, and (12.33) [in conjunction with (12.32)] is the corrector. These formulas do not require prior data values, and therefore can be started without prior function values. Because of this fact, the modified Euler's method can be used to obtain a number of values of a function, and then these can be used in getting a more sophisticated method started.

12.5.1 Adams' Method

The exact value of the function y evaluated at the end of one interval of integration is given by

$$y_{k+1} = y_k + \int_{x_k}^{x_{k+1}} y'(x)\,dx \tag{12.19}$$

Since y' is dependent on y, we must approximate the above integral. In the predictor–corrector methods, the integral is approximated by passing a polynomial in x through known points of the function $y'(x)$, then integrating over the limits x_k to x_{k+1}. Since initially no points in this interval are known, the predictor is formed by passing a polynomial through the derivatives of known points and then integrating analytically over one interval of the independent variable to obtain a formula for an approximate ordinate value at the end of the interval. The upper part of Fig. 12.4 illustrates the relationship between the points used in the polynomial that is integrated and the predicted ordinate. The required polynomial is easily formed by using the Gregory–Newton backwards interpolation formula. We can write by Eq. (4.41)

$$\begin{aligned} y'(x) = y'_k + \nabla y'_k u + \nabla^2 y'_k u(u+1)/2! + \nabla^3 y'_k u(u+1)(u+2)/3! + \cdots \\ + \nabla^N y'_k u(u+1)(\cdots)(u+N-1) \end{aligned} \tag{12.36}$$

where $u = (x-x_k)/h$, and hence

$$h\,du = dx \tag{12.37}$$

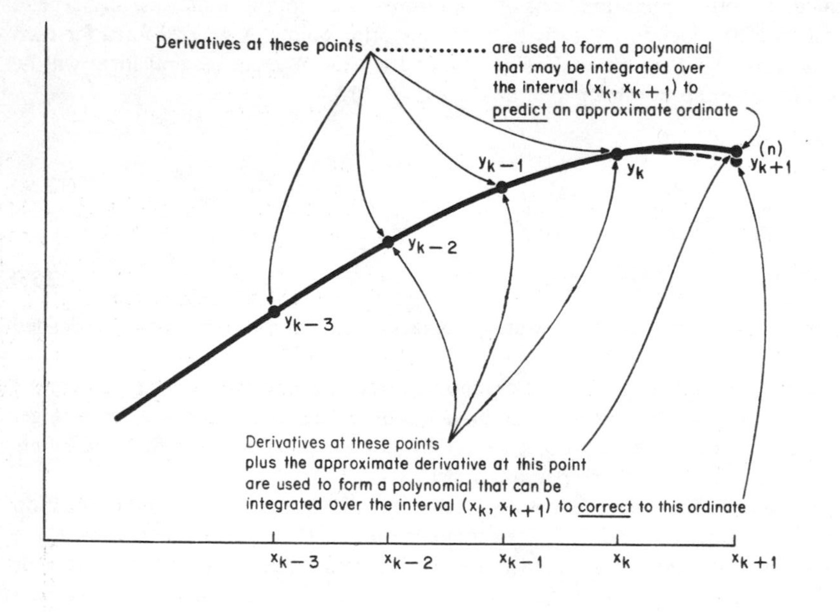

Fig. 12.4. Graphical description of a predictor–corrector method.

Since $u = 0$ for x_k and $u = 1$ for x_{k+1},

$$\int_{x_k}^{x_{k+1}} y'(x)dx = h\int_0^1 (y'_k + \nabla y'_k u + \nabla^2 y'_k u(u+1)/2 + \nabla^3 y'_k u(u+1)(u+2)/6 + \cdots)du \tag{12.38}$$

Upon integrating and substituting in Eq. (12.19),

$$y_{k+1} = y_k + h(y'_k + \nabla y'_k/2 + 5\nabla^2 y'_k/12 + 3\nabla^3 y'_k/8 + \cdots \tag{12.39}$$

Truncating this after the third difference and expressing in terms of ordinate values,

$$y_{k+1} = y_k + h(55y'_k - 59y'_{k-1} + 37y'_{k-2} - 9y'_{k-3})/24 \tag{12.40}$$

This formula can be used by itself as the basis for an integration method. This would be comparable, in technique, with Euler's method (Sec. 12.3). However, a method based on Eq. (12.40) alone can be improved, as Euler's method was improved, by using a value obtained by use of (12.40) to start an iterative procedure.

To form an iterative procedure, a corrector formula is needed. This formula is obtained by first passing a polynomial through the derivatives of known points and the derivative of the predicted point. Then this polynomial is substituted in Eq. (12.19), and the indicated integration is performed. The relationship between the points used in the polynomial that is integrated and the corrected point is illustrated in the lower part of Fig. 12.4. The required polynomial can also be obtained by the use of the Gregory–Newton backward interpolation formula. However, it is necessary in this case to expand the formula backwards from the end point of the interval rather than the starting point of the interval.

$$y'(x) = y'_{k+1} + \nabla y'_{k+1}u + \nabla^2 y'_{k+1}u(u+1)/2! + \nabla^2 y'_{k+1}u(u+1)(u+2)/3! + \cdots \tag{12.41}$$

Integrating as before but now with limits $u = -1$ and $u = 0$,

$$y_{k+1} = y_k + h(y'_{k+1} - \nabla y'_{k+1}/2 - \nabla^2 y'_{k+1}/12 - \nabla^3 y'_{k+1}/24 - \cdots) \tag{12.42}$$

Again, using only the first three differences and expressing in terms of ordinates,

$$y_{k+1} = y_k + h(9y'_{k+1} + 19y'_k - 5y'_{k-1} + y'_{k-2})/24 \tag{12.43}$$

This is called Adams' four-point formula because it is based on Eq. (12.42), which was used by J. C. Adams as early as 1855. Adams also developed Eq. (12.39) but did not use it as the predictor equation as we do today. This is because other methods can be used to obtain the first approximation to y_{k+1}. However, the method used should be a reasonable compromise between ease of computing and accuracy. The method employed by Adams was particularly well suited to hand computing.†

To use Eqs. (12.40) and (12.43) in a predictor–corrector method, (12.40) [which is derived from (12.39)] is used as the predictor in the form

$$y^{(0)}_{k+1} = y_k + h(55y'_k - 59y'_{k-1} + 37y'_{k-2} - 9y'_{k-3})/24 \tag{12.44}$$

and (12.43) is used as the corrector in the form

$$y^{(n+1)}_{k+1} = y_k + h(9y'^{(n)}_{k+1} + 19y'_k - 5y'_{k-1} + y'_{k-2})/24 \tag{12.45}$$

---

†See *Numerical Mathematical Analysis*, 6th edition, by J. B. Scarborough (Baltimore, Md.: The John Hopkins Press, 1966) for a complete description of Adams' method.

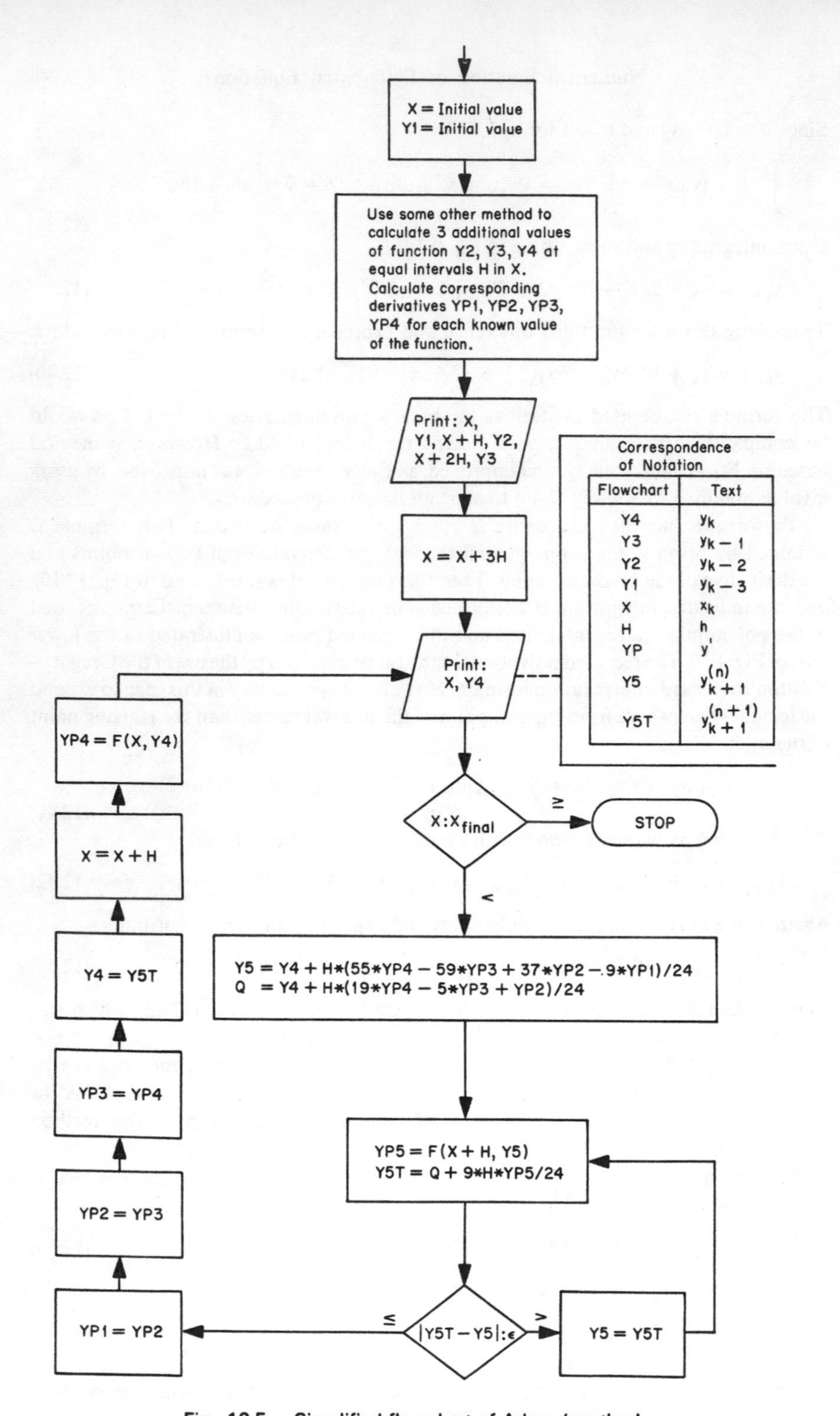

Fig. 12.5. Simplified flowchart of Adams' method.

In (12.44) and (12.45), y'_{k-1}, y'_{k-2}, and y'_{k-3} are values retained from previous steps. The value of y'_k is calculated once for the current step by

$$y'_k = F(x_k, y_k) \tag{12.46}$$

and the value of $y'^{(n)}_{k+1}$ is calculated for each iteration by

$$y'^{(n)}_{k+1} = F(x_{k+1}, y^{(n)}_{k+1}) \tag{12.47}$$

To improve computational efficiency, (12.45) can be put in the form

$$y^{(n+1)}_{k+1} = Q_1 + Q_2 y'^{(n)}_{k+1} \tag{12.48}$$

by lumping together quantities that remain constant during the iteration.

Applying the method of Eqs. (12.44) and (12.45) to the previous example, the results are those shown in Table 12.3. Note that for $h = 0.001$, the results are less accurate than for $h = 0.01$. This is due to buildup of rounding error. In our example, the solution was started by using the exact values of the derivative of the function for $y'(-3h)$, $y'(-2h)$, and $y'(-h)$. A simplified flowchart for Adams' method is shown in Fig. 12.5.

TABLE 12.3 ADAMS' METHOD SOLUTIONS OF $y' = xe^y$ FOR VARIOUS INTEGRATION INTERVALS

x	*Exact*	*h = 0.1*	*h = 0.01*	*h = 0.001*
0.0	0.00000000	0.00000000	0.00000000	0.00000000
0.2	0.02020271	0.02020306	0.02020271	0.02020271
0.4	0.08338161	0.08338649	0.08338161	0.08338159
0.6	0.19845094	0.19847057	0.19845093	0.19845088
0.8	0.38566248	0.38573466	0.38566247	0.38566229
1.0	0.69314718	0.69348488	0.69314720	0.69314668

12.5.2 Milne's Method

Other predictor–corrector methods can be devised based on the foregoing analysis, but using more or fewer points in the formulas. In addition, methods can be developed based on a more general form of Eq. (12.19), namely,

$$y_{k+1} = y_{k-j} + \int_{x_{k-j}}^{x_{k+1}} y'(x)\,dx \tag{12.49}$$

These methods are called Milne's methods. As before, $y'(x)$ can be approximated by the Newton–Gregory backward interpolation formula expanded about x_k. Expressed in terms of u, Eq. (12.49) can be written as

$$y_{k+1} = y_{k-j} + h\int_{-j}^{1} (y'_k + \nabla y'_k u + \nabla^2 y'_k u(u+1)/2 + \nabla^3 y'_k u(u+1)(u+2)/6 + \cdots)\,du \tag{12.50}$$

Integrating yields

$$y_{k+1} = y_{k-j} + h\,[(1+j)y'_k + (1-j^2)\nabla y'_k/2 + (5-3j^2+2j^3)\nabla^2 y'_k/12 + (9-4j^2+4j^3-j^4)\nabla^3 y'_k/24 + \cdots] \tag{12.51}$$

This equation can be evaluated for any $j > 0$. [For $j = 0$, this equation reduces to (12.39).] By selecting a value for j and truncating the series after some finite number of terms, (12.51) can be used as a predictor equation. The corresponding corrector equation can be developed by expanding the series for $y'(x)$ about x_{k+1} and integrating from $u = -j-1$ to $u = 0$.

$$y_{k+1} = y_{k-j} + h\int_{-j-1}^{0} (y'_{k+1} + \nabla y'_{k+1}u + \nabla^2 y'_{k+1}u(u+1)/2 + \nabla^3 y'_{k+1}u(u+1)(u+2)/6 + \cdots)\,du \tag{12.52}$$

Integrating,

$$y_{k+1} = y_{k-j} + h\,[(1 + j)y'_{k+1} - (1+2j+j^2)\nabla y'_{k+1}/2 - (1-3j^2-2j^3)\nabla^2 y'_{k+1}/12 - (1-2j^2+j^4)\nabla^3 y'_{k+1}/24 - \cdots] \tag{12.53}$$

One factor to consider in selecting a value for j is that the interpolating polynomials pass exactly through the table points being used; and the error of the polynomial between table points has opposite signs on adjacent intervals. This is illustrated in Figs. 10.5 and 11.2. Because of this feature, it is better to integrate over an even number of intervals so there is a chance for some of the error to cancel out. In the current application, in order to have an even number of intervals, j must be odd. Furthermore, when j is odd, some term in both (12.51) and (12.53) will vanish.

In Milne's second-difference method, for $j = 3$, the third-order difference term vanishes from Eq. (12.51) yielding

$$y_{k+1} = y_{k-3} + h(4y'_k - 4\nabla y'_k + 8\nabla^2 y'_k/3 + (0)\nabla^3 y'_k) \tag{12.54}$$

or

$$y_{k+1} = y_{k-3} + 4h(2y'_k - y'_{k-1} + 2y'_{k-2})/3 \tag{12.55}$$

Similarly, in (12.53), the third-order difference term vanishes when $j = 1$.

$$y_{k+1} = y_{k-1} + h(2y'_{k+1} - 2\nabla y'_{k+1} + \nabla^2 y'_{k+1}/3 + (0)\nabla^3 y'_{k+1}) \tag{12.56}$$

or

$$y_{k+1} = y_{k-1} + h(y'_{k+1} + 4y'_k + y'_{k-1})/3 \tag{12.57}$$

Equation (12.55) can be used as a predictor and (12.57) can be written in the iterative form

$$y_{k+1}^{(n+1)} = y_{k-1} + h(y'^{(n)}_{k+1} + 4y'_k + y'_{k-1})/3 \tag{12.58}$$

to use as a corrector. Equation (12.57) is essentially Simpson's rule applied in an iterative fashion to solve a differential equation.

In using this method, even though Eq. (12.58) determines the change in y for two steps in x, x is only advanced one interval at a time. That is, y_3 is determined by applying a two-step correction to y_1; then y_4 is found by applying a two-step correction to y_2, etc. Figure 12.6 gives a simplified flowchart of this method.

Table 12.4 gives comparative results for various interval sizes using Milne's second-difference method applied to the previous example. In this example, as in the previous one, the solution was started by using exact values. Note that the solution for $h = 0.001$ is less accurate than for $h = 0.01$. This is due to rounding errors.

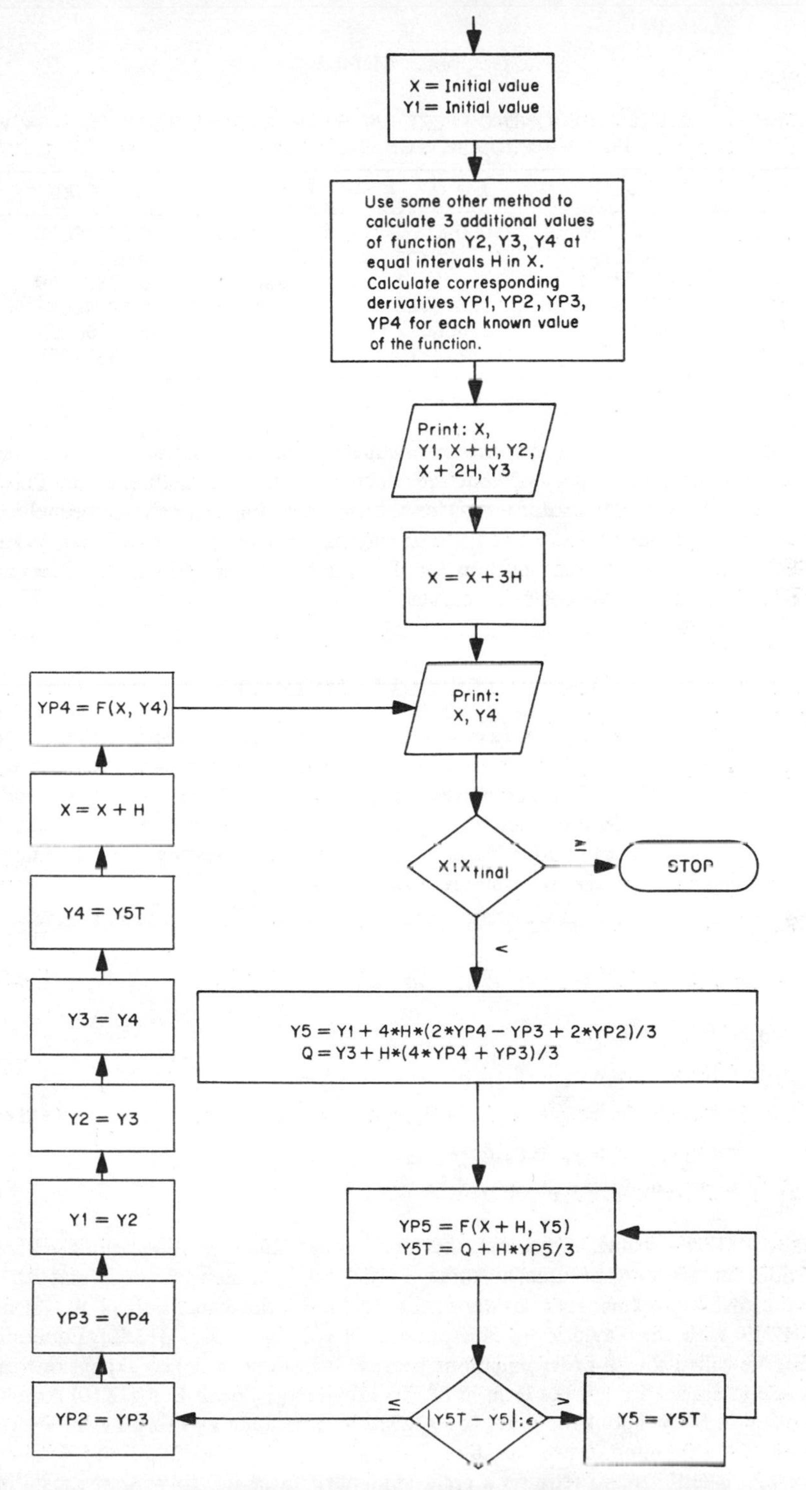

Fig. 12.6. Simplified flowchart of Milne's second difference method.

TABLE 12.4 MILNE'S SECOND-DIFFERENCE METHOD SOLUTIONS OF $y' = xe^y$ FOR VARIOUS INTEGRATION INTERVALS

x	*Exact*	*h = 0.1*	*h = 0.01*	*h = 0.001*
0.0	0.00000000	0.00000000	0.00000000	0.00000000
0.2	0.02020271	0.02020306	0.02020271	0.02020271
0.4	0.08338161	0.08338341	0.08338161	0.08338159
0.6	0.19845094	0.19845732	0.19845093	0.19845088
0.8	0.38566248	0.38568612	0.38566247	0.38566229
1.0	0.69314718	0.69326826	0.69314715	0.69314668

In the examples given in this section, adequate accuracy was attained (for at least one integration interval size) without introducing too much rounding error. This is not always true. Therefore, some means for reducing rounding error should be included in the integration method. Also, the largest integration interval that can be tolerated should be used. The formulas given in Sec. 12.7 can be used to estimate the necessary interval size to obtain the desired accuracy.

12.6 RUNGE–KUTTA METHODS

The Runge–Kutta methods are a group of methods based on one devised by C. Runge and modified by W. Kutta and others. In these methods, the integration interval is divided into subintervals, the derivatives are found at each of the subintervals, and a weighted average is taken of these multiplied by the integration interval in order to determine an increment to add to the dependent variable. Expressing this mathematically for a single, first-order equation given by

$$y'(x) = F(x,y) \tag{12.59}$$

then

$$\begin{aligned}
\delta_1 &= hF(x_k, y_k) \\
\delta_2 &= hF(x_k + \alpha_1 h, y_k + \beta_{11}\delta_1) \\
\delta_3 &= hF(x_k + \alpha_2 h, y_k + \beta_{21}\delta_1 + \beta_{22}\delta_2) \\
\delta_4 &= hF(x_k + \alpha_3 h, y_k + \beta_{31}\delta_1 + \beta_{32}\delta_2 + \beta_{33}\delta_3) \\
y_{k+1} &= y_k + w_1\delta_1 + w_2\delta_2 + w_3\delta_3 + w_4\delta_4
\end{aligned} \tag{12.60}$$

Equations (12.60) define, in general, the fourth-order Runge–Kutta methods. These equations contain many parameters, some of which can be chosen arbitrarily and others determined by equating, term by term, the Taylor's series expansion of y_{k+1} from Eq. (12.19) with the Taylor's series expansion of y_{k+1} from Eqs. (12.60). Equations (12.60) are called fourth-order equations because the aforementioned expansions can be made to agree through the term in h^4. To expand y_{k+1} from Eqs. (12.60) requires expansion of F(x,y) about the point (x_k, y_k). This can be done by using Taylor's series expansion of a function of two variables (Sec. 5.4). Although this expansion is straightforward, it is tedious and requires a great amount of space just to write the resulting series. Therefore, the detailed mathematics will be omitted here. Instead, we will just present the results obtained by others.†

Before examining particular fourth-order methods, it is interesting to compare a second-order method with the modified Euler's method. By a proper choice of parameters, we can derive the following second-order method:

$$\begin{aligned} \delta_1 &= hF(x_k, y_k) \\ \delta_2 &= hF(x_k + h, y_k + \delta_1) \\ y_{k+1} &= y_k + (\delta_1 + \delta_2)/2 \end{aligned} \tag{12.61}$$

This method is essentially the modified Euler's method but *without* iteration.

The original fourth-order method is

$$\begin{aligned} \delta_1 &= hF(x_k, y_k) \\ \delta_2 &= hF(x_k + h/2, y_k + \delta_1/2) \\ \delta_3 &= hF(x_k + h/2, y_k + \delta_2/2) \\ \delta_4 &= hF(x_k + h, y_k + \delta_3) \\ y_{k+1} &= y_k + (\delta_1 + 2\delta_2 + 2\delta_3 + \delta_4)/6 \end{aligned} \tag{12.62}$$

This method can be interpreted geometrically as follows:

1. The derivative is evaluated at the starting point of the interval.
2. The above derivative is used to obtain an approximate ordinate to determine an approximate derivative for the midpoint of the interval.
3. The above derivative is used to obtain a second approximation of the ordinate to determine a second approximation to the derivative at the midpoint of the interval.
4. The above derivative is used to obtain an approximate ordinate to determine an approximate derivative for the end point of the interval.
5. A weighted average of the above four derivatives is taken to determine a total increment in the ordinate for the whole interval.

Since δ_2/h and δ_3/h are approximations to the derivative at the midpoint, we may average these as follows:

$$\delta_{23} = (\delta_2 + \delta_3)/2 \tag{12.63}$$

which would allow us to write

$$2\delta_2 + 2\delta_3 = 4\delta_{23} \tag{12.64}$$

On substituting in the last of Eqs. (12.62), we have

$$y_{k+1} = y_k + (\delta_1 + 4\delta_{23} + \delta_4)/6 \tag{12.65}$$

This can be recognized as being analogous to Simpson's rule when we make adjustments because the interval is defined differently and because certain approximations must be made to the derivatives.

†Table 12.10 in Sec. 12.10 summarizes the parameters of four different variations of the fourth-order Runge–Kutta method.

The Kutta variation is defined as follows:

$$\begin{aligned}
\delta_1 &= hF(x_k, y_k) \\
\delta_2 &= hF(x_k + h/3, y_k + \delta_1/3) \\
\delta_3 &= hF(x_k + 2h/3, y_k - \delta_1/3 + \delta_2) \\
\delta_4 &= hF(x_k + h, y_k + \delta_1 - \delta_2 + \delta_3) \\
y_{k+1} &= y_k + (\delta_1 + 3\delta_2 + 3\delta_3 + \delta_4)/8
\end{aligned} \tag{12.66}$$

This method is analogous to the third-order integration method given in Table 11.4. Again, adjustments must be made because the interval size is defined differently and because only approximations to the derivatives are available.

Many other Runge–Kutta methods can be developed. Methods using iteration to improve the approximations can be devised, and higher-order methods can also be derived. The primary attraction of the Runge–Kutta methods is that they are self-starting; that is, they do not require prior values. Sometimes, a Runge–Kutta method is used to get a predictor–corrector method started.

Figure 12.7 gives a simplified flowchart of the Runge–Kutta method defined by Eqs. (12.62) with provision for solving several equations simultaneously, where

$$\begin{aligned}
D1 &= Y1' = F1(X, Y1, Y2, \ldots, YN) \\
D2 &= Y2' = F2(X, Y1, Y2, \ldots, YN) \\
&\vdots \\
DN &= YN' = FN(X, Y1, Y2, \ldots, YN)
\end{aligned} \tag{12.67}$$

Applying this method to the sample problem used to illustrate previous methods yields the results shown in Table 12.5. The results are tabulated for interval sizes of h = 0.2, h = 0.02, and h = 0.002 instead of h = 0.1, h = 0.01, and h = 0.001 as in the earlier examples because the method used actually calculates at intervals of h/2. The results for h = 0.02 are nearly exact and, in fact, are better than for h = 0.002. This latter fact is due to buildup of rounding error.

TABLE 12.5 RUNGE–KUTTA SOLUTIONS OF $y' = xe^y$ FOR VARIOUS INTEGRATION INTERVALS

x	*Exact*	*h = 0.2*	*h = 0.02*	*h = 0.002*
0	0.00000000	0.00000000	0.00000000	0.00000000
0.2	0.02020271	0.02020304	0.02020271	0.02020270
0.4	0.08338161	0.08338316	0.08338161	0.08338159
0.6	0.19845094	0.19845566	0.19845093	0.19845087
0.8	0.38566248	0.38567641	0.38566246	0.38566228
1.0	0.69314718	0.69319566	0.69314714	0.69314669

12.7 ERROR ANALYSIS

The inherent error of the numerical methods discussed in Secs. 12.3 through 12.6 can be derived by use of Taylor's series. It must be remembered that other errors, namely computational, can also occur.

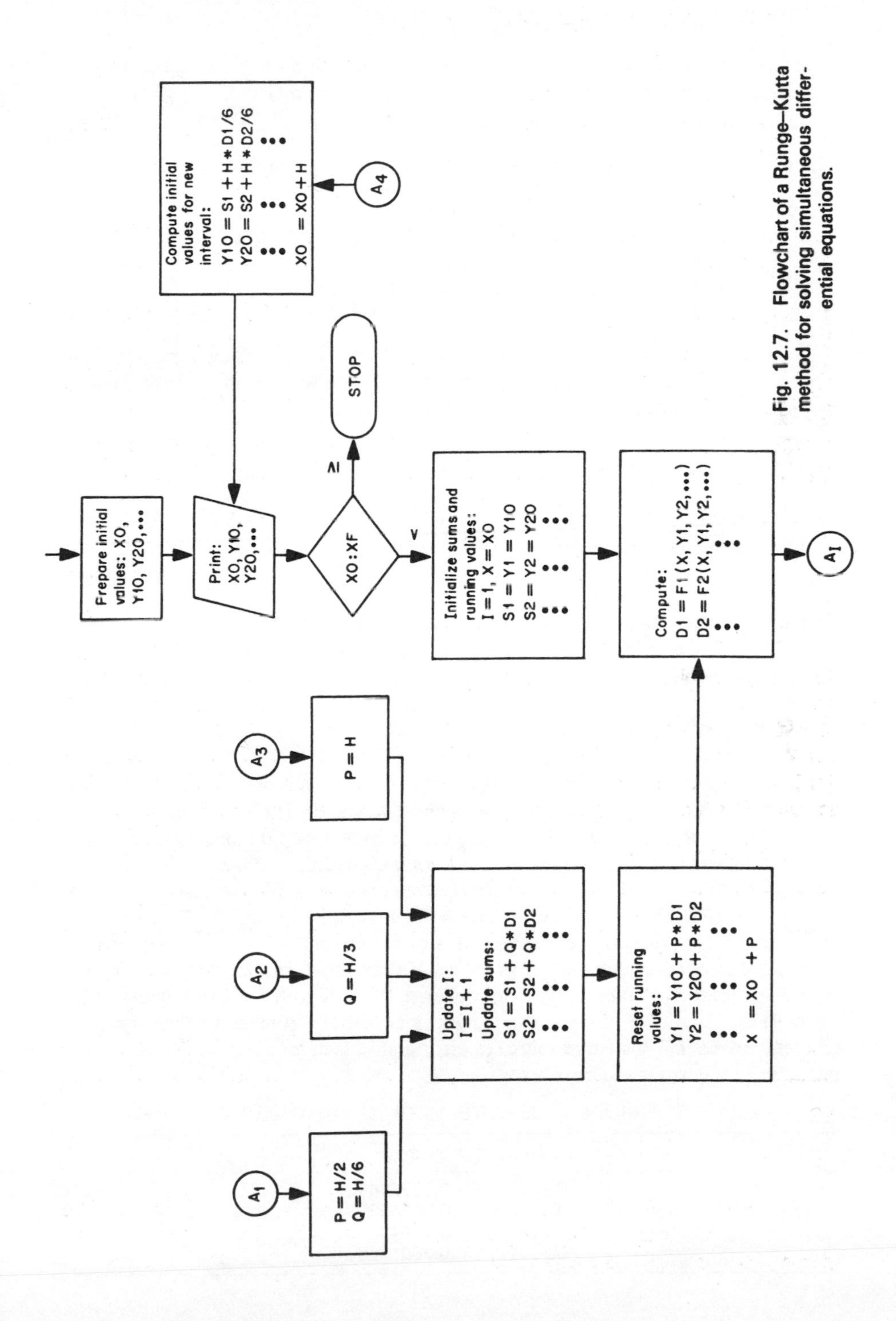

Fig. 12.7. Flowchart of a Runge–Kutta method for solving simultaneous differential equations.

12.7.1 Inherent Error of Euler's Method

Equation (12.19) can be approximated by a Taylor's series by first expanding y'(x) in a Taylor's series. Retaining only two terms in the series, we have

$$y'(x) = y'(x_k) + (x-x_k)y''(\xi) \qquad (12.68)$$

where $x_k \leqq \xi \leqq x_{k+1}$. Substituting in Eq. (12.19) and integrating,

$$y(x_{k+1}) = y(x_k) + hy'(x_k) + h^2y''(\xi)/2 \qquad (12.69)$$

Equation (12.23), Euler's method, agrees with (12.69) through the first two terms. The inherent error, therefore, for a single interval is

$$E = h^2y''(\xi)/2 \qquad (12.70)$$

In order to determine the total error, we must sum the individual errors for each interval. In the example, for x varying from 0 to 1, there will be 1/h intervals. Hence, the total error will be given by

$$E_T = \sum_{i=1}^{1/h} h^2y''(\xi_i)/2 \qquad (12.71)$$

where $x_{i-1} \leqq \xi_i \leqq x_i$. For the example, y'' varies from 1 to 6 as x varies from 0 to 1. Therefore, $y''(\xi_i) \leqq 6$, and we can say from (12.71)

$$E_T \leqq 3h \qquad (12.72)$$

Actually, this is a very pessimistic estimate. For the example given, y'' increases monotonically over the entire range 0 to 1. This then implies that on any one interval, $y''(\xi_i) \leqq y_i''$. Therefore,

$$E_T \leqq \sum_{i=1}^{1/h} h^2y_i''/2 \qquad (12.73)$$

For the example problem, since we know the analytic solution, we can obtain an analytic expression for y_i'' in terms of x = ih and evaluate Eq. (12.73) for any value of h. Doing this, we obtain, for h = 0.1, that E_T should be less than 0.11365353. Table 12.6 gives the actual values of the errors at x = 1. for various values of h as computed from the results given in Table 12.1. All of these errors are well within the pessimistic limit set by Eq. (12.72). However, the error at h = 0.1 is slightly worse than the theoretical upper limit of the truncation error calculated by the use of (12.73). The reason for this is the buildup of computational error. In fact, we can show that, for this example, the theoretical total truncation error is approximately equal to h for small values of h. From Table 12.6, this is apparently not true in practice. This emphasizes the point that inherent error is not the entire source of error and in fact, for small values of h, may not even be the major source of error.

TABLE 12.6 ERRORS IN EULER'S METHOD SOLUTIONS OF $y' = xe^y$

h	*Error*
0.1	0.12775492
0.01	0.01466378
0.001	0.00149398
0.0001	0.00018566

12.7.2 Inherent Error of the Modified Euler's Method

The analysis of the inherent error of the modified Euler's method is complicated by the fact that an iteration is involved. However, we can still obtain a series expansion of the method to any desired degree of accuracy. We have first, by Eq. (12.31),

$$y_{k+1}^{(0)} = y_k + hy_k' \tag{12.74}$$

Then, by (12.33), we can write successively

$$y_{k+1}^{(1)} = y_k + h(y_k' + y_{k+1}'^{(0)})/2 \tag{12.75a}$$

$$y_{k+1}^{(2)} = y_k + h(y_k' + y_{k+1}'^{(1)})/2 \tag{12.75b}$$

$$y_{k+1}^{(3)} = y_k + h(y_k' + y_{k+1}'^{(2)})/2 \tag{12.75c}$$

We can differentiate (12.74) to obtain

$$y_{k+1}'^{(0)} = y_k' + hy_k'' \tag{12.76}$$

which can be substituted in Eq. (12.75a) to give

$$y_{k+1}^{(1)} = y_k + hy_k' + h^2y_k''/2 \tag{12.77}$$

Differentiating this gives

$$y_{k+1}'^{(1)} = y_k' + hy_k'' + h^2y_k'''/2 \tag{12.78}$$

which can be substituted in (12.75b) to give

$$y_{k+1}^{(2)} = y_k + hy_k' + h^2y_k''/2 + h^3y_k'''/4 \tag{12.79}$$

Then this can be differentiated in a similar fashion and substituted in (12.75c) to give

$$y_{k+1}^{(3)} = y_k + hy_k' + h^2y_k''/2 + h^3y_k'''/4 + h^4y_k^{iv}/8 \tag{12.80}$$

An inspection of Eqs. (12.74), (12.77), (12.79), and (12.80) shows that the only difference between these equations is that at each iteration, a term of the form $h^n y_k^{(n)}/2^{n-1}$ is added. (The superscript n on y_k means the nth derivative of y with respect to the independent variable, evaluated at the point k.)

A Taylor's series expansion of Eq. (12.19) is

$$y_{k+1} = y_k + h\,y_k' + h^2y_k''/2 + h^3y'''(\xi)/6 \tag{12.81}$$

which agrees, through the first three terms, with (12.80). Ignoring terms containing powers of h greater than 3, the error for one interval is approximately

$$E \doteq h^3[y'''(\xi)/6 - y_k'''/4] \doteq -h^3y'''(\xi)/12 \tag{12.82}$$

The total error can be obtained, as was done in the previous subsection, by totaling over all intervals. In the example, there were 1/h intervals, therefore,

$$E_T \doteq -\sum_{i=1}^{1/h} h^3y'''(\xi_i)/12 \tag{12.83}$$

Since y''', for the example, varies from 0 to 28 as x varies from 0 to 1,

$$|E_T| \leqq 2.333h^2 \tag{12.84}$$

From the results given in Table 12.2, we can tabulate the total errors at $x = 1$. These

TABLE 12.7 ERRORS IN THE MODIFIED EULER'S METHOD SOLUTIONS OF $y' = xe^y$

h	Error
0.1	−0.00558197
0.01	−0.00005507
0.001	−0.00000005

are given in Table 12.7. These errors are considerably less than the maximum predicted by the inequality (12.84).

The individual errors vary as h^3 varies, but the total error will vary approximately as h^2. That is, if h is reduced by a factor of 10, then the total error should be reduced by a factor of 100. This is verified by the example, except that a reduction in interval from $h = 0.01$ to $h = 0.001$ improved the error by a factor of 1000 instead of 100. A further reduction of the interval to $h = 0.0001$ actually resulted in a larger error. This is due to computational error buildup because of the large number of integration intervals. It is obvious that a method with a smaller inherent error is more desirable than a method with a larger inherent error. An additional benefit is that it reduces the number of computations, thereby reducing computational error buildup.

12.7.3 Inherent Error of Predictor–Corrector Methods

The Adams method, given by Eqs. (12.44) and (12.45), can be analyzed in a manner similar to that given in Sec. 12.7.2. Expanding $y'_{k-1}, y'_{k-2}, y'_{k-3}$ by Taylor's series gives

$$\begin{aligned} y'_{k-1} &= y'_k - hy''_k + h^2y'''_k/2 - h^3y^{iv}_k/6 + h^4y^v(\xi_1)/24 \\ y'_{k-2} &= y'_k - 2hy''_k + 2h^2y'''_k - 4h^3y^{iv}_k/3 + 16h^4y^v(\xi_2)/24 \\ y'_{k-3} &= y'_k - 3hy''_k + 9h^2y'''_k/2 - 9h^3y^{iv}_k/2 + 81h^4y^v(\xi_3)/24 \end{aligned} \quad (12.85)$$

Substituting the values obtained in Eqs. (12.85) into (12.44) gives

$$\begin{aligned} y^{(0)}_{k+1} &= y_k + hy'_k + h^2y''_k/2 + h^3y'''_k/6 + h^4y^{iv}_k/24 \\ &\quad + h^5[-59y^v(\xi_1) + 592y^v(\xi_2) - 729y^v(\xi_3)]/576 \end{aligned} \quad (12.86)$$

The Taylor's series expansion of y_{k+1} is

$$(y_{k+1})_T = y_k + hy'_k + h^2y''_k/2 + h^3y'''_k/6 + h^4y^{iv}_k/24 + h^5y^v(\xi_4)/120 \quad (12.87)$$

The error in the predictor is found by subtracting (12.86) from (12.87):

$$E_P = h^5\{y^v(\xi_4)/120 + [59y^v(\xi_1) - 592y^v(\xi_2) + 729y^v(\xi_3)]/576\} \quad (12.88)$$

A single value of ξ can be found such that (12.88) can be simplified to Eq. (12.89) without changing the value of the right-hand side:

$$E_P = 251\,h^5y^v(\xi)/720 \quad (12.89)$$

The error in the corrector can be found in a similar fashion. Since Eq. (12.45) is in an iterative form, it must be analyzed for particular values of n. Thus if $n = 0$, (12.45) becomes

$$y^{(1)}_{k+1} = y_k + h(9y'^{(0)}_{k+1} + 19y'_k - 5y'_{k-1} + y'_{k-2})/24 \quad (12.90)$$

By differentiating (12.86) and neglecting derivatives above the fifth, an approximation for $y'^{(0)}_{k+1}$ can be obtained which can be substituted in (12.90). Also, expressions for y'_{k-1} and y'_{k-2}, given by (12.85), can be substituted in (12.90). Therefore,

$$y^{(1)}_{k+1} = y_k + hy'_k + h^2y''_k/2 + h^3y'''_k/6 + h^4y^{iv}_k/24 + h^5[9y^v_k - 5y^v(\xi_1) + 16y^v(\xi_2)]/576 \tag{12.91}$$

It is not necessary to determine values of $y^{(n+1)}_{k+1}$ for other values of n because the terms through h^4 in Eqs. (12.91) and (12.87) agree. After simplifying the difference, (12.87) minus (12.91), the error in the corrector can be written as

$$E_C = -19h^5y^v(\xi)/720 \tag{12.92}$$

In the example, the fifth derivative ranges from 0 to 2400. Therefore, no single error will be greater in absolute value than $190h^5/3$, and the total error for all 1/h intervals must be less than

$$|E_T| \leqq 63.3333h^4 \tag{12.93}$$

From the results given in Table 12.3, we can determine the total errors at x = 1. These are given in Table 12.8. Inequality (12.93) is satisfied for h = 0.1 and h = 0.01, but is not satisfied for h = 0.001. This is probably due to buildup of computational errors. It is important to note that for the same interval size, this method requires considerably more computations than the modified Euler's method.

TABLE 12.8 ERRORS IN ADAMS' METHOD SOLUTIONS OF $y' = xe^y$

h	*Error*
0.1	−0.00033770
0.01	−0.00000002
0.001	0.00000050

The inherent error of Milne's second-difference method [Eqs. (12.55) and (12.58)] can be derived in a similar way. The errors in the predictor and corrector are, respectively,

$$E_P = 14h^5y^v(\xi)/45 \tag{12.94}$$

$$E_C = -h^5y^v(\xi)/90 \tag{12.95}$$

In Sec. 12.5, a few predictor–corrector integration methods were presented. Each of these methods was derived by essentially the same process. To obtain the formulas for a given method, Eq. (12.49) [or Eq. (12.19)] is integrated analytically using a polynomial approximation for y′(x). Different methods result, depending on certain parameters that can be selected arbitrarily. These parameters are 1) the number of derivative values through which the polynomial is passed, and 2) in Eq. (12.49), j, which determines the number of intervals over which the integration is made. The formula will be a predictor if the approximating polynomial passes only through the derivatives of known points. It will be a corrector if the derivative value of the point being computed is also required.

In each of these methods, the parameters can be selected, the formulas developed, and then the truncation error determined. It was noted in Sec. 12.5 that a more general form for the predictor–corrector methods would be given in this section. This form is

$$y_{k+1} = \sum_{i=0}^{N} A_i y_{k-i} + h \sum_{i=-1}^{M} B_i y'_{k-i} \tag{12.96}$$

If $B_{-1} = 0$, the formula is a predictor; if $B_{-1} \neq 0$, the formula is a corrector and would be written in iterative form.

Each of the methods given in Sec. 12.5 is in the form of (12.96). The point in introducing this generalized form in this section is to emphasize that another method of developing predictor–corrector formulas is possible. This method consists of starting with the general form, Eq. (12.96), selecting N and M and perhaps some values of the A's and B's. The remaining A's and B's are determined by expanding the right side of (12.96) in a Taylor's series, and then equating it term by term with the Taylor's expansion of y_{k+1} and solving the resulting set of equations for the unknown parameters. This method provides a technique for specifying the order of the truncation error prior to developing the formulas.

Formulas of the form of Eq. (12.96) can also be derived by taking linear combinations of other predictor–corrector methods. An example proposed by Milne and Reynolds is a linear combination of Eqs. (12.45) and (12.58). Taking one eighth of the former and adding seven eighths of the latter yields

$$y_{k+1}^{(n+1)} = (y_k + 7y_{k-1})/8 + h(65y'^{(n)}_{k+1} + 243y'_k + 51y'_{k-1} + y'_{k-2})/192 \tag{12.97}$$

The inherent error of this formula is

$$E_C = -5h^5 y^{v}(\xi)/384 \tag{12.98}$$

This corrector can be used with any predictor, but the authors of the method suggest

$$y_{k+1}^{(0)} = y_{k-1} + h(8y'_k - 5y'_{k-1} + 4y'_{k-2} - y'_{k-3})/3 \tag{12.99}$$

This formula results from using $j = 1$ in Eq. (12.51) and truncating all terms involving fourth or higher-order differences. The inherent error of the predictor is given by

$$E_P = 29h^5 y^{v}(\xi)/90 \tag{12.100}$$

The inherent errors of these formulas are slightly better than those for Adams' method, but not as good as those for Milne's second-difference method. In view of this fact, it would seem to be more desirable to use Milne's second-difference method rather than the Milne–Reynolds method. Furthermore, by taking the linear combination $-8/11$ of Eq. (12.45) plus $19/11$ of (12.58), the resulting formula agrees with the Taylor's series expansion through the term involving the fifth derivative. This means that the inherent error is of the order of h^6. However, the Milne–Reynolds method has greater stability than either Adams' method or Milne's second-difference method. The stability of methods for the numerical solution of differential equations will be discussed briefly in Sec. 12.7.5.

12.7.4 The Inherent Error of the Runge–Kutta Methods

It was stated in Sec. 12.6 that the parameters of the general Eqs. (12.60) can be selected such that the Runge–Kutta methods will agree with the Taylor's series expansion for y_{k+1} through the term containing h^4. To do this involves a very long algebraic analysis. For the sake of brevity, that analysis is omitted. Suffice it to say that the order of the error is $O(h^5)$. The error results of the example problem at $x = 1$. as calculated from Table 12.5 are shown in Table 12.9. For the example, much better results are obtained by the Runge–Kutta method than any of the previous methods. *However*, this is not always the case. Some problems would give better results using a predictor–corrector method. Several of the predictor-corrector methods and most of the Runge–Kutta methods given have inherent errors of the order $0(h^5)$. The selection of which method to use must be based on a variety of factors. Although inherent error is an important factor, complexity of computation (which is directly related to computational error) and the stability of the integration method should also be considered.

TABLE 12.9 ERRORS IN THE RUNGE–KUTTA SOLUTIONS OF $y' = xe^y$

h	*Error*
0.2	−0.00004848
0.02	0.00000004
0.002	0.00000049

12.7.5 Stability of Numerical Solutions of Differential Equations

In this subsection, stability will be defined and a few comments about stability will be made. It is beyond the scope of this text to discuss stability rigorously.

The numerical solution of a differential equation is *stable* if the propagated error is bounded. This definition implies that instability is caused by the buildup of error. This is true. It has been emphasized in the error analyses that error other than the inherent error of the method being used can occur. This error is the computational error or roundoff error. Because we must work with numbers having finite precision, we usually make small errors at each computational step of the procedure. Although the absolute errors are usually small, the relative errors, for some operations, can be quite large. Furthermore, if there are numerous steps in the procedure, even small errors may accumulate into large errors. It was noted in Sec. 3.2 that taking the difference between two numbers of nearly equal magnitude may result in a large relative error. Therefore, such operations should be avoided if possible.

The stability of a numerical solution of a differential equation depends on the numerical method employed *and* on the nature of the differential equation being solved. Both inherent error and computational error contribute to the total buildup of error. Generally, as the step size is decreased, the inherent error of the method decreases; but this causes the number of steps to increase, which causes the roundoff error to increase. Therefore, some optimum step size will exist for which the total error is minimum. If the numerical method is complex or if the differential equation is complex, then there will be many computations and many opportunities for error.

Equations whose solutions are oscillatory are especially prone to stability problems. Numerical methods that combine terms of nearly equal magnitude but of opposite sign may be unstable. This explains why, as noted in Sec. 12.7.3, that the use of one method may be preferable to the use of another even though the first method may have a larger inherent error than the second.

The stability of predictor–corrector methods can sometimes be improved by decreasing the amount of "correcting." In the derivation of the inherent error of Adams' method, we saw by Eq. (12.91) that the Taylor's series expansion for $y_{k+1}^{\prime(1)}$ would be the same as for the first five terms of the Taylor's series expansion of $y'^{(0)}_{k+1}$. Therefore, the Taylor's series expansion of $y^{(n+1)}_{k+1}$ would be the same (for the same number of terms) as Eq. (12.91) for any n greater than zero. Therefore, after the first application of the corrector, the expression for the inherent error does not change. This is also true of Milne's second-difference method and the Milne–Reynolds method. Hence, repeated applications of the corrector can only alter the rounding error. Several authors of predictor–corrector methods recommend that, in order to improve stability, the corrector should *not* be iterated.

12.8 SEMI-ANALYTIC SOLUTIONS OF DIFFERENTIAL EQUATIONS

In Sec. 12.2, semi-numerical integration methods were given. These methods consisted of finding analytic approximations to the solutions of differential equations. Here we will discuss another method of obtaining approximate (or even exact) solutions by analytic means.

Although the problem analyst will usually attempt to determine an analytic solution to his problem, he may overlook the possibility of approximating a solution with a set of analytic solutions. This concept is best described with an example whose solution is given by a set of exact solutions. Suppose that the differential equation to be solved is

$$y' = y + f(t) \tag{12.101}$$

where f(t) is a function with finite discontinuities. For example, let $f(t) = k$ for $k-1 \leqq t \leqq k$. At first it may appear that no analytic solution exists for this problem. Technically this is true, because the derivative of the function being sought has discontinuities. However, the problem can be solved in segments yielding a solution with only finite discontinuities.

On any one interval (k–1, k), we can write Eq. (12.101) in the form

$$y' = y + k \tag{12.102}$$

This equation can be solved analytically to give

$$y = C_k e^t - k \tag{12.103}$$

To obtain the particular solutions needed on each interval, we need the initial values for each interval to evaluate the C_k. If $y = 0$ when $t = 0$, then $C_1 = 1$. Then for the interval (0,1),

$$y = e^t - 1 \tag{12.104}$$

The initial values for the second interval are obtained by noting that on the first interval as t approaches 1, y approaches e-1. These initial values give $C_2 = (e+1)/e$. Therefore, on the second interval (1, 2),

$$y = \left(\frac{e+1}{e}\right)e^t - 2 \tag{12.105}$$

Although the literal expressions for the C_k become increasingly complex, in an actual computer implementation only the numerical values of the C_k would be needed.

Therefore, additional formulas similar to Eqs. (12.104) and (12.105) can be developed quite easily, and any number of numerical values of the solution can be obtained. The only limitations to the accuracy of the values computed in this manner will be the roundoff error in each computation and the possibility of some buildup of error in the computation of the C_k.

Although it may appear that this technique would have few applications, there are a surprising number of real problems where this technique or some variation of it may be used. For example, some control systems make only discrete corrections at discrete intervals of time. The closed loop response of many such systems can be analyzed by this method.

In the example given, a set of exact solutions can be obtained. Suppose that f(t) is a continuous function, which is given only by a finite set of tabulated values. The foregoing technique can still be applied to obtain solutions; or it may be possible to approximate the function in whole or in part with a form that would still permit analytic integration. In either of these events, the solutions would only be approximate. However, such approximate solutions may be more accurate than those obtained by a numerical integration method. Furthermore, the problem of instability is avoided.

Even though this technique may not yield an exact solution to the given problem, the solution is, at least, an exact solution to a problem which was substituted for the original problem. This same statement cannot be made for the other integration techniques of this chapter. A numerical integration method may give the exact solution of some problem, but we may not know what that problem is.

12.9 SOLUTION OF HIGHER ORDER DIFFERENTIAL EQUATIONS

In this chapter, numerical methods for solving first-order differential equations in a single unknown have been presented. These methods can easily be expanded to solve sets of first-order equations in several unknowns. A flowchart for solving a set of first-order differential equations by a Runge–Kutta process was given in Fig. 12.7. For a set of m equations, instead of the single general equation given by (12.3), we will have the general set of equations

$$
\begin{aligned}
&G_1(x, y_1, y_2, \ldots, y_m, y'_1, y'_2, \ldots, y'_m) = 0 \\
&G_2(x, y_1, y_2, \ldots, y_m, y'_1, y'_2, \ldots, y'_m) = 0 \\
&\vdots \\
&G_m(x, y_1, y_2, \ldots, y_m, y'_1, y'_2, \ldots, y'_m) = 0
\end{aligned}
\tag{12.106}
$$

Each of these equations can be solved for the derivative of a different dependent variable resulting in the set.

$$
\begin{aligned}
y'_1 &= F_1(x, y_1, y_2, \ldots, y_m, \quad y'_2, y'_3, \ldots, y'_{m-1}, y'_m) \\
y'_2 &= F_2(x, y_1, y_2, \ldots, y_m, y'_1, \quad y'_3, \ldots, y'_{m-1}, y'_m) \\
&\vdots \\
y'_m &= F_m(x, y_1, y_2, \ldots, y_m, y'_1, y'_2, y'_3, \ldots, y'_{m-1} \quad)
\end{aligned}
\tag{12.107}
$$

Once the equations have been written in this form, any of the methods presented can be applied to each of these equations, and the set solved simultaneously.†

The reason for discussing solving sets of first-order equations in a section on higher-order methods is because differential equations of second or higher order can be reduced to sets of first-order equations. Consider the kth-order differential equation given by Eq. (12.1). Solving this for the highest order derivative, we can write

$$y^{(k)} = F(x, y, y', y'', \ldots, y^{(k-1)}) \tag{12.108}$$

Now making the substitutions,

$$y_j = y^{(j)} \tag{12.109}$$

(The superscript here in parentheses means the jth derivative of y with respect to x.) Noting that this implies

$$y_j' = y_{j+1} \tag{12.110}$$

we can write (12.108) as the set of equations

$$\begin{aligned} y' &= y_1 \\ y_1' &= y_2 \\ y_2' &= y_3 \\ &\vdots \\ y_{k-2}' &= y_{k-1} \\ y_{k-1}' &= F(x, y, y_1, y_2, \ldots, y_{k-1}) \end{aligned} \tag{12.111}$$

Once the equations are in this form [which is the same as the general form given by Eqs. (12.107)], then any one of the methods discussed can be applied to solve these simultaneously.

Even if we have a *set* of higher-order differential equations, the same technique can be applied to reduce that set to a set of first-order differential equations. The general form will not be given here, since to do so would require double subscripting. In actual practice, the analyst would most likely use unique symbols for each variable rather than the same symbol with different subscripts. An example of this was given in Eqs. (12.2), where v was used to denote velocity, which is the first time derivative of the position coordinate y.

As an example, let us generalize the problem of Eqs. (12.2) to the case of a particle moving in two dimensions subject only to the forces of gravity and the retarding force of the medium through which it moves:

$$\begin{aligned} Mx'' + x'R(x,y) &= 0 \\ My'' + y'R(x,y) - Mg &= 0 \end{aligned} \tag{12.112}$$

†When solving Eqs. (12.106) for the individual derivatives, (12.107), it is desirable that in each equation, the derivative on the left of the equals sign does not appear on the right side also. Otherwise, an iterative solution would be required.

where the derivatives are taken with respect to time, which is the independent variable. Letting $v_x = x'$ and $v_y = y'$, this set of second-order equations can be written as the following set of first-order equations:

$$\begin{aligned} x' &= v_x \\ y' &= v_y \\ v_x' &= -v_x R(x,y)/M \\ v_y' &= -v_y R(x,y)/M + g \end{aligned} \qquad (12.113)$$

12.10 SUMMARY

In this chapter a variety of methods for the numerical solution of differential equations have been presented. The user must make his own judgment as to which method to use for a particular problem as there is no single best method for all problems.

The methods presented can be divided into two general classes: those involving some analytic techniques (Secs. 12.2 and 12.8) and those using only numerical techniques (Secs. 12.3 through 12.6). The methods of Secs. 12.2 and 12.8 are more dependent on the differential equations being solved than are the methods of Secs. 12.3 through 12.6. Therefore it is difficult to give a general summary of the methods of Secs. 12.2 and 12.8. This section only gives general summaries of the methods of Secs. 12.3 through 12.6. These methods can be grouped into two major types: predictor–corrector (Secs. 12.3 through 12.5) and Runge–Kutta (Sec. 12.6).

In each of the following subsections, the symbol y_k' is the value of the derivative defined by

$$y_k' = F(x_k, y_k) \qquad (12.114)$$

and h is the increment in the independent variable. The inherent error of the method is given by the letter E, and $y^{(n)}(\xi)$ is the nth derivative of the dependent variable evaluated at some value ξ of the independent variable. The value ξ satisfies the inequalities

$$x_{k-j} \leqq \xi \leqq x_{k+1} \qquad (12.115)$$

where $k-j$ is the lowest subscript used in the formula.

12.10.1 Summary of Predictor–Corrector Methods

Predictor–corrector methods are summarized in Table 12.10. The formulas of each of the methods can be written in the form of Eq. (12.96), which is repeated below. Therefore, a tabulation of the values of the A's and B's along with (12.96) completely defines each equation.

$$y_{k+1} = \sum_{i=0}^{N} A_i y_{k-i} + h \sum_{i=-1}^{M} B_i y_{k-i}' \qquad (12.96)$$

The name of the method is also given in the table, followed by the letter P or C to identify the predictor or the corrector, respectively. The equation number used earlier

TABLE 12.10 COEFFICIENTS FOR FOUR PREDICTOR–CORRECTOR METHODS

Method	*P/C*	*Eq. No.*	A_0	A_1	A_3	B_{-1}	B_0	B_1	B_2	B_3	C	n
Modified	P	(12.31)	1				1				1/2	2
Euler's	C	(12.33)	1			1/2	1/2				−1/12	3
Adams'	P	(12.44)	1				55/24	−59/24	37/24	−9/24	251/720	5
	C	(12.45)	1			9/24	19/24	−5/24	1/24		−19/720	5
Milne's												
Second-	P	(12.55)			1		8/3	−4/3	8/3		14/45	5
Difference	C	(12.58)		1		1/3	4/3	1/3			−1/90	5
Milne-	P	(12.99)		1			8/3	−5/3	4/3	−1/3	29/90	5
Reynolds	C	(12.97)	1/8	7/8		65/192	243/192	51/192	1/192		−5/384	5

in the text is given as a cross reference. In addition, the inherent error for each formula can be written in the form

$$E = Ch^n y^{(n)}(\xi) \tag{12.116}$$

Also tabulated in Table 12.10 are the values of C and n for each formula.

12.10.2 Fourth-Order Runge–Kutta Methods

Equations (12.60) defined the fourth-order Runge–Kutta methods in general. These equations are repeated below. The parameters of four of these methods are given in Table 12.11.

$$\begin{aligned}
\delta_1 &= hF(x_k, y_k) \\
\delta_2 &= hF(x_k + \alpha_1 h, y_k + \beta_{11}\delta_1) \\
\delta_3 &= hF(x_k + \alpha_2 h, y_k + \beta_{21}\delta_1 + \beta_{22}\delta_2) \\
\delta_4 &= hF(x_k + \alpha_3 h, y_k + \beta_{31}\delta_1 + \beta_{32}\delta_2 + \beta_{33}\delta_3) \\
y_{k+1} &= y_k + w_1\delta_1 + w_2\delta_2 + w_3\delta_3 + w_4\delta_4
\end{aligned} \tag{12.60}$$

For this form, it is also necessary that

$$\alpha_i = \sum_{j=1}^{i} \beta_{ij} \tag{12.117}$$

and

$$\sum_{i=1}^{4} w_i = 1 \tag{12.118}$$

The inherent error of the fourth-order Runge–Kutta methods is of the order $O(h^5)$. The coefficients of the Ralston variation have been selected to minimize this error.

TABLE 12.11 PARAMETERS FOR FOUR RUNGE–KUTTA METHODS

Parameter	*Runge*	*Kutta*	*Gill*	*Ralston*
α_1	0.50000000	0.33333333	0.50000000	0.40000000
α_2	0.50000000	0.66666667	0.50000000	0.45573725
α_3	1.00000000	1.00000000	1.00000000	1.00000000
β_{11}	0.50000000	0.33333333	0.50000000	0.40000000
β_{21}	0.00000000	−0.33333333	0.20710678	0.29697761
β_{22}	0.50000000	1.00000000	0.29289322	0.15875964
β_{31}	0.00000000	1.00000000	0.00000000	0.21810040
β_{32}	0.00000000	−1.00000000	−0.70710678	−3.05096516
β_{33}	1.00000000	1.00000000	1.70710678	3.83286476
w_1	0.16666667	0.12500000	0.16666667	0.17476028
w_2	0.33333333	0.37500000	0.09763107	−0.55148066
w_3	0.33333333	0.37500000	0.56903559	1.20553560
w_4	0.16666667	0.12500000	0.16666667	0.17118478

EXERCISES

1. The equation of motion of a simple pendulum is given by the differential equation

$$\ddot{\theta} = -k \sin \theta$$

For $k = 11.22243167$ radians/sec^2 and the initial values $\theta(0) = 1$ radian and $\dot{\theta}(0) = 0$, obtain Taylor's series expansions for θ and $\dot{\theta}$. Omit terms involving derivatives higher than the sixth.

2. Prepare a flowchart and write a FORTRAN program to solve a first-order differential equation by Euler's method. Provide some method for printing results less frequently than once per integration interval.

3. Execute the program in Exercise 2 for the differential equation

$$y' = y + \sqrt{y}$$

with the initial value $y(0) = 1$. Integrate the differential equation from $x = 0$ to $x = 1$. Do this for each of the following integration intervals: 0.1, 0.01, and 0.001. Print results at intervals of 0.2 in x.

4. Prepare a flowchart and write a FORTRAN program to solve a first-order differential equation by the modified Euler's method. Provide also for printing results less frequently than once per integration interval.

5. Repeat Exercise 3 using the program in Exercise 4.

6. The equation in Exercise 1 can be replaced by the two first-order equations

$$\dot{\theta} = D$$
$$\dot{D} = -11.22243167 \sin \theta$$

Modify the program in Exercise 4 to solve these two first-order equations by the modified Euler's method.

7. Execute the program in Exercise 6 for the initial values given in Exercise 1. Integrate from $t = 0$ to $t = 1$ with each of the following integration intervals: 0.1, 0.01, and 0.001. Print results at intervals of 0.1 in t.

8. Prepare a flowchart and write a FORTRAN program to solve the differential equations of Exercise 6 by the predictor–corrector equations (12.44) and (12.45). Assume that functions are available for calculating the necessary starting values.

9. Execute the program in Exercise 8. Use the Taylor's series approximations found for Exercise 1 to obtain the necessary starting values. Integrate from $t = 0$ to $t = 1$ using each of the following integration intervals: 0.1, 0.01, and 0.001. Print results at intervals of 0.1 in t.

10. Modify the program in Exercise 8 to use the predictor–corrector equations (12.55) and (12.58).

11. Execute the program in Exercise 10 for the conditions specified in Exercise 9.

12. Modify the program in Exercise 8 to use the predictor–corrector equations (12.99) and (12.97).

13. Execute the program in Exercise 12 for the conditions specified in Exercise 9.

14. Prepare a flowchart and write a FORTRAN program that will solve the equations in Exercise 6 by any fourth-order Runge–Kutta method. [See Eq. (12.60).] Note that since the independent variable t does not appear explicitly in the two differential equations, the parameters α_1, α_2, and α_3 are not needed.

15. Execute the program in Exercise 14 using the Runge coefficients. (See Table 12.11) Integrate from $t = 0$ to $t = 1$ using each of the following integration intervals: 0.1, 0.02, and 0.002. Print results at intervals of 0.1 in t.

16. Repeat Exercise 15 using the Kutta coefficients.

17. Repeat Exercise 15 using the Gill coefficients.

18. Repeat Exercise 15 using the Ralston coefficients.

Bibliography

Most of the references listed here have been used in the preparation of this text. The majority of these are general references covering a variety of numerical methods. Other more specialized references that have been consulted for particular topics have also been included. The reader who is interested in additional references will find that several of the references in this list contain extensive bibliographies. In addition, new numerical methods and variations on old methods may be found in current technical literature such as the various publications of The Association for Computing Machinery. Publications of other professional organizations, such as The Institute of Electrical and Electronic Engineers, also frequently contain papers on numerical methods.

AMERICAN NATIONAL STANDARDS INSTITUTE. *American NATIONAL STANDARD: Flowchart Symbols and Their Usage in Information Processing.* ANSI X3.5, New York, 1970.

BECKETT, Royce and James Hurt. *Numerical Calculations and Algorithms.* New York: McGraw-Hill, 1967.

BENNETT, Albert A., William E. Milne, and Harry Bateman. *Numerical Integration of Differential Equations.* New York: Dover Publications, 1956.

BURKE, A. W., H. H. Goldstein, and J. von Neuman. "Preliminary Discussion of the Logical Design of the Electronic Computing Instrument." A report prepared for the Ordnance Department, U. S. Army, at The Institute for Advanced Study, Princeton, N. J. Reprinted, in part, in *Datamation,* vol. 8, nos. 9 and 10, 1962.

BUSINESS EQUIPMENT MANUFACTURERS ASSOCIATION. *American Standard Flowchart Symbols for Information Processing.* New York: American National Standards Institute, 1965.

CARNAHAN, Brice, H. A. Luther, and James O. Wilkes. *Applied Numerical Methods.* New York: Wiley, 1969.

CHURCHILL, Ruel V. *Modern Operational Mathematics In Engineering,* 2nd ed. New York: McGraw-Hill, 1958.

CLENSHAW, C. W. "Chebyshev Series for Mathematical Functions." In *Mathematical Tables,* vol. 5. London: Her Majesty's Stationery Office, 1962.

CONTE, Samuel D. *Elementary Numerical Analysis: An Algorithmic Approach,* New York: McGraw-Hill, 1965.

EKSTROM, Ralph E. "A Basic Course In Numerical Methods." Reprinted from *Machine Design,* October 26, 1967 through June 20, 1968. Cleveland: Penton Publishing, 1967–1968.

FREEBURG, H.A. *A History of Mathematics.* New York: Macmillan, 1961.

FRY, C. R. "An Experimental Comparison of Runge–Kutta and Predictor–Corrector Numerical Integration Processes," *CARDE Technical Note* 1713/66. Valcartier, Quebec: Canadian Armament Research and Development Establishment, 1966.

GOLD, Bernard and Charles M. Rader. *Digital Processing of Signals.* New York: McGraw-Hill, 1969.

HAMMING, Richard W. *Numerical Methods for Scientists and Engineers.* New York· McGraw-Hill, 1962.

HARTREE, Douglas R. *Numerical Analysis,* 2nd ed. London: Oxford University Press, 1958.

HILDEBRAND, Francis B. *Introduction to Numerical Analysis.* New York: McGraw-Hill, 1956.

HOUSEHOLDER, Alston S. *Principles of Numerical Analysis.* New York: McGraw-Hill, 1953.

KUNZ, Kaiser S. *Numerical Analysis.* New York: McGraw-Hill, 1957.

KUO, Shan S. *Numerical Methods and Computers.* Reading, Mass.: Addison-Wesley, 1965.

LEE, John A. *Numerical Analysis for Computers.* New York: Van Nostrand Reinhold, 1966.

LEVY, H. and E. A. Baggott. *Numerical Solutions of Differential Equations.* New York: Dover Publications, 1950.

McCALLA, Thomas R. *Introduction to Numerical Methods and FORTRAN Programming.* New York: Wiley, 1967.

McCORMICK, John M. and Mario G. Salvadori. *Numerical Methods in FORTRAN.* Englewood Cliffs, N. J.: Prentice-Hall, 1964.

McCRACKEN, Daniel D. and William S. Dorn. *Numerical Methods and FORTRAN Programming.* New York: Wiley, 1964.

MILNE, William E. *Numerical Solution of Differential Equations.* New York: Dover Publications, 1970. Originally published in 1953.

NEUGEBAUER, Otto. *The Exact Sciences in Antiquity.* Providence, R. I.: Brown University Press, 1957.

NIELSEN, Kaj L. *Methods in Numerical Analysis,* 2nd ed. New York: Macmillan, 1964.

PENNINGTON, Ralph H. *Introductory Computer Methods and Numerical Analysis,* 2nd ed. New York: Macmillan, 1970.

REDDICK, Harry W. and Frederick H. Miller. *Advanced Mathematics for Engineers,* 3rd ed. New York: Wiley, 1955.

REX, Frederick J., Jr. "Herman Hollerith, the First 'Statistical Engineer'." *Computers and Automation,* vol. 10, August 1961.

SALVADORI, Mario G. and M. L. Baron. *Numerical Methods in Engineering,* 2nd ed. Englewood Cliffs, N. J.: Prentice-Hall, 1961.

SCARBOROUGH, James B. *Numerical Mathematical Analysis,* 6th ed. Baltimore, Md.: The Johns Hopkins Press, 1966.

SOUTHWORTH, Raymond W. and Samuel L. De Leeuw. *Digital Computation and Numerical Methods.* New York: McGraw-Hill, 1965.

STIBITZ, George R. and Evelyn Loveday. "The Relay Computers at Bell Labs." *Datamation,* vol. 13, April, May 1967.

THOMAS, George B., Jr. *Calculus and Analytic Geometry,* 4th ed. Reading, Mass.: Addison-Wesley, 1968.

Appendix: Correspondence between Standard Flowchart Symbols and FORTRAN Statements

Symbol	Symbol Name	Standard FORTRAN Statement
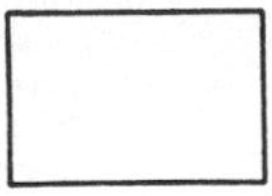	Process	Arithmetic assignment Assigned GO TO Computed GO TO CONTINUE GO TO assignment Logical assignment Optional use: Arithmetic statement function FUNCTION subprogram SUBROUTINE subprogram
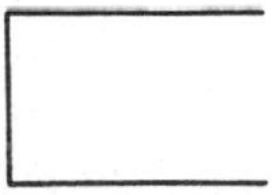	Annotation	Comments COMMON DATA DIMENSION EQUIVALENCE EXTERNAL FORMAT TYPE
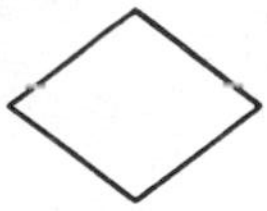	Decision	Arithmetic IF Logical IF
	Terminal	PAUSE RETURN STOP
	Connector	GO TO
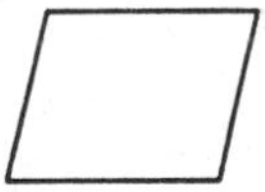	Generalized Input/Output	May include any of the following:
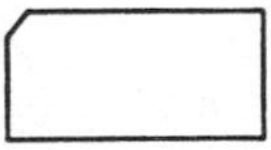	Punched Card Input/Output	READ PUNCH (WRITE)
	Punched Tape Input/Output	READ WRITE
	Magnetic Tape Input/Output	BACKSPACE ENDFILE READ REWIND WRITE
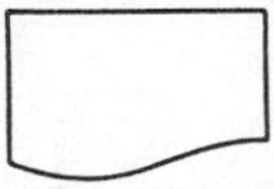	Document Output	PRINT (WRITE)
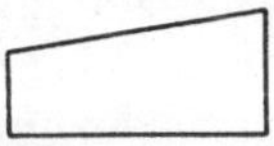	Manual Input	READ

Answer Section

This answer section provides results for all *Exercises* given in the text. The format used herein, however, is not a reproduction of the actual computer print-outs of the associated programs, due to space limitations.

CHAPTER 2

2.1

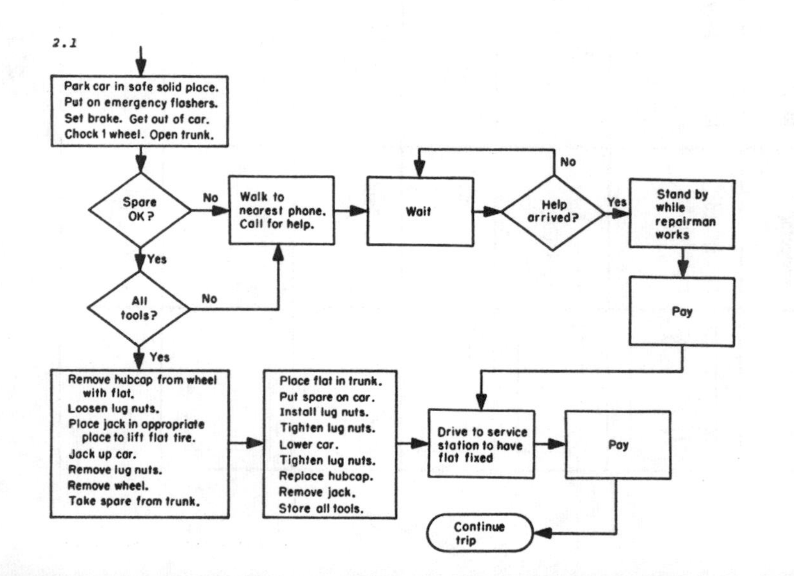

2.3

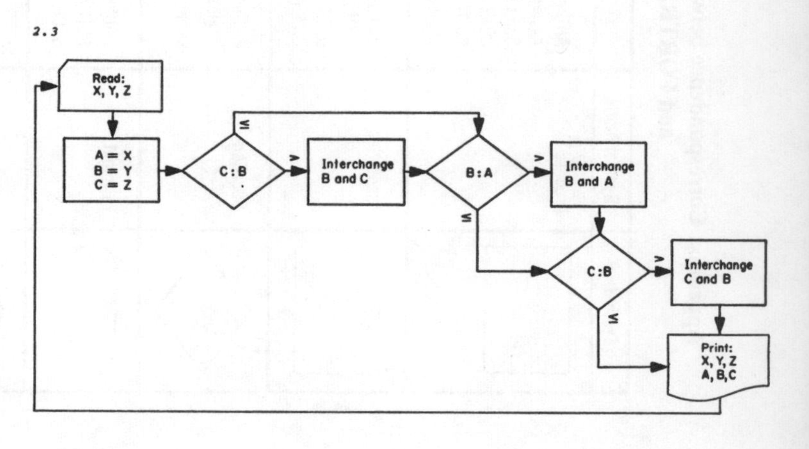

2.2 How to prepare spaghetti

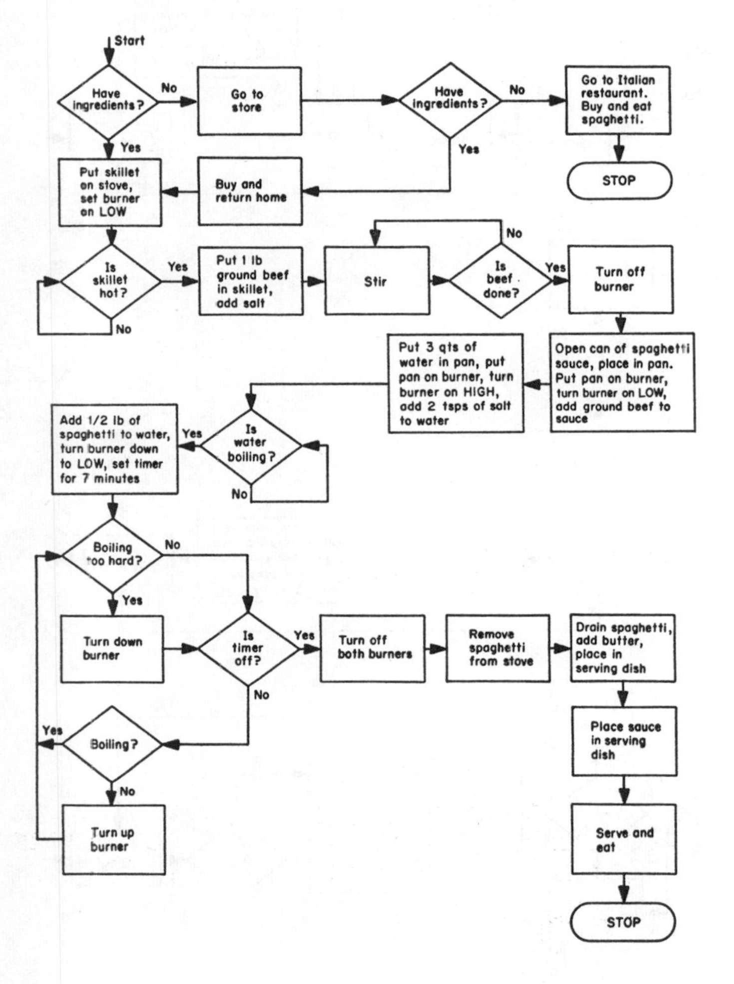

2.4

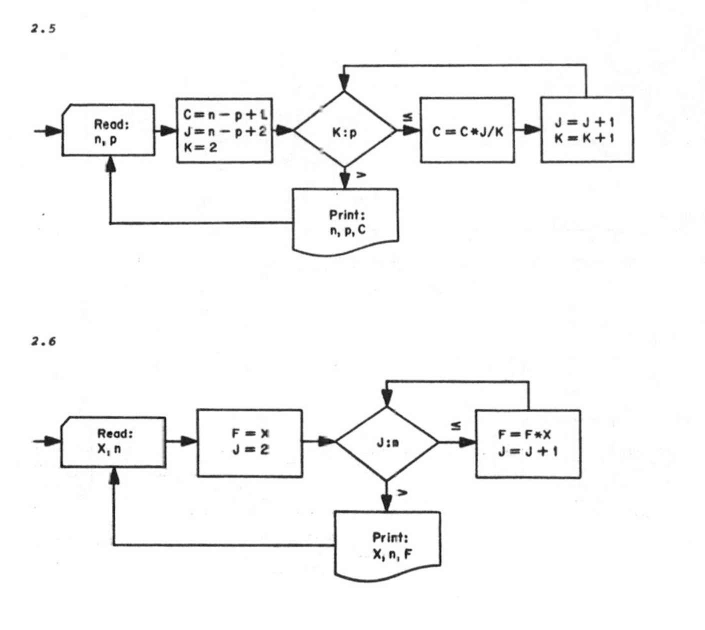

2.5

2.6

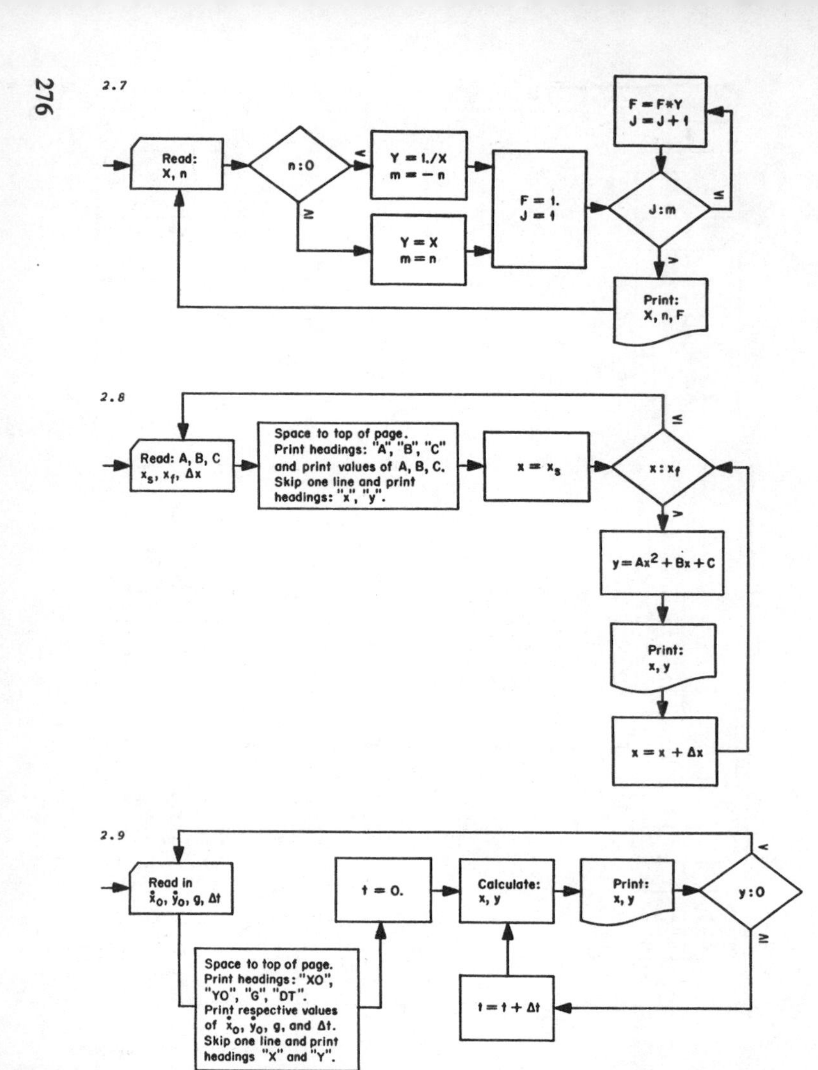
2.7
Read:
X, n
n:0
<
≥
Y = 1./X
m = −n
Y = X
m = n
F = 1.
J = 1
F = F*Y
J = J+1
J:m
≤
>
Print:
X, n, F
2.8
Read: A, B, C
x_s, x_f, Δx
Space to top of page.
Print headings: "A", "B", "C"
and print values of A, B, C.
Skip one line and print
headings: "x", "y".
$x = x_s$
$x : x_f$
≤
>
$y = Ax^2 + Bx + C$
Print:
x, y
x = x + Δx
2.9
Read in
$\dot{x}_0$, $\dot{y}_0$, g, Δt
Space to top of page.
Print headings: "XO",
"YO", "G", "DT".
Print respective values
of $\dot{x}_0$, $\dot{y}_0$, g, and Δt.
Skip one line and print
headings "X" and "Y".
t = 0.
Calculate:
x, y
t = t + Δt
Print:
x, y
y:0
<
≥

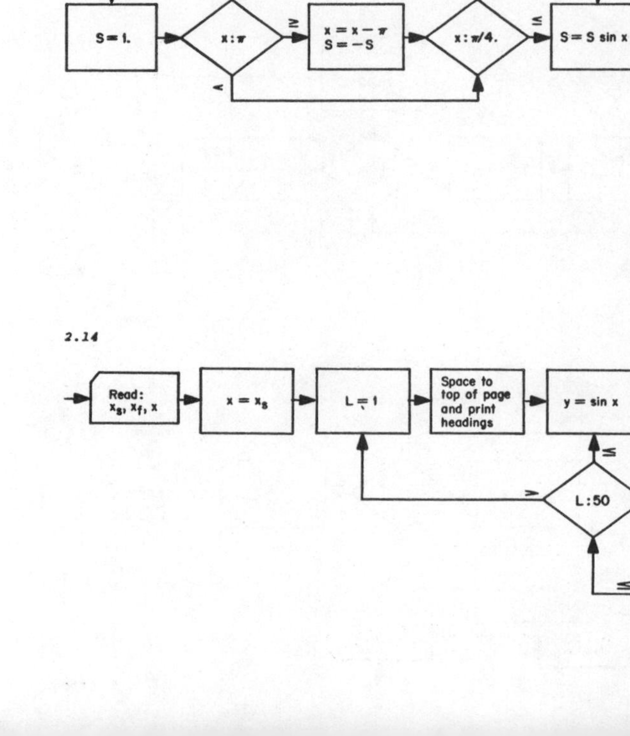
2.13
Set x =
remainder
of A/2π
A:0.
<
≥
S = −1.
x = −x
S = 1.
x:π
≥
<
x = x − π
S = −S
x:π/4.
≤
>
x:3π/4.
<
≥
x = π/2. − x
S = S cos x
x = π − x
S = S sin x
2.14
Read:
x_s, x_f, x
$x = x_s$
L = 1
Space to
top of page
and print
headings
y = sin x
Print:
x, y
x = x + x
L = L + 1
$x : x_f$
≤
>
L:50
≤
>
STOP

2.10

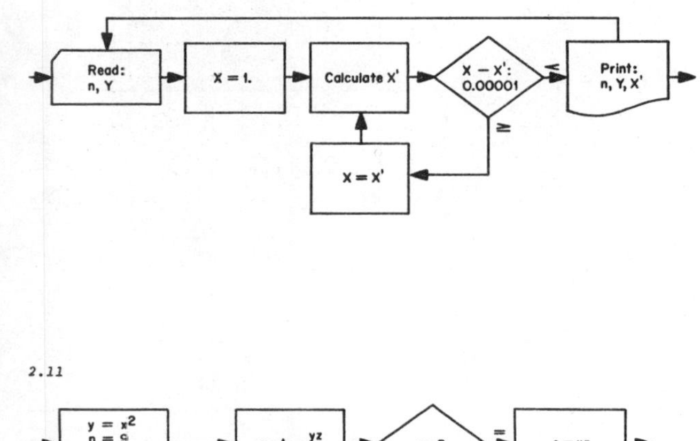

2.11

2.12

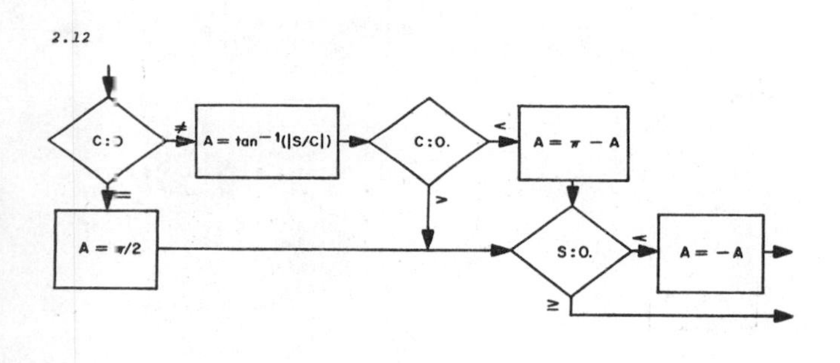

2.15

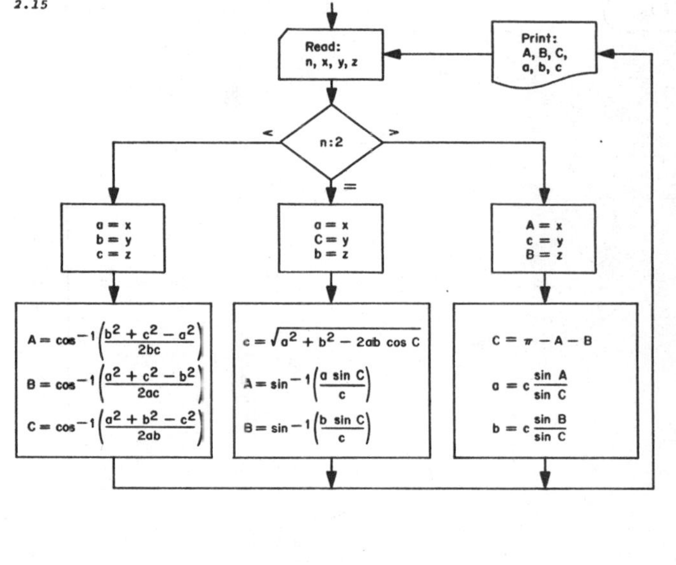

CHAPTER 3

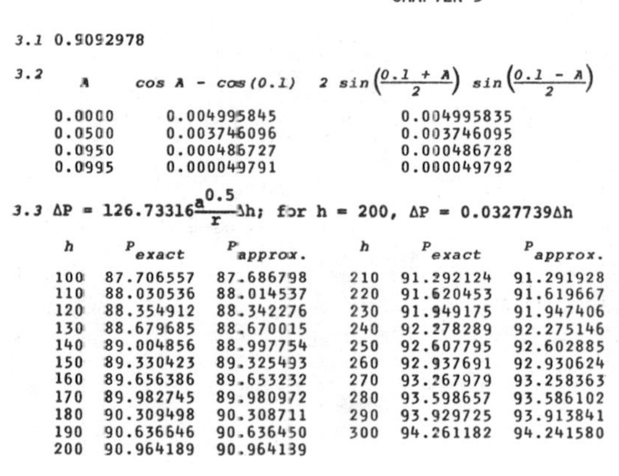

3.1 0.9092978

3.2

A	$\cos A - \cos(0.1)$	$2 \sin\left(\frac{0.1 + A}{2}\right) \sin\left(\frac{0.1 - A}{2}\right)$
0.0000	0.004995845	0.004995835
0.0500	0.003746096	0.003746095
0.0950	0.000486727	0.000486728
0.0995	0.000049791	0.000049792

3.3 $\Delta P = 126.73316\frac{a^{0.5}}{r}\Delta h$; for $h = 200$, $\Delta P = 0.0327739\Delta h$

h	P_{exact}	$P_{approx.}$	h	P_{exact}	$P_{approx.}$
100	87.706557	87.686798	210	91.292124	91.291928
110	88.030536	88.014537	220	91.620453	91.619667
120	88.354912	88.342276	230	91.949175	91.947406
130	88.679685	88.670015	240	92.278289	92.275146
140	89.004856	88.997754	250	92.607795	92.602885
150	89.330423	89.325493	260	92.937691	92.930624
160	89.656386	89.653232	270	93.267979	93.258363
170	89.982745	89.980972	280	93.598657	93.586102
180	90.309498	90.308711	290	93.929725	93.913841
190	90.636646	90.636450	300	94.261182	94.241580
200	90.964189	90.964139			

3.4 $\lambda = 3.801440$

x	$\Delta\lambda_{exact}$	$\Delta\lambda_{approx.}$
180	2.481049	1.506045
220	1.361166	1.004030
260	0.577907	0.502015
300	0.000000	0.000000
340	-0.443673	-0.502015
380	-0.794886	-1.004030
420	-1.079749	-1.506045

3.5 Flowchart

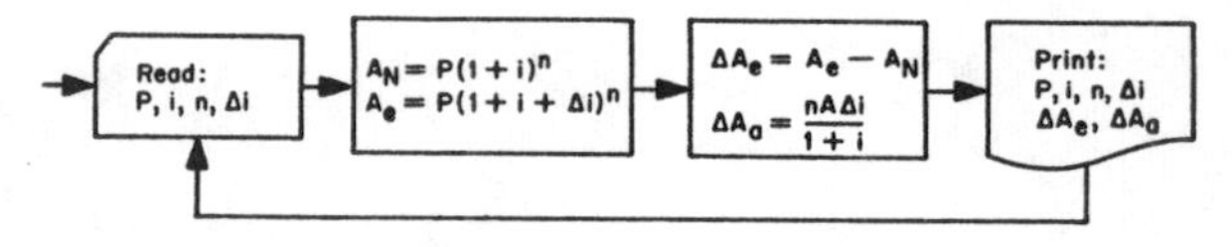

Program

```
 1 READ 10, P,X,N,DX
10 FORMAT(F10.0, F10.3, I4, F10.3)
   AN = P*(1. + X)**N
   AE = P*(1. + X + DX)**N
   EN = N
   DAE = AE - AN
   DAA = EN*AN*DX/(1. + X)
   PRINT 20, P,X,N,DX,DAE,DAA
20 FORMAT(1H , F7.0, F7.3, I5, F7.3, 2F10.2)
   GO TO 1
   END
```

Results

P	i	n	Δi	ΔA_{exact}	$\Delta A_{approx.}$
10000.	0.005	240	0.001	8923.70	7904.96
5000.	0.010	300	0.002	80169.29	58777.62
5000.	0.015	48	0.005	2717.96	2415.93

3.6 Flowchart

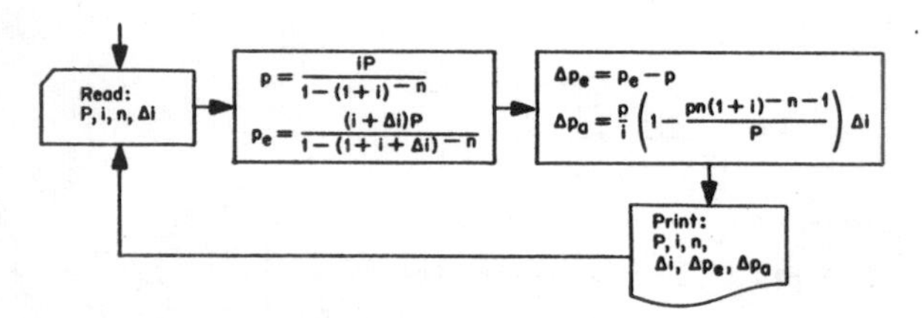

3.8 and *3.9*

			$n = 1.615$, $r_1 = 20$, $r_2 = 20$		$n = 1.360$, $r_1 = -5$, $r_2 = 10$	
Δn	Δr_1	Δr_2	Δf_{direct}	$\Delta f_{differ.}$	Δf_{direct}	$\Delta f_{differ.}$
0.000	0.00	0.00	0.000000	0.000000	0.000000	0.000000
-0.005	0.00	0.00	0.133280	0.132196	-0.391236	-0.385802
0.005	0.00	0.00	-0.131131	-0.132196	0.380518	0.385802
0.000	-0.01	0.00	-0.004066	-0.004065	-0.111338	-0.111111
-0.005	-0.01	0.00	0.129180	0.128131	-0.504138	-0.496913
0.005	-0.01	0.00	-0.135164	-0.136261	0.270709	0.274691
0.000	0.01	0.00	0.004064	0.004065	0.110890	0.111111
-0.005	0.01	0.00	0.137377	0.136261	-0.278785	-0.274691
0.005	0.01	0.00	-0.127100	-0.128131	0.489888	0.496913
0.000	0.00	-0.01	-0.004066	-0.004065	-0.027833	-0.027778
-0.005	0.00	-0.01	0.129180	0.128131	-0.419462	-0.413580
0.005	0.00	-0.01	-0.135164	-0.136261	0.353066	0.358024
0.000	-0.01	-0.01	-0.008130	-0.008130	-0.139391	-0.138889
-0.005	-0.01	-0.01	0.125083	0.124066	-0.532590	-0.524691
0.005	-0.01	-0.01	-0.139195	-0.140326	0.243036	0.246913
0.000	0.01	-0.01	-0.000004	0.000000	0.083278	0.083333
-0.005	0.01	-0.01	0.133276	0.132196	-0.306785	-0.302469
0.005	0.01	-0.01	-0.131135	-0.132196	0.462655	0.469135
0.000	0.00	0.01	0.004064	0.004065	0.027723	0.027778
-0.005	0.00	0.01	0.137377	0.136261	-0.363123	-0.358024
0.005	0.00	0.01	-0.127100	-0.128131	0.407860	0.413580
0.000	-0.01	0.01	-0.000004	0.000000	-0.083389	-0.083333
-0.005	-0.01	0.01	0.133276	0.132196	-0.475799	-0.469135
0.005	-0.01	0.01	-0.131135	-0.132196	0.298271	0.302469
0.000	0.01	0.01	0.008130	0.008130	0.138391	0.138889
-0.005	0.01	0.01	0.141476	0.140326	-0.250896	-0.246913
0.005	0.01	0.01	-0.123066	-0.124066	0.517013	0.524691

3.10 $\Delta Y = (3Ax^2 \cos\theta + Bye^{-x} \sin\phi)\,\Delta x - Ax^3 \sin\theta\,\Delta\theta$
$- Be^{-x} \sin\phi\,\Delta y - Bye^{-x} \cos\phi\,\Delta\phi$

3.11 $\Delta Y = 0.00561507$

3.12 Sample results

$N(1)$	$N(2)$	$N(3)$	$N(4)$	$N(5)$
10	27	31	23	9
8	26	28	30	8
5	22	43	19	11
13	29	30	22	6
10	20	33	28	9
15	18	46	14	7

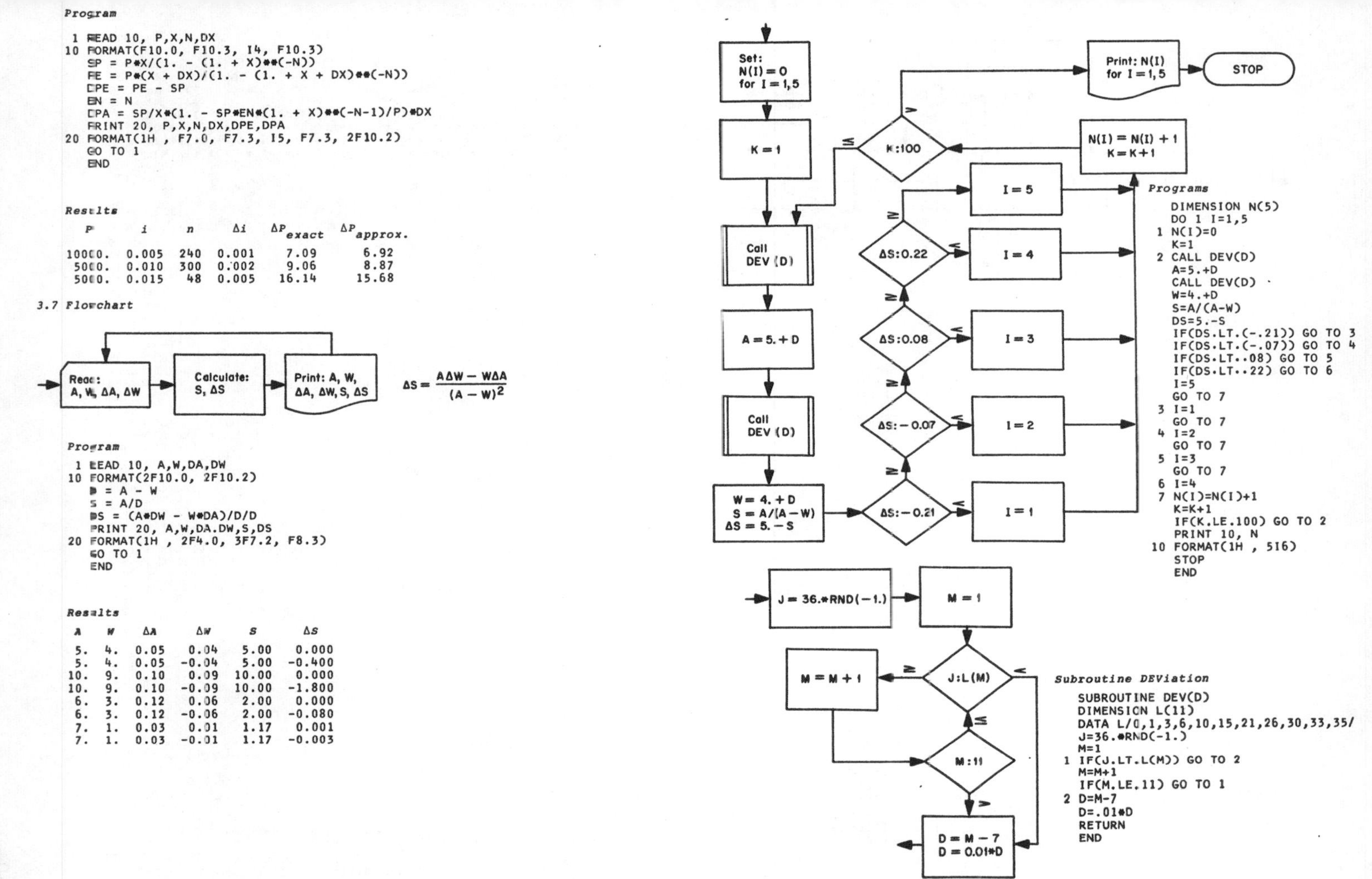

Program

```
 1 READ 10, P,X,N,DX
10 FORMAT(F10.0, F10.3, I4, F10.3)
   SP = P*X/(1. - (1. + X)**(-N))
   PE = P*(X + DX)/(1. - (1. + X + DX)**(-N))
   DPE = PE - SP
   EN = N
   DPA = SP/X*(1. - SP*EN*(1. + X)**(-N-1)/P)*DX
   PRINT 20, P,X,N,DX,DPE,DPA
20 FORMAT(1H , F7.0, F7.3, I5, F7.3, 2F10.2)
   GO TO 1
   END
```

Results

P	i	n	Δi	ΔP_{exact}	$\Delta P_{approx.}$
10000.	0.005	240	0.001	7.09	6.92
5000.	0.010	300	0.002	9.06	8.87
5000.	0.015	48	0.005	16.14	15.68

3.7 Flowchart

$$\Delta S = \frac{A\Delta W - W\Delta A}{(A - W)^2}$$

Program

```
 1 READ 10, A,W,DA,DW
10 FORMAT(2F10.0, 2F10.2)
   D = A - W
   S = A/D
   DS = (A*DW - W*DA)/D/D
   PRINT 20, A,W,DA.DW,S,DS
20 FORMAT(1H , 2F4.0, 3F7.2, F8.3)
   GO TO 1
   END
```

Results

A	W	ΔA	ΔW	S	ΔS
5.	4.	0.05	0.04	5.00	0.000
5.	4.	0.05	-0.04	5.00	-0.400
10.	9.	0.10	0.09	10.00	0.000
10.	9.	0.10	-0.09	10.00	-1.800
6.	3.	0.12	0.06	2.00	0.000
6.	3.	0.12	-0.06	2.00	-0.080
7.	1.	0.03	0.01	1.17	0.001
7.	1.	0.03	-0.01	1.17	-0.003

Programs

```
   DIMENSION N(5)
   DO 1 I=1,5
 1 N(I)=0
   K=1
 2 CALL DEV(D)
   A=5.+D
   CALL DEV(D)
   W=4.+D
   S=A/(A-W)
   DS=5.-S
   IF(DS.LT.(-.21)) GO TO 3
   IF(DS.LT.(-.07)) GO TO 4
   IF(DS.LT..08) GO TO 5
   IF(DS.LT..22) GO TO 6
   I=5
   GO TO 7
 3 I=1
   GO TO 7
 4 I=2
   GO TO 7
 5 I=3
   GO TO 7
 6 I=4
 7 N(I)=N(I)+1
   K=K+1
   IF(K.LE.100) GO TO 2
   PRINT 10, N
10 FORMAT(1H , 5I6)
   STOP
   END
```

Subroutine DEViation

```
   SUBROUTINE DEV(D)
   DIMENSION L(11)
   DATA L/0,1,3,6,10,15,21,26,30,33,35/
   J=36.*RND(-1.)
   M=1
 1 IF(J.LT.L(M)) GO TO 2
   M=M+1
   IF(M.LE.11) GO TO 1
 2 D=M-7
   D=.01*D
   RETURN
   END
```

3.13

Case	Least	Greatest
1 (Direct)	-0.139195	0.141476
1 (Differential)	-0.140326	0.140326
2 (Direct)	-0.532590	0.517013
2 (Differential)	-0.524691	0.524691

Yes

3.14 $Y_{least} = -0.60925291$

$Y_{greatest} = -0.59802286$

3.15

ΔP_{upper}	ΔP_{lower}
1.964	-1.832
2.259	-2.106
3.345	-3.118

Program

```
  DIMENSION X(4), H(4), L(4)
  REAL L
  POW(A,B,C,D) = .000029*(1. + .7854*A*B*B/C)*A*B*B*D
1 READ 2, X
2 FORMAT(3F10.3, F10.0)
  PN = POW(X(1),X(2),X(3),X(4))
  DO 9 I = 1,4
  D = .01*X(I)
  X(I) = X(I) + D
  PH = POW(X(1),X(2),X(3),X(4))
  X(I) = X(I) - 2.*D
  PL = POW(X(1),X(2),X(3),X(4))
  X(I) = X(I) + D
  IF(PH.GT.PL) GO TO 3
  IF(PH.GT.PN) GO TO 4
  IF(PN.GT.PL) GO TO 5
  H(I) = -1.
  L(I) = 1.
  GO TO 9
3 IF(PH.GT.PN) GO TO 6
  H(I) = 0.
  L(I) = -1.
  GO TO 9
6 IF(PN.GT.PL) GO TO 7
  H(I) = 1.
  L(I) = 0.
  GO TO 9
7 H(I) = 1.
  L(I) = -1.
  GO TO 9
4 H(I) = -1.
  L(I) = 0.
  GO TO 9
5 H(I) = 0.
  L(I) = 1.
9 CONTINUE
  PH = POW(X(1)*(1. + .01*H(1)),X(2)*(1. + .01*H(2)),
 1       X(3)*(1. + .01*H(3)),X(4)*(1. + .01*H(4)))
  PL = POW(X(1)*(1. + .01*L(1)),X(2)*(1. + .01*L(2)),
 1       X(3)*(1. + .01*L(3)),X(4)*(1. + .01*L(4)))
  DH = PN - PH
  DL = PN - PL
  PRINT 8, X, DH, DL
8 FORMAT(1H , 4F8.1, 2F8.3)
  GO TO 1
  END
```

3.16 $R_{nom} = 200$ (all cases)

ΔR_{upper}	ΔR_{lower}
10.4	-10.5
10.7	-10.9
11.2	-11.3
14.0	-14.1
13.1	-13.1
11.2	-11.3
15.5	-15.5
16.0	-16.0
17.5	-17.5

CHAPTER 4

4.1

x	sin x
0.000	0.00000000
0.200	0.19866933
0.324	0.31836098
0.428	0.41505204
0.521	0.49774771
0.607	0.57040594
0.688	0.63499342
0.765	0.69253700
0.839	0.74397529
0.911	0.79011709
0.981	0.83105398
1.049	0.86692522
1.116	0.89835052
1.182	0.92536601
1.247	0.94803438
1.311	0.96644232
1.375	0.98089306
1.438	0.99119552
1.501	0.99756523
1.564	0.99997691
1.627	0.99842099

4.2 *Flowchart*

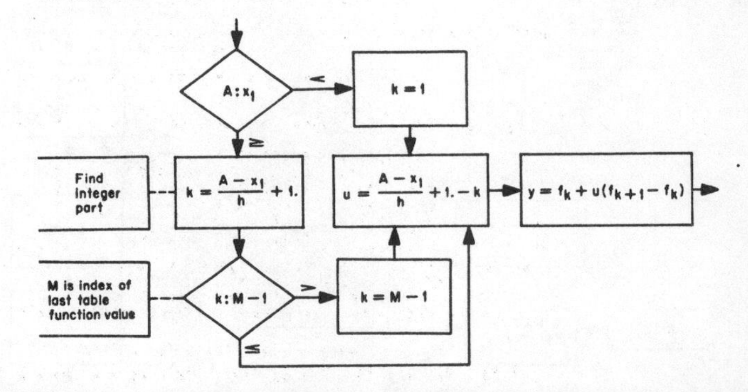

Flowchart

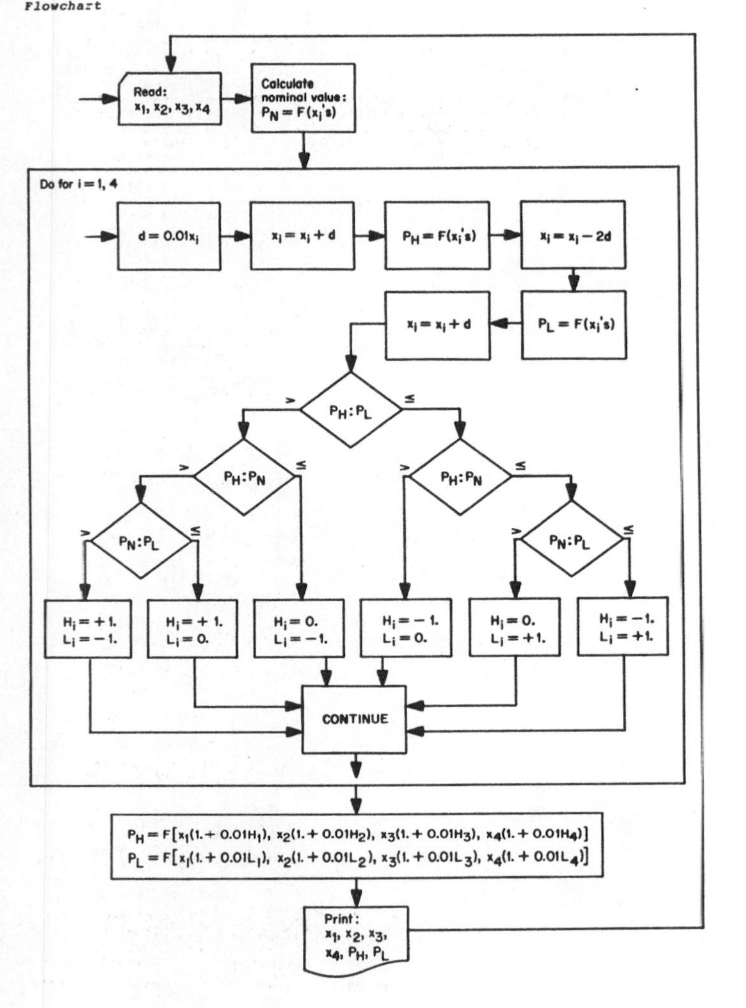

4.3 Let $u = (x - x_{k-1})/h$. Then $y = y_{k-1} + (y_k - y_{k-1})u + (y_{k+2} - 3y_k + 2y_{k-1})u(u - 1)/6$

4.4 $y = 0.19739556\ u - 0.01428474\ u(u - 1)/2$
$- 0.00891296\ u(u - 1)(u - 2)/6$
$+ 0.00651898\ u(u - 1)(u - 2)(u - 3)/24$
$- 0.00219754\ u(u - 1)(u - 2)(u - 3)(u - 4)/120$

4.5

x	y_{exact}	$y_{approx.}$	error
0.1	0.09966865	0.09961158	0.00005707
0.3	0.29145679	0.29147216	-0.00001537
0.5	0.46364761	0.46363931	0.00000083
0.7	0.61072596	0.61073425	-0.00000829
0.9	0.73281510	0.73279839	0.00001671

4.6 $y = 0.99748845\ A + 0.03044044\ A^2 - 0.46952344\ A^3 + 0.28422031\ A^4 - 0.05722760\ A^5$

4.7 Answers are same as for Exercise 4.5

4.8 $y = 0.38050638 + 0.15991312\ u$
$- 0.02319770\ u(u - 1)/2 - 0.00239398\ u(u^2 - 1)/6$
$+ 0.00651898\ u(u^2 - 1)(u - 2)/24$
$- 0.00219754\ u(u^2 - 1)(u^2 - 4)/120$

4.9 Answers are same as for Exercise 4.5

4.10 *Flowchart*

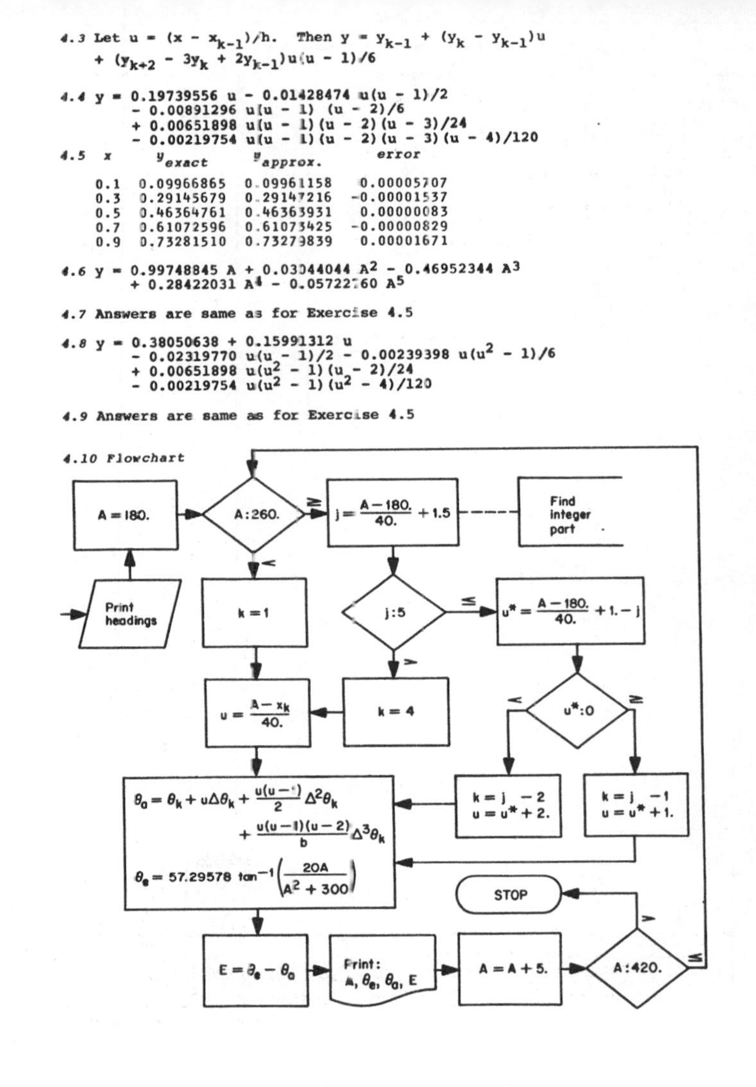

4.10 Program

```
      DIMENSION X(4),TH(4),D1(4),D2(4),D3(4)
      DATA X/180.,220.,260.,300./
      DATA TH/6.282489,5.162606,4.379347,3.801440/
      DATA D1/-1.119883,-.783259,-.577907,-.443673/
      DATA D2/.336624,.205352,.134234,.092460/
      DATA D3/-.131272,-.071118,-.041774,-.026110/
      PRINT 10
   10 FORMAT( 33H1  X     EXACT     APPROX.     ERROR//)
      A=180.
    1 IF(A.LT.260.) GO TO 2
      J=(A-180.)/40.+1.5
      IF(J.GT.5) GO TO 3
      XJ=J
      US=(A-180.)/40.+1.-XJ
      IF(US) 4,5,5
    4 K=J-2
      U=US+2.
      GO TO 6
    5 K=J-1
      U=US+1.
      GO TO 6
    2 K=1
      GO TO 7
    3 K=4
    7 U=(A-X(K))/40.
    6 THAP=TH(K)+U*(D1(K)+(U-1.)/2.*(D2(K)+(U-2.)/3.*D3(K)))
      THEX=57.29578*ATAN(20.*A/(A*A+300.))
      ERR=THEX-THAP
      PRINT 20, A, THEX, THAP, ERR
   20 FORMAT(1H , F5.0, 3F10.6)
      A=A+5.
      IF(A.LE.420.) GO TO 1
      STOP
      END
```

Results

X	EXACT	APPROX.	ERROR
180.	6.282492	6.282489	0.000003
185.	6.116969	6.119608	-0.002639
190.	5.959839	5.963781	-0.003942
195.	5.810489	5.814752	-0.004263
200.	5.668360	5.672265	-0.003905
205.	5.532948	5.536063	-0.003115
210.	5.403791	5.405890	-0.002099
215.	5.280471	5.281490	-0.001019
220.	5.162604	5.162606	-0.000002
225.	5.049839	5.048982	0.000857
230.	4.941855	4.940361	0.001495
235.	4.838357	4.836487	0.001870
240.	4.739072	4.737103	0.001969
245.	4.643752	4.641954	0.001798
250.	4.552163	4.550782	0.001381
255.	4.464094	4.463332	0.000762
260.	4.379346	4.379347	-0.000001
265.	4.297737	4.297337	0.000400
270.	4.219095	4.218397	0.000699
275.	4.143264	4.142387	0.000877
280.	4.070095	4.069169	0.000926
285.	3.999452	3.998605	0.000847
290.	3.931207	3.930554	0.000653
295.	3.865241	3.864879	0.000362
300.	3.801441	3.801440	0.000001
305.	3.739703	3.739497	0.000206
310.	3.679930	3.679569	0.000361
315.	3.622029	3.621576	0.000453
320.	3.565914	3.565435	0.000479
325.	3.511505	3.511066	0.000439
330.	3.458724	3.458385	0.000339
335.	3.407501	3.407313	0.000188
340.	3.357767	3.357767	-0.000000
345.	3.309458	3.309344	0.000114
350.	3.262515	3.262316	0.000200
355.	3.216881	3.216629	0.000252
360.	3.172501	3.172235	0.000267
365.	3.129326	3.129081	0.000245
370.	3.087306	3.087117	0.000189
375.	3.046396	3.046292	0.000104
380.	3.006553	3.006554	-0.000001
385.	2.967735	2.967853	-0.000118
390.	2.929905	2.930138	-0.000233
395.	2.893024	2.893357	-0.000334
400.	2.857057	2.857461	-0.000403
405.	2.821972	2.822397	-0.000425
410.	2.787735	2.788114	-0.000379
415.	2.754317	2.754563	-0.000245
420.	2.721689	2.721691	-0.000002

4.12 Flowchart

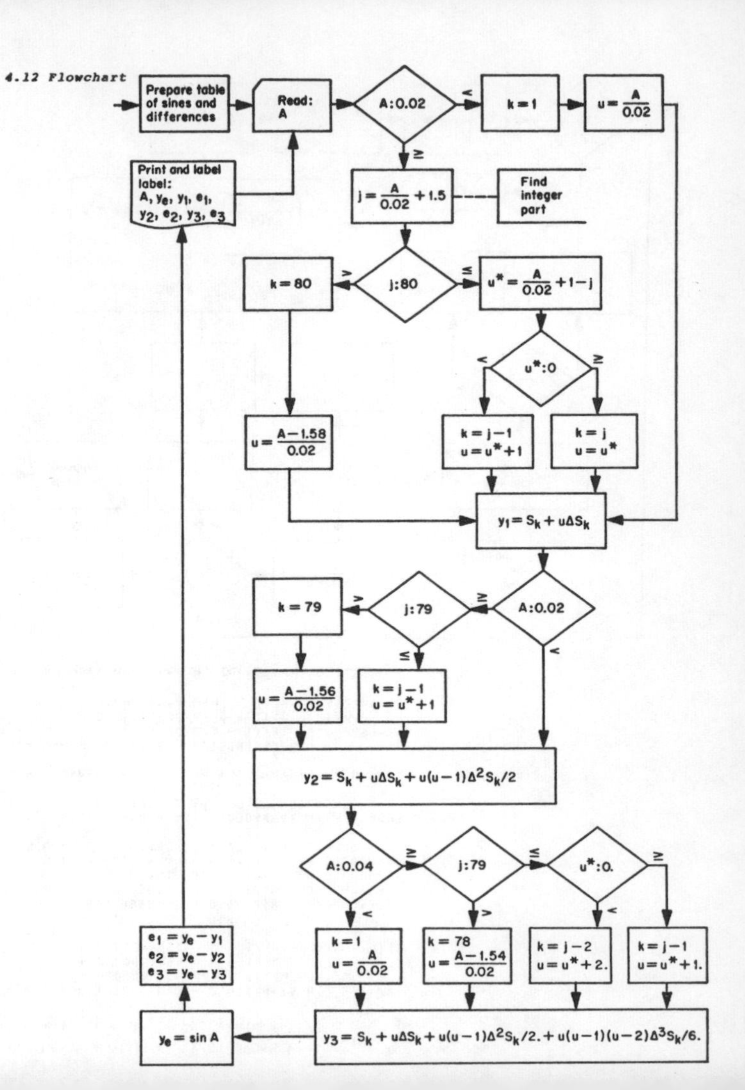

4.11 Flowchart

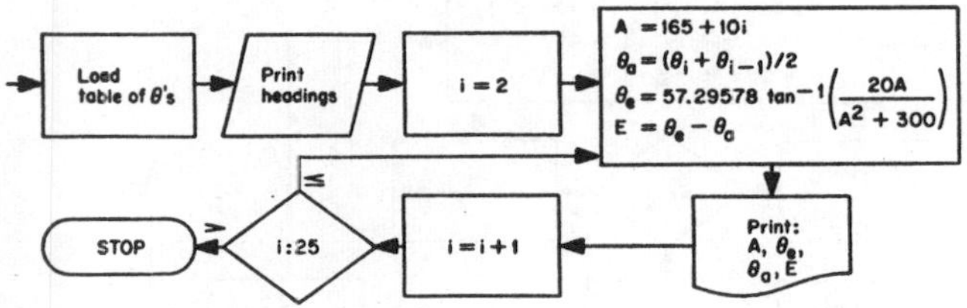

Load table of θ's → Print headings → i = 2 → A = 165 + 10i; $\theta_a = (\theta_i + \theta_{i-1})/2$; $\theta_e = 57.29578 \tan^{-1}\left(\frac{20A}{A^2 + 300}\right)$; $E = \theta_e - \theta_a$ → Print: A, θ_e, θ_a, E → i = i + 1 → i:25 (≤ back to A = 165 + 10i; > STOP)

Program

```
    DIMENSION TH(25)
    DATA TH/6.282489,5.963781,5.672265,5.405890,5.162606,
  1         4.940361,4.737103,4.550782,4.379347,4.218397,
  2         4.069169,3.930554,3.801440,3.679569,3.565435,
  3         3.458385,3.357767,3.262316,3.172235,3.087117,
  4         3.006554,2.930138,2.857461,2.788114,2.721691/
    PRINT 10
 10 FORMAT( 33H1  X     EXACT     APPROX.     ERROR//)
    I=2
  1 XI=I
    A=165.+10.*XI
    THAP=(TH(I)+TH(I-1))/2.
    THEX=57.29578*ATAN(20.*A/(A*A+300.))
    ERR=THEX-THAP
    PRINT 20, A, THEX, THAP, ERR
 20 FORMAT(1H , F5.0, 3F10.6)
    I=I+1
    IF(I.LE.25) GO TO 1
    STOP
    END
```

Results

X	EXACT	APPROX.	ERROR
185.	6.116969	6.123135	-0.006167
195.	5.810489	5.818023	-0.007534
205.	5.532948	5.539078	-0.006130
215.	5.280471	5.284248	-0.003777
225.	5.049839	5.051484	-0.001644
235.	4.838357	4.838732	-0.000375
245.	4.643752	4.643943	-0.000191
255.	4.464094	4.465065	-0.000970
265.	4.297737	4.298872	-0.001135
275.	4.143264	4.143783	-0.000519
285.	3.999452	3.999862	-0.000409
295.	3.865241	3.865997	-0.000757
305.	3.739703	3.740505	-0.000801
315.	3.622029	3.622502	-0.000473
325.	3.511505	3.511910	-0.000405
335.	3.407501	3.408076	-0.000575
345.	3.309458	3.310041	-0.000583
355.	3.216881	3.217276	-0.000394
365.	3.129326	3.129676	-0.000350
375.	3.046396	3.046836	-0.000440
385.	2.967735	2.968346	-0.000611
395.	2.893024	2.893800	-0.000776
405.	2.821972	2.822788	-0.000816
415.	2.754317	2.754903	-0.000585

Program

```
    DIMENSION S(81),D1(80),D2(79),D3(78)
    X=0.
    DO 1 I=1,81
    S(I)=SIN(X)
  1 X=X+.02
    DO 2 I=1,80
  2 D1(I)=S(I+1)-S(I)
    DO 3 I=1,79
  3 D2(I)=D1(I+1)-D1(I)
    DO 4 I=1,78
  4 D3(I)=D2(I+1)-D2(I)
  5 READ 10, A
 10 FORMAT(F11.0)
    IF(A.LT..02) GO TO 6
    J=A/.02+1.5
    IF(J.GT.80) GO TO 7
    XJ=J
    US=A/.02+1.-XJ
    IF(US) 8,9,9
  8 K=J-1
    U=US+1.
    GO TO 11
  9 K=J
    U=US
    GO TO 11
  6 K=1
    U=A/.02
    GO TO 11
  7 K=80
    U=(A-1.58)/.02
 11 Y1=S(K)+U*D1(K)
    IF(A.LT..02) GO TO 12
    IF(J.GT.79) GO TO 13
    K=J-1
    U=US+1.
    GO TO 12
 13 K=79
    U=(A-1.56)/.02
 12 Y2=S(K)+U*(D1(K)+(U-1.)/2.*D2(K))
    IF(A.LT..04) GO TO 14
    IF(J.GT.79) GO TO 15
    IF(US) 16,17,17
 14 K=1
    U=A/.02
    GO TO 18
 15 K=78
    U=(A-1.54)/.02
    GO TO 18
 16 K=J-2
    U=US+2.
    GO TO 18
 17 K=J-1
    U=US+1.
 18 Y3=S(K)+U*(D1(K)+(U-1.)/2.*(D2(K)+(U-2.)/3.*D3(K)))
    YE=SIN(A)
    E1=YE-Y1
    E2=YE-Y2
    E3=YE-Y3
    PRINT 20,A,YE,Y1,E1,Y2,E2,Y3,E3
 20 FORMAT(3H A =,F12.8,8H, EXACT=,F12.8/
  1 9X, 13HFIRST  ORDER=,F12.8, 8H, ERROR=,F12.8/
  2 9X, 13HSECOND ORDER=,F12.8, 8H, ERROR=,F12.8/
  3 9X, 13HTHIRD  ORDER=,F12.8, 8H, ERROR=,F12.8)
    GO TO 5
    END
```

4.12 Results

```
A= -0.01000000, EXACT= -0.00999983
        FIRST  ORDER= -0.00999933, ERROR= -0.00000050
        SECOND ORDER= -0.01000233, ERROR=  0.00000250
        THIRD  ORDER= -0.00999983, ERROR=  0.00000000
A=  0.00500000, EXACT=  0.00499998
        FIRST  ORDER=  0.00499967, ERROR=  0.00000031
        SECOND ORDER=  0.00500042, ERROR= -0.00000044
        THIRD  ORDER=  0.00499998, ERROR= -0.00000000
A=  0.46000000, EXACT=  0.44394810
        FIRST  ORDER=  0.44394810, ERROR=  0.00000001
        SECOND ORDER=  0.44394810, ERROR=  0.00000001
        THIRD  ORDER=  0.44394810, ERROR=  0.00000001
A=  0.55000000, EXACT=  0.52268723
        FIRST  ORDER=  0.52266109, ERROR=  0.00002614
        SECOND ORDER=  0.52268679, ERROR=  0.00000043
        THIRD  ORDER=  0.52268722, ERROR=  0.00000001
A=  0.78539816, EXACT=  0.70710678
        FIRST  ORDER=  0.70707885, ERROR=  0.00002792
        SECOND ORDER=  0.70710657, ERROR=  0.00000021
        THIRD  ORDER=  0.70710680, ERROR= -0.00000002
A=  1.01000001, EXACT=  0.84683184
        FIRST  ORDER=  0.84678954, ERROR=  0.00004230
        SECOND ORDER=  0.84683214, ERROR= -0.00000030
        THIRD  ORDER=  0.84683188, ERROR= -0.00000004
A=  1.23450001, EXACT=  0.94398332
        FIRST  ORDER=  0.94394573, ERROR=  0.00003760
        SECOND ORDER=  0.94398344, ERROR= -0.00000012
        THIRD  ORDER=  0.94398332, ERROR=  0.00000000
A=  1.57079633, EXACT=  1.00000000
        FIRST  ORDER=  0.99995032, ERROR=  0.00004968
        SECOND ORDER=  0.99999999, ERROR=  0.00000001
        THIRD  ORDER=  0.99999999, ERROR=  0.00000001
A=  1.65000001, EXACT=  0.99686503
        FIRST  ORDER=  0.99861350, ERROR= -0.00174847
        SECOND ORDER=  0.99686367, ERROR=  0.00000136
        THIRD  ORDER=  0.99686372, ERROR=  0.00000131
A=  1.67000000, EXACT=  0.99508335
        FIRST  ORDER=  0.99822946, ERROR= -0.00314611
        SECOND ORDER=  0.99507976, ERROR=  0.00000358
        THIRD  ORDER=  0.99507987, ERROR=  0.00000348
```

4.13 Flowchart

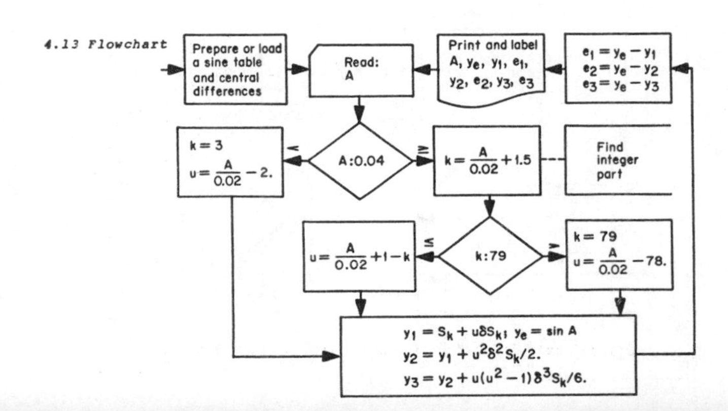

4.13 Results (continued)

```
A=  0.55000000, EXACT=  0.52268723
        FIRST  ORDER=  0.52271250, ERROR= -0.00002527
        SECOND ORDER=  0.52268679, ERROR=  0.00000043
        THIRD  ORDER=  0.52268723, ERROR=  0.00000000
A=  0.78539816, EXACT=  0.70710678
        FIRST  ORDER=  0.70711681, ERROR= -0.00001004
        SECOND ORDER=  0.70710657, ERROR=  0.00000021
        THIRD  ORDER=  0.70710681, ERROR= -0.00000003
A=  1.01000001, EXACT=  0.84683184
        FIRST  ORDER=  0.84687475, ERROR= -0.00004291
        SECOND ORDER=  0.84683215, ERROR= -0.00000031
        THIRD  ORDER=  0.84683189, ERROR= -0.00000004
A=  1.23450001, EXACT=  0.94398332
        FIRST  ORDER=  0.94399774, ERROR= -0.00001442
        SECOND ORDER=  0.94398344, ERROR= -0.00000011
        THIRD  ORDER=  0.94398332, ERROR=  0.00000000
A=  1.57079633, EXACT=  1.00000000
        FIRST  ORDER=  1.00005828, ERROR= -0.00005828
        SECOND ORDER=  1.00000000, ERROR=  0.00000000
        THIRD  ORDER=  1.00000000, ERROR=  0.00000000
A=  1.65000001, EXACT=  0.99686503
        FIRST  ORDER=  1.00091331, ERROR= -0.00404828
        SECOND ORDER=  0.99686362, ERROR=  0.00000141
        THIRD  ORDER=  0.99686243, ERROR=  0.00000259
A=  1.67000000, EXACT=  0.99508335
        FIRST  ORDER=  1.00112921, ERROR= -0.00604586
        SECOND ORDER=  0.99507967, ERROR=  0.00000367
        THIRD  ORDER=  0.99507748, ERROR=  0.00000587
```

4.14 Flowchart

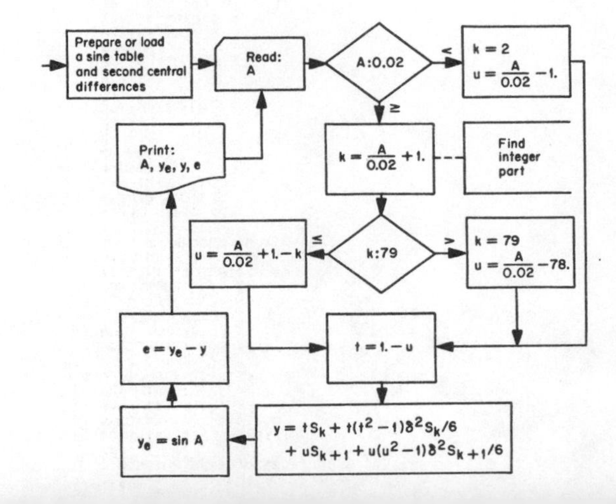

Program

```
   DIMENSION S(81),D1(80),D2(79),D3(78),CD1(80),CD2(80),CD3(79)
   X=0.
   DO 1 I=1,81
   S(I)=SIN(X)
 1 X=X+.02
   DO 2 I=1,80
 2 D1(I)=S(I+1)-S(I)
   DO 3 I=1,79
 3 D2(I)=D1(I+1)-D1(I)
   DO 4 I=1,78
 4 D3(I)=D2(I+1)-D2(I)
   DO 11 I=2,80
11 CD1(I)=(D1(I)+D1(I-1))/2.
   DO 12 I=2,80
12 CD2(I)=D2(I-1)
   DO 13 I=3,79
13 CD3(I)=(D3(I-1)+D3(I-2))/2.
 5 READ 10, A
10 FORMAT(F11.8)
   IF(A.LT..04) GO TO 6
   K=A/.02+1.5
   IF(K.GT.79) GO TO 7
   XK=K
   U=A/.02+1.-XK
   GO TO 8
 6 K=3
   U=A/.02-2.
   GO TO 8
 7 K=79
   U=A/.02-78.
 8 Y1=S(K)+U*CD1(K)
   Y2=Y1+U*U*CD2(K)/2.
   Y3=Y2+U*(U*U-1.)*CD3(K)/6.
   YE=SIN(A)
   E1=YE-Y1
   E2=YE-Y2
   E3=YE-Y3
   PRINT 20,A,YE,Y1,E1,Y2,E2,Y3,E3
20 FORMAT(3H A=,F12.8,8H, EXACT=,F12.8/
  1 9X, 13HFIRST  ORDER=,F12.8, 8H, ERROR=,F12.8/
  2 9X, 13HSECOND ORDER=,F12.8, 8H, ERROR=,F12.8/
  3 9X, 13HTHIRD  ORDER=,F12.8, 8H, ERROR=,F12.8)
   GO TO 5
   END
```

Results

```
A= -0.01000000, EXACT= -0.00999983
         FIRST  ORDER= -0.00996734, ERROR= -0.00003249
         SECOND ORDER= -0.01001733, ERROR=  0.00001749
         THIRD  ORDER= -0.00999984, ERROR=  0.00000001
A=  0.00500000, EXACT=  0.00499998
         FIRST  ORDER=  0.00501966, ERROR= -0.00001968
         SECOND ORDER=  0.00499517, ERROR=  0.00000481
         THIRD  ORDER=  0.00499998, ERROR=  0.00000000
A=  0.46000000, EXACT=  0.44394810
         FIRST  ORDER=  0.44394810, ERROR=  0.00000001
         SECOND ORDER=  0.44394810, ERROR=  0.00000001
         THIRD  ORDER=  0.44394810, ERROR=  0.00000001
```

Program

```
   DIMENSION S(81),D2(80)
   X=0.
   DO 1 I=1,81
   S(I)=SIN(X)
 1 X=X+.02
   DO 2 I=2,80
 2 D2(I)=S(I+1)-2.*S(I)+S(I-1)
 5 READ 10, A
10 FORMAT(F11.8)
   IF(A.LT..02) GO TO 6
   K=A/.02+1.
   IF(K.GT.79) GO TO 7
   XK=K
   U=A/.02+1.-XK
   GO TO 8
 6 K=2
   U=A/.02-1.
   GO TO 8
 7 K=79
   U=A/.02-78.
 8 T=1.-U
   Y=T*(S(K)+(T*T-1.)*D2(K)/6.)
  1 +U*(S(K+1)+(U*U-1.)*D2(K+1)/6.)
   YE=SIN(A)
   E=YE-Y
   PRINT 20, A,YE,Y,E
20 FORMAT(1H , 4F12.8)
   GO TO 5
   END
```

Results

A	y_{exact}	$y_{Everett}$	*Error*
-0.01000000	-0.00999983	-0.00999983	0.00000000
0.00500000	0.00499998	0.00499998	-0.00000000
0.46000000	0.44394810	0.44394810	0.00000001
0.55000000	0.52268723	0.52268723	0.00000000
0.78539816	0.70710678	0.70710680	-0.00000002
1.01000001	0.84683184	0.84683188	-0.00000004
1.23450001	0.94398332	0.94398332	0.00000000
1.57079633	1.00000000	0.99999999	0.00000001
1.65000001	0.99686503	0.99686372	0.00000131
1.67000000	0.99508335	0.99507987	0.00000348

4.15 $$y = -f_k\frac{(u-1)(u-2)(u-3)}{6} + f_{k+1}\frac{u(u-2)(u-3)}{2}$$

$$-f_{k+2}\frac{u(u-1)(u-3)}{2} + f_{k+3}\frac{u(u-1)(u-2)}{6}$$

4.16 Flowchart

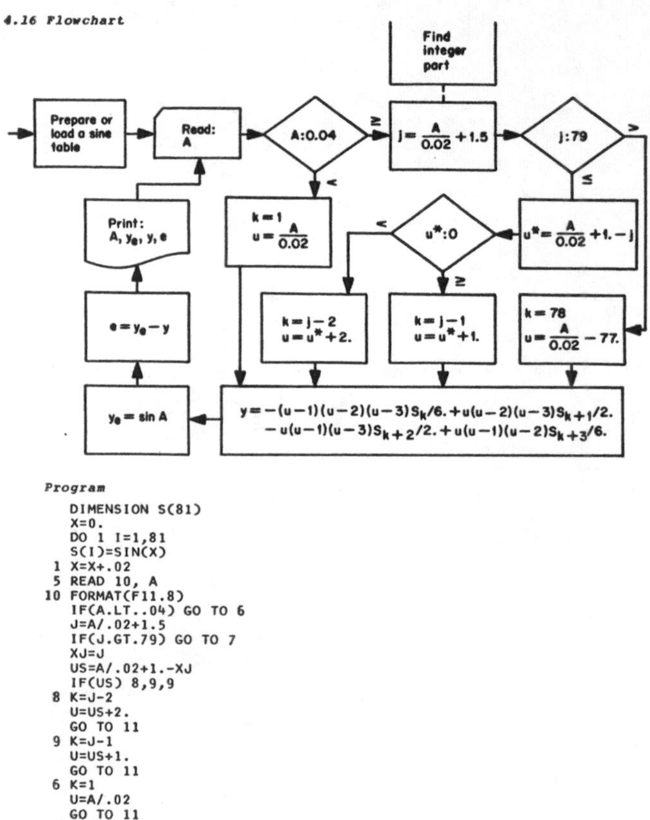

Program

```
      DIMENSION S(81)
      X=0.
      DO 1 I=1,81
      S(I)=SIN(X)
    1 X=X+.02
    5 READ 10, A
   10 FORMAT(F11.8)
      IF(A.LT..04) GO TO 6
      J=A/.02+1.5
      IF(J.GT.79) GO TO 7
      XJ=J
      US=A/.02+1.-XJ
      IF(US) 8,9,9
    8 K=J-2
      U=US+2.
      GO TO 11
    9 K=J-1
      U=US+1.
      GO TO 11
    6 K=1
      U=A/.02
      GO TO 11
    7 K=78
      U=A/.02-77.
   11 U1=U-1.
      U2=U-2.
      U3=U-3.
      Y=-S(K)*U1*U2*U3/6.+S(K+1)*U*U2*U3/2.
     1 -S(K+2)*U*U1*U3/2.+S(K+3)*U*U1*U2/6.
      YE=SIN(A)
      E=YE-Y
      PRINT 20, A,YE,Y,E
   20 FORMAT(1H , 4F12.8)
      GO TO 5
      END
```

5.8 Flowchart

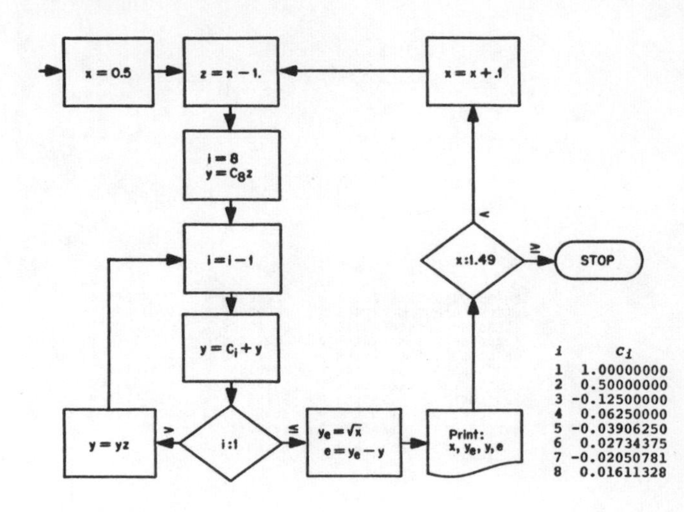

i	C_i
1	1.00000000
2	0.50000000
3	-0.12500000
4	0.06250000
5	-0.03906250
6	0.02734375
7	-0.02050781
8	0.01611328

Program

```
   DIMENSION C(8)
   DATA C/1.0000000,.50000000,-.12500000,.06250000,
  1 -.3906250,.02734375,-.02050781,.01611328/

   X = .5
 1 Z = X - 1.
   I = 8
   Y = C(8)*Z
 2 I = I - 1
   Y = C(I) + Y
   IF(I - 1) 3,3,4
 4 Y = Y*Z
   GO TO 2
 3 YE = SQRT(X)
   E = YE - Y
   PRINT 5, X, YE, Y, E
 5 FORMAT(1H , F4.1, 3F10.6)
   IF(X.GT.1.49) GO TO 6
   X = X + .1
   GO TO 1
 6 STOP
   END
```

Results

x	y_{exact}	y_{series}	error
0.5	0.707107	0.707195	-0.000088
0.6	0.774597	0.774610	-0.000013
0.7	0.836660	0.836661	-0.000001
0.8	0.894427	0.894427	0.000000
0.9	0.948683	0.948683	0.000000
1.0	1.000000	1.000000	0.000000
1.1	1.048809	1.048809	0.000000
1.2	1.095445	1.095445	0.000000
1.3	1.140175	1.140176	-0.000001
1.4	1.183216	1.183222	-0.000006
1.5	1.224745	1.224781	-0.000036

5.9 $$P = K\left[1 + \frac{3}{2}x + \frac{3}{2}e + \frac{3}{8}x^2 + \frac{9}{4}xe + \frac{15}{8}e^2 - \frac{1}{16}x^3 + \frac{9}{16}x^2e + \frac{45}{16}xe^2 + \frac{35}{16}e^3\right]$$

Results

A	y_{exact}	$y_{Lagrange}$	*Error*
-0.01000000	-0.00999983	-0.00999983	0.00000000
0.00500000	0.00499998	0.00499998	-0.00000000
0.46000000	0.44394810	0.44394810	0.00000001
0.55000000	0.52268723	0.52268722	0.00000001
0.78539816	0.70710678	0.70710680	-0.00000002
1.01000001	0.84683184	0.84683188	-0.00000004
1.23450001	0.94398332	0.94398332	0.00000000
1.57079633	1.00000000	0.99999999	0.00000001
1.65000001	0.99686503	0.99686372	0.00000131
1.67000001	0.99508335	0.99507987	0.00000348

CHAPTER 5

5.1 $f(t) = \sum_{n=0}^{\infty} \frac{t^{2n}}{(2)^n n!}$

5.2 $R_4 = \frac{f^V(\xi)t^5}{5!}$; $R_4(1) = -0.0303265$,

Actual error = -0.0184693

5.3 $y = x + x^2 + \frac{x^3}{3} - \frac{x^5}{30} - \frac{x^6}{90} - \frac{x^7}{630}$

5.4 $R_7 = -16e^{\xi} \sin \xi \frac{x^8}{8!}$; $R_7 \Big]_{x = \xi = \pi/2} = -0.07075329$

5.5 $y = \frac{\pi}{4} + \frac{(x-1)}{2} - \frac{(x-1)^2}{4} + \frac{(x-1)^3}{12} - \frac{(x-1)^5}{40}$

$y(1.1) = 0.83298125$; $\tan^{-1}(1.1) = 0.83298127$

5.6 $\theta = 3.801440 - 0.01255037(x - 300) + 0.00004123402(x - 300)^2 - 0.0000001348050(x - 300)^3$

$\theta(280)$ = 4.070019 (series)
4.071003 (third-order interpolation)
4.070097 (exact)

$\theta(350)$ = 3.260156 (series)
3.262316 (third-order interpolation)
3.262520 (exact)

5.7 $y = 1 + \frac{x}{2} + \sum_{n=2}^{\infty} \frac{(-1)^{n-1}(2n-3)!}{2^{2n-2}(n-2)!n!} x^n$

5.10 $F(R_1,R_2) = \frac{R_0}{2} + \frac{R_1 - R_0}{4} + \frac{R_2 - R_0}{4} = \frac{R_1 + R_2}{4}$

5.11 Flowchart

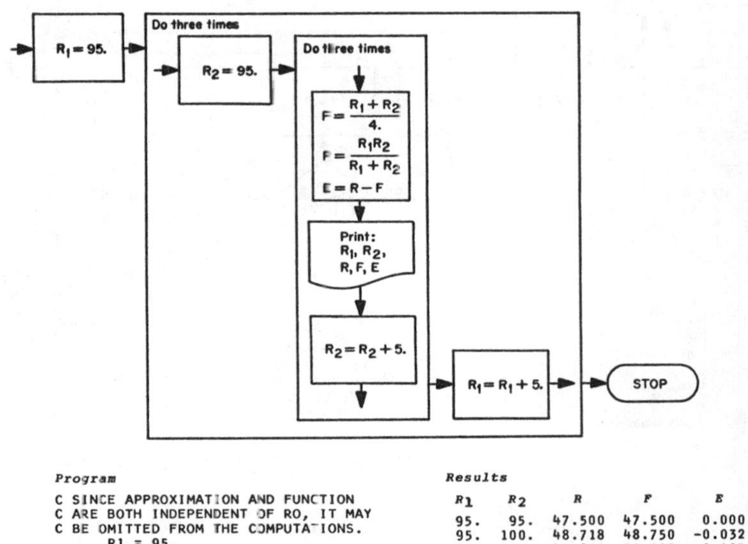

Program

```
C SINCE APPROXIMATION AND FUNCTION
C ARE BOTH INDEPENDENT OF RO, IT MAY
C BE OMITTED FROM THE COMPUTATIONS.
      R1 = 95.
      DO 1 I = 1,3
      R2 = 95.
      DO 2 J = 1,3
      F = (R1 + R2)/4.
      R = R1*R2/(R1 + R2)
      E = R - F
      PRINT 3, R1, R2, R, F, E
    3 FORMAT(1H , 2F5.0, 3F8.3)
    2 R2 = R2 + 5.
    1 R1 = R1 + 5.
      STOP
      END
```

Results

R_1	R_2	*R*	*F*	*E*
95.	95.	47.500	47.500	0.000
95.	100.	48.718	48.750	-0.032
95.	105.	49.875	50.000	-0.125
100.	95.	48.718	48.750	-0.032
100.	100.	50.000	50.000	0.000
100.	105.	51.220	51.250	-0.030
105.	95.	49.875	50.000	-0.125
105.	100.	51.220	51.250	-0.030
105.	105.	52.500	52.500	0.000

5.12 $\left|E_{GN}^{(2)}\right| \le \frac{\sqrt{3}\, h^3 f'''(\xi)}{27}$

5.13 $\left|E_S^{(2)}\right| \le \frac{h^3 f'''(\xi)}{9\sqrt{3}}$

5.14 $\left|E_S^{(3)}\right| \le \dfrac{h^4 f^{iv}(\xi)}{2}$

5.15 $$y_{GN}^{(3)}(4/3) = -\frac{5}{81}f_k + \frac{20}{27}f_{k+1} + \frac{10}{27}f_{k+2} - \frac{4}{81}f_{k+3}$$

$$y_{GN}^{(3)}(5/3) = -\frac{4}{81}f_k + \frac{10}{27}f_{k+1} + \frac{20}{27}f_{k+2} - \frac{5}{81}f_{k+3}$$

$$\bar{y} = -\frac{1}{18}f_k + \frac{10}{9}f_{k+1} + \frac{10}{9}f_{k+2} - \frac{1}{18}f_{k+3}$$

$$\left|E(\bar{y})\right| \le \frac{h^2 f''(\xi)}{72}$$

CHAPTER 6

6.1 See flowchart of Fig. 2.5

```
      PRINT 1
    1 FORMAT(1H1,7X,1HA,13X,1HB,13X,1HC,12X,2HR1,12X,2HX1,
   1          12X,2HR2,12X,2HX2,7X,7HREMARKS)
    2 READ 3, A,B,C
    3 FORMAT(3F10.1)
      IF(A) 4,5,4
    5 IF(B) 6,7,6
    7 PRINT 8, A,B,C
    8 FORMAT(1H ,1P3E14.7,57X,21HNO SOLUTION POSSIBLE.)
      GO TO 2
    6 R1 = - C/B
      PRINT 9, A,B,C,R1
    9 FORMAT(1H ,1P4E14.7,43X,31HTHIS IS A LINEAR EQUATION ONLY.)
      GO TO 2
    4 R = B*B - 4.*A*C
      D = - B/2./A
      IF(R) 10,11,11
   10 X1 = SQRT(- R)/2./A
      X2 = - X1
      R1 = D
      R2 = D
      GO TO 12
   11 E = SQRT(R)/2./A
      R1 = D + E
      R2 = D - E
      X1 = 0.
      X2 = 0.
   12 PRINT 13, A,B,C,R1,X1,R2,X2
   13 FORMAT(1H ,1P7E14.7)
      GO TO 2
      END
```

○THIS IS A LINEAR EQUATION ONLY.

●NO SOLUTION POSSIBLE.

6.2

A	*B*	*C*	*R1*	*X1*	*R2*	*X2*
14.7	-3.2	8.7	0.10884353	0.76157020	0.10884353	-0.76157020
14.7	-3.2	-8.7	0.88581435	0.0	-0.66812728	0.0
○0.0	○-5.6	○4.3	○0.76785714			
-9.8	0.0	6.3	-0.80178373	0.0	0.80178373	0.0
●0.0	●0.0	●8.5				
-16.2	14.7	0.0	0.00000000	0.0	0.90740741	0.0

6.3 Program (continued)

```
    5 BP = -B0/2. - D
      B = (ABS(BP))**.33333333
      IF(BP) 6,7,7
    6 B = -B
    7 Y1 = A + B
      Y2 = -Y1/2.
      Y3 = Y2
      X2 = .86602543*(A - B)
      X3 = -X2
    3 E = A2/3.
      R1 = Y1 - E
      R2 = Y2 - E
      R3 = Y3 - E
      X2 = 0.
      X3 = 0.
      RETURN
      END
```

6.4

A	*B*	*C*	*R1*	*R2*	*X2*	*R3*	*X3*
6.0	9.0	4.0	-4.0	-1.0	0.0	-1.0	0.0
9.7	3.0	-29.7	1.5	-9.0	0.0	-2.2	0.0
4.6	6.85	-21.35	1.4	-3.0	2.5	-3.0	-2.5

6.5 Flowchart (see Exercise 6.3 for subroutine CUBIC)

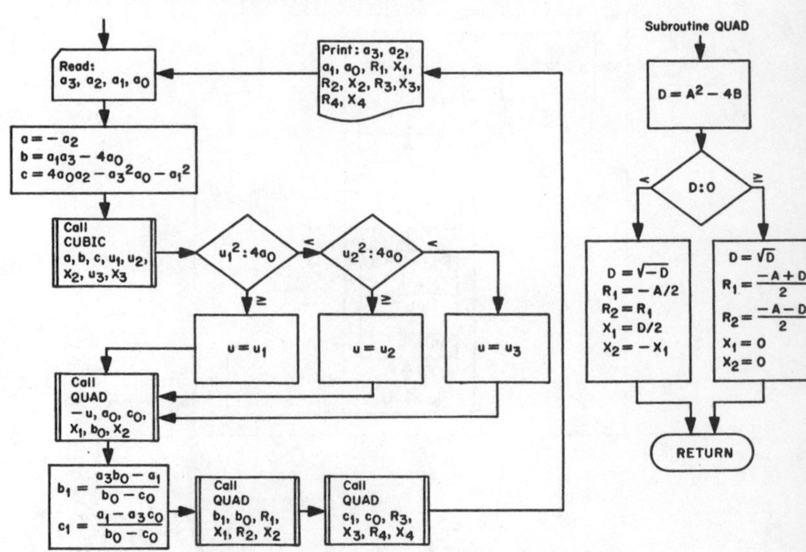

6.3 Flowchart

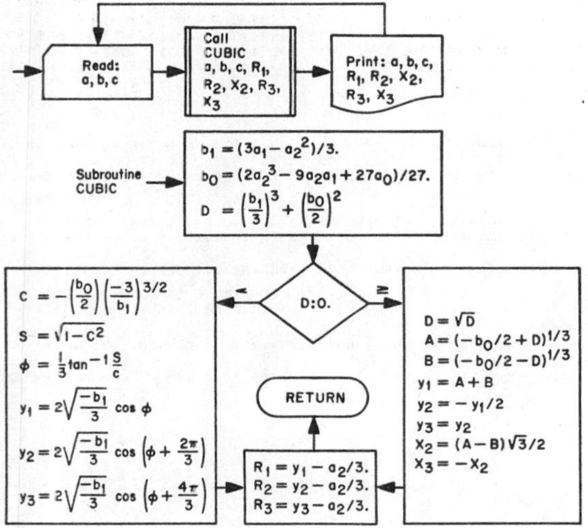

Read: a, b, c → Call CUBIC a, b, c, R_1, R_2, X_2, R_3, X_3 → Print: a, b, c, R_1, R_2, X_2, R_3, X_3

Subroutine CUBIC →

$b_1 = (3a_1 - a_2^2)/3.$

$b_0 = (2a_2^3 - 9a_2a_1 + 27a_0)/27.$

$D = \left(\frac{b_1}{3}\right)^3 + \left(\frac{b_0}{2}\right)^2$

D:O.

<:

$C = -\left(\frac{b_0}{2}\right)\left(\frac{-3}{b_1}\right)^{3/2}$

$S = \sqrt{1-C^2}$

$\phi = \frac{1}{3}\tan^{-1}\frac{S}{c}$

$y_1 = 2\sqrt{\frac{-b_1}{3}}\cos\phi$

$y_2 = 2\sqrt{\frac{-b_1}{3}}\cos\left(\phi + \frac{2\pi}{3}\right)$

$y_3 = 2\sqrt{\frac{-b_1}{3}}\cos\left(\phi + \frac{4\pi}{3}\right)$

≥:

$D = \sqrt{D}$

$A = (-b_0/2 + D)^{1/3}$

$B = (-b_0/2 - D)^{1/3}$

$y_1 = A + B$

$y_2 = -y_1/2$

$y_3 = y_2$

$X_2 = (A - B)\sqrt{3}/2$

$X_3 = -X_2$

$R_1 = y_1 - a_2/3.$

$R_2 = y_2 - a_2/3.$

$R_3 = y_3 - a_2/3.$

RETURN

Programs

```
1 READ 2, A,B,C
2 FORMAT(3F10.0)
  CALL CUBIC(A,B,C,R1,R2,X2,R3,X3)
  PRINT 3, A,B,C,R1,R2,X2,R3,X3
3 FORMAT(1H ,1P8E14.7)
  GO TO 1
  END

  SUBROUTINE CUBIC(A2,A1,A0,R1,R2,X2,R3,X3)
  B1 = (3.*A1 - A2*A2)/3.
  B0 = (2.*A2*A2*A2 - 9.*A2*A1 + 27.*A0)/27.
  D = B1*B1*B1/27. + B0*B0/4.
  IF(D) 1,2,2
1 T = -3./B1
  E = SQRT(T)
  C = - B0/2.*T*E
  S = SQRT(1. - C*C)
  P = ATAN2(S,C)/3
  E = 2./E
  Y1 = E*COS(P)
  Y2 = E*COS(P + 2.0943951)
  Y3 = E*COS(P + 4.1887902)
  GO TO 3
2 D = SQRT(D)
  AP = -B0/2. + D
  A = (ABS(AP))**.33333333
  IF(AP) 4,5,5
4 A = -A
```

Programs

```
1 READ 2, A3,A2,A1,A0
2 FORMAT(4F10.0)
  A = -A2
  B = A1*A3 - 4.*A0
  C = 4.*A0*A2 - A3*A3*A0 - A1*A1
  CALL CUBIC(A,B,C,U1,U2,X2,U3,X3)
  IF(U1*U1 - 4.*A0) 3,4,4
3 IF(U2*U2 - 4.*A0) 5,6,6
5 U = U3
  GO TO 7
6 U = U2
  GO TO 7
4 U = U1
7 CALL QUAD(-U,A0,C0,X1,B0,X2)
  D = B0 - C0
  B1 = (A3*B0 - A1)/D
  C1 = (A1 - A3*C0)/D
  CALL QUAD(B1,B0,R1,X1,R2,X2)
  CALL QUAD(C1,C0,R3,X3,R4,X4)
  PRINT 8, A3,A2,A1,A0,R1,X1,R2,X2,R3,X3,R4,X4
8 FORMAT(1H ,1P4E14.7/1H ,1P8E14.7)
  GO TO 1
  END

  SUBROUTINE QUAD(A,B,R1,X1,R2,X2)
  D = A*A - 4.*B
  IF(D) 1,2,2
1 D = SQRT(-D)
  R1 = -A/2.
  R2 = R1
  X1 = D/2.
  X2 = -X1
  RETURN
2 D = SQRT(D)
  R1 = (-A + D)/2.
  R2 = (-A - D)/2.
  X1 = 0.
  X2 = 0.
  RETURN
  END
```

6.6

a_3	a_2	a_1	a_0	R_1, X_1	R_2, X_2	R_3, X_3	R_4, X_4
8.0	19.25	-17.5	-122.0	2.00000000	-4.00000000	-3.00000000	-3.00000000
				0.00000000	0.00000000	2.50000000	-2.50000000
-2.0	-13.00	14.0	24.0	-1.00000000	-3.00000000	4.00000000	2.00000000
				0.00000000	0.00000000	0.00000000	0.00000000

6.7 **See answers to Exercise 6.4**

6.8 **See answers to Exercise 6.4**

6.9 **See answers to Exercise 6.4**

6.10 **±1.1141571, ±2.7726047, ±6.4391172**

6.11 Flowchart

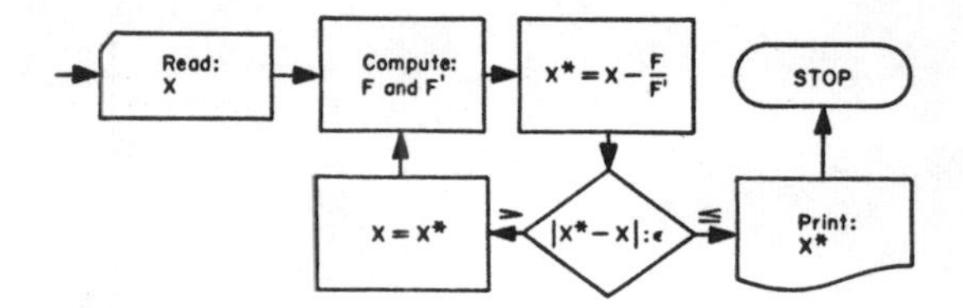

Program

```
C THE FOLLOWING TWO ARITHMETIC STATEMENT FUNCTIONS MAY
C BE CHANGED TO PERMIT THE SOLUTION OF OTHER PROBLEMS.
C THE ONES GIVEN ALLOW THE SOLUTION OF EXERCISE 6.12.
      FUNC(U) = 1./U - SIN(U)
      DERIV(U) = -1./U/U - COS(U)
      READ 1, X
    1 FORMAT(F10.6)
    2 F = FUNC(X)
      FP = DERIV(X)
      XS = X - F/FP
      IF(ABS(XS - X) - .0000001) 4,4,3
    3 X = XS
      GO TO 2
    4 PRINT 5, XS
    5 FORMAT(1H ,1PE14.7)
      STOP
      END
```

6.12 See answers to Exercise 6.10

6.13 Change the arithmetic statement functions of Exercise 6.11 to:

```
FUNC(U) = SIN(U) - EXP(-U)
DERIV(U) = COS(U) + EXP(-U)
```

Answers: 2.9907352, 6.2948049, 56.5486680

6.14 To change the program of Exercise 6.11 to solve for complex roots, it is necessary to add the statement

```
COMPLEX FUNC,U,DERIV,X,F,FP,XS
```

at the beginning of the program and change the ABS function to CABS. Also, the output FORMAT should be changed to allow printing of both real and imaginary parts.

6.15 Case 1: 1.4, -3.0, +2.5i, -3.0 - 2.5i
Case 2: -1.0, i, -i

6.16 Case 1: 2.0, -4.0, -3.0 + 2.5i, -3.0 -2.5i
Case 2: 0.3090170 + 0.951056651i,
0.3090170 - 0.951056651i,
-0.8090170 + 0.587785251i,
-0.8090170 - 0.587785251i

6.18 See answers to Exercise 6.4

6.19 See answers to Exercise 6.6

6.20 Flowchart

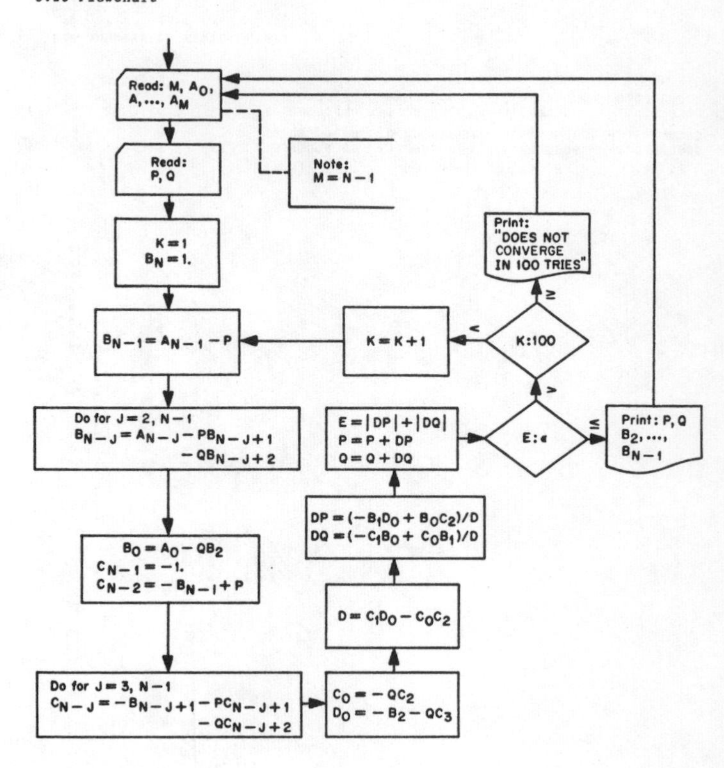

6.17 Flowchart

Program

```
      DIMENSION A(20),B(20)
    1 READ 2, N,(A(I), I = 1,N)
    2 FORMAT(I3/(8F10.0))
      READ 3, P,Q
    3 FORMAT(2F10.0)
      K = 1
    4 J = N - 2
      B(N) = A(N) - P
      B(N - 1) = A(N - 1) - P*B(N) - Q
    5 IF(J.EQ.2) GO TO 6
      L = N - J
      B(L) = A(L) - P*B(L + 1) - Q*B(L + 2)
      J = J - 1
      GO TO 5
    6 QS = A(1)/B(3)
      PS = (A(2) - Q*B(4))/B(3)
      E1 = ABS(QS - Q)
      E2 = ABS(PS - P)
      IF(E1 + E2 - .0000001) 7,7,8
    7 PRINT 9, PS,QS,(B(I), I = 3,N)
    9 FORMAT(1H ,1P2E14.7/(1H ,8E14.7))
      GO TO 1
    8 IF(K.LT.100) GO TO 10
      PRINT 11
   11 FORMAT(32H DOES NOT CONVERGE IN 100 TRIES.)
      GO TO 1
   10 K = K + 1
      P = PS
      Q = QS
      GO TO 4
      END
```

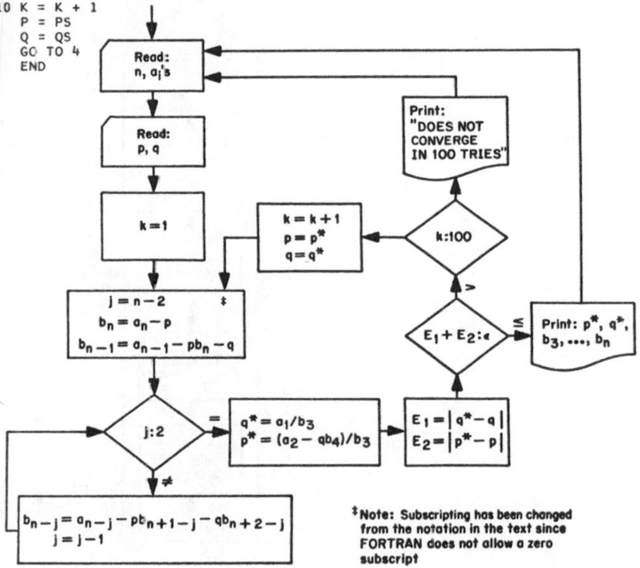

Read: n, a_i's

Read: p, q

$k = 1$

$j = n - 2$ ‡
$b_n = a_n - p$
$b_{n-1} = a_{n-1} - pb_n - q$

$j:2$

$=$: $q^* = a_1/b_3$
$p^* = (a_2 - qb_4)/b_3$

$E_1 = |q^* - q|$
$E_2 = |p^* - p|$

$E_1 + E_2 : \epsilon$

$\leq$: Print: p^*, q^*, $b_3, \ldots, b_n$

$>$: $k:100$

$k = k + 1$
$p = p^*$
$q = q^*$

Print: "DOES NOT CONVERGE IN 100 TRIES"

$\neq$: $b_{n-j} = a_{n-j} - pb_{n+1-j} - qb_{n+2-j}$
$j = j - 1$

‡Note: Subscripting has been changed from the notation in the text since FORTRAN does not allow a zero subscript

Program

```
      DIMENSION A(20),B(20),C(20)
    1 READ 2, M,AZ, (A(I), I = 1,M)
    2 FORMAT(I3/(8F10.6))
      N = M + 1
      READ 3, P,Q
    3 FORMAT(2F10.6)
      K = 1
      B(N) = 1.
    4 B(M) = A(M) - P
      DO 5 J = 2,M
      L = N - J
    5 B(L) = A(L) - P*B(L + 1) - Q*B(L + 2)
      BZ = AZ - Q*B(2)
      C(M) = -1.
      C(N - 2) = -B(N - 1) + P
      DO 6 J = 3,M
      L = N - J
    6 C(L) = -B(L + 1) - P*C(L + 1) - Q*C(L + 2)
      CZ = -Q*C(2)
      DZ = -B(2) - Q*C(3)
      D = C(1)*DZ - CZ*C(2)
      DP = (-B(1)*DZ + BZ*C(2))/D
      DQ = (-C(1)*BZ + CZ*B(1))/D
      E = ABS(DP) + ABS(DQ)
      P = P + DP
      Q = Q + DQ
      IF(E.GT..0000001) GO TO 7
      PRINT 8, P,Q,(B(I), I = 2,M)
    8 FORMAT(1H ,1P2E14.7/(1H ,8E14.7))
      GO TO 1
    7 IF(K.LT.100) GO TO 9
      PRINT 10
   10 FORMAT(32H DOES NOT CONVERGE IN 100 TRIES.)
      GO TO 1
    9 K = K + 1
      GO TO 4
      END
```

6.21 See answers to Exercise 6.6

6.22 $x^2 - 0.34729636x + 1.$
$x^2 + 1.87938525x + 1.$
$x^2 + 1.00000000x + 1.$
$x^2 - 1.53208889x + 1.$

CHAPTER 7

7.1

```
      SUBROUTINE DET4(A, D)
      DIMENSION A(4,4)
      B1 = A(3,3)*A(4,4) - A(4,3)*A(3,4)
      B2 = A(3,2)*A(4,4) - A(3,4)*A(4,2)
      B3 = A(4,3)*A(3,2) - A(3,3)*A(4,2)
      B4 = A(3,1)*A(4,4) - A(3,4)*A(4,1)
      B5 = A(4,3)*A(3,1) - A(3,3)*A(4,1)
      B6 = A(4,2)*A(3,1) - A(3,2)*A(4,1)
      C1 =  A(2,2)*B1 - A(2,3)*B2 + A(2,4)*B3
      C2 = -A(2,1)*B1 + A(2,3)*B4 - A(2,4)*B5
      C3 =  A(2,1)*B2 - A(2,2)*B4 + A(2,4)*B6
      C4 = -A(2,1)*B3 + A(2,2)*B5 - A(2,3)*B6
      D = A(1,1)*C1 + A(1,2)*C2 + A(1,3)*C3 + A(1,4)*C4
      RETURN
      END
```

7.2 Flowchart

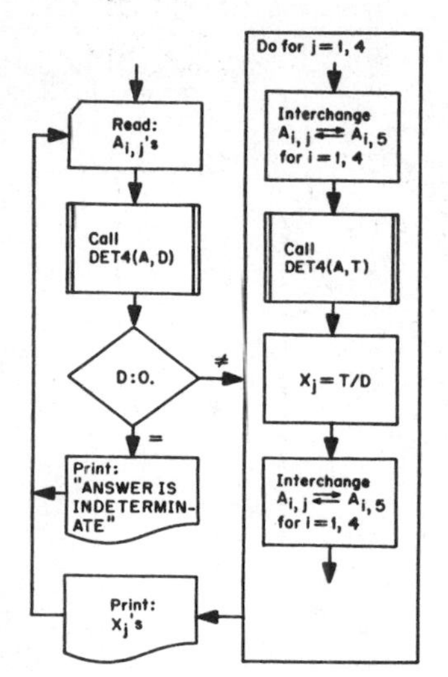

Program

```
  DIMENSION A(4,5), X(4)
1 READ 2, ((A(I,J), J = 1,5), I = 1,4
2 FORMAT(5F10.6)
  CALL DET4(A,D)
  IF(D.EQ.0.) GO TO 7
  DO 3 J = 1,4
  DO 4 I = 1,4
  T = A(I,J)
  A(I,J) = A(I,5)
4 A(I,5) = T
  CALL DET4(A,T)
  X(J) = T/D
  DO 5 I = 1,4
  T = A(I,J)
  A(I,J) = A(I,5)
5 A(I,5) = T
3 CONTINUE
  PRINT 6, X
6 FORMAT(1H ,4E14.7)
  GO TO 1
7 PRINT 8
8 FORMAT(25H ANSWER IS INDETERMINATE.)
  GO TO 1
  END
```

7.4 Program (continued)

```
 3 FORMAT(F10.6)
   DO 4 K = 1,NM1
   P = A(K,K)
   IF(P.NE.0.) GO TO 5
   PRINT 6
 6 FORMAT(42H ZERO DIVISOR.  TRY REARRANGING EQUATIONS.)
   GO TO 1
 5 DO 7 J = K,NP1
 7 A(K,J) = A(K,J)/P
   KP1 = K + 1
   DO 4 I = KP1,N
   R = A(I,K)
   DO 4 J = K,NP1
 4 A(I,J) = A(I,J) - R*A(K,J)
   X(N) = A(N,NP1)/A(N,N)
   I = N - 1
 8 X(I) = A(I,NP1)
   IP1 = I + 1
   DO 9 J = IP1,N
 9 X(I) = X(I) - A(I,J)*X(J)
   I = I - 1
   IF(I.NE.0) GO TO 8
   PRINT 10, (X(J), J = 1,N)
10 FORMAT(1H ,1PE15.7)
   GO TO 1
   END
```

7.5 See answers to Exercise 7.3

7.6 Flowchart

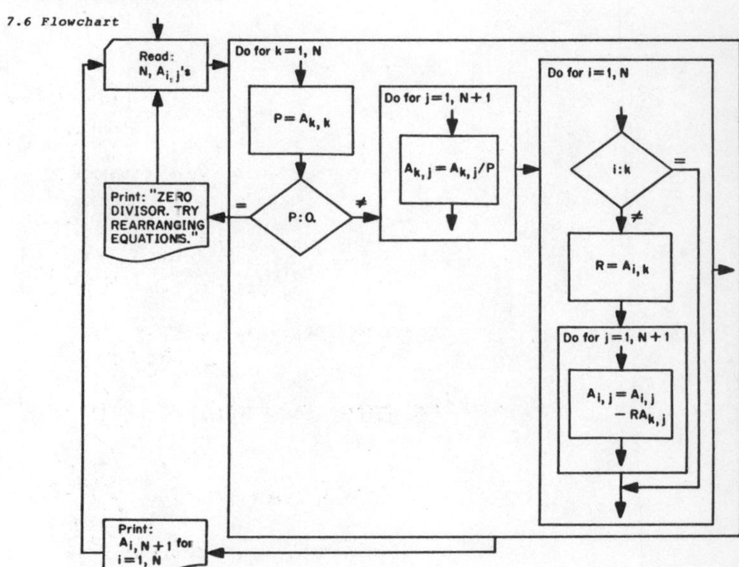

```
7.3  4.000000  -3.000000   2.000000  -1.000000
    ANSWER IS INDETERMINATE
    -1.732051   3.968119  -0.504017  -1.732051
     1.500000   0.500000  -1.250000   0.250000
    -2.000000   7.000000  10.000000   0.000000
```

7.4 Flowchart

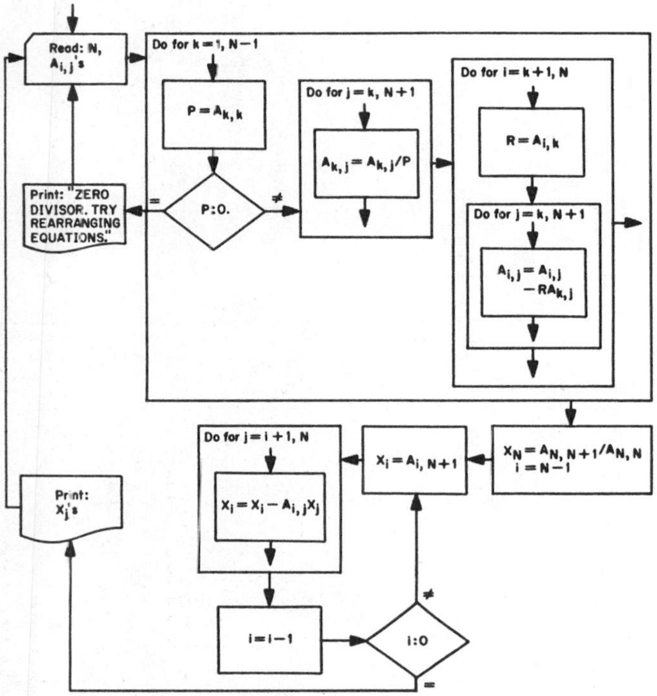

Read: N, $A_{i,j}$'s

Do for $k = 1, N-1$

$P = A_{k,k}$

P:0. (= / ≠)

Print: "ZERO DIVISOR. TRY REARRANGING EQUATIONS."

Do for $j = k, N+1$

$A_{k,j} = A_{k,j}/P$

Do for $i = k+1, N$

$R = A_{i,k}$

Do for $j = k, N+1$

$A_{i,j} = A_{i,j} - RA_{k,j}$

$X_N = A_{N,N+1}/A_{N,N}$; $i = N-1$

$X_i = A_{i,N+1}$

Do for $j = i+1, N$

$X_i = X_i - A_{i,j}X_j$

$i = i-1$

i:0 (≠ / =)

Print: X_j's

Program

```
   DIMENSION A(8,9), X(9)
 1 READ 2, N
 2 FORMAT(I3)
   NP1 = N + 1
   NM1 = N - 1
   READ 3, ((A(I,J),J = 1,NP1), I = 1,N)
```

Program

```
   DIMENSION A(8,9)
 1 READ 2, N
 2 FORMAT(I3)
   NP1 = N + 1
   READ 3, ((A(I,J),J = 1,NP1), I = 1,N)
 3 FORMAT(F10.6)
   DO 4 K = 1,N
   P = A(K,K)
   IF(P.NE.0.) GO TO 5
   PRINT 6
 6 FORMAT(42H ZERO DIVISOR.  TRY REARRANGING EQUATIONS.)
   GO TO 1
 5 DO 7 J = 1, NP1
 7 A(K,J) = A(K,J)/P
   DO 8 I = 1,N
   IF(I.EQ.K) GO TO 8
   R = A(I,K)
   DO 9 J = 1,NP1
 9 A(I,J) = A(I,J) - R*A(K,J)
 8 CONTINUE
 4 CONTINUE
   PRINT 10, (A(I,NP1), I = 1,N)
10 FORMAT(1H ,1PE15.7)
   GO TO 1
   END
```

7.7 See answers to Exercise 7.3

7.8 $x_1 = -3.$, $x_2 = 2.$, $x_3 = -1.$, $x_4 = 0.$, $x_5 = 1.$, $x_6 = -2.$, $x_7 = 3.$, $x_8 = -4.$

7.9 Flowchart

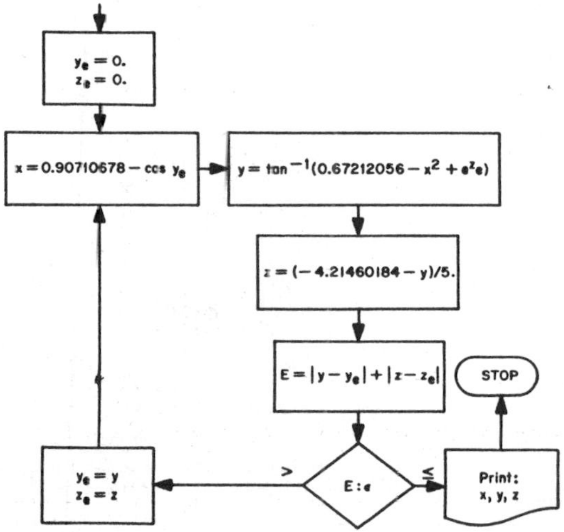

$y_e = 0.$; $z_e = 0.$

$x = 0.90710678 - \cos y_e$

$y = \tan^{-1}(0.67212056 - x^2 + e^{z_e})$

$z = (-4.21460184 - y)/5.$

$E = |y - y_e| + |z - z_e|$

E:ε (> / ≤)

$y_e = y$; $z_e = z$

Print: x, y, z

STOP

7.9 Program

```
      YE = 0.
      ZE = 0.
    1 X = .90710678 - COS(YE)
      Y = ATAN(.67212056 - X*X + EXP(ZE))
      Z = (-4.21460184 - Y)/5.
      IF(ABS(Y - YE) + ABS(Z - ZE) .LE. .00000002) GO TO 2
      YE = Y
      ZE = Z
      GO TO 1
    2 PRINT 3, X,Y,Z
    3 FORMAT(1H ,1P3E15.7)
      STOP
      END
```

Results: x = 0.20000000, y = 0.78539816, z = -1.00000000

7.10 Flowchart

$x = 0.$, $y = 0.$, $z = 0.$

$F_1 = x + \cos y - 0.90710678$
$F_2 = x^2 + \tan y - e^z - 0.67212056$
$F_3 = y + 5z + 4.21460184$
$D = 5\sec^2 y + 10x \sin y + e^z$

$\Delta x = [-F_1(5\sec^2 y + e^z) - 5F_2 \sin y - F_3 e^z \sin y]/D$
$\Delta y = (10xF_1 - 5F_2 - e^z F_3)/D$
$\Delta z = [-2xF_1 + F_2 - F_3(\sec^2 y + 2x \sin y)]/D$

$x = x + \Delta x$
$y = y + \Delta y$
$z = z + \Delta z$

$|\Delta x| + |\Delta y| + |\Delta z| : \epsilon$ ($>$ / $\leq$)

Print: x, y, z

STOP

Program

```
      X = 0.
      Y = 0.
      Z = 0.
    1 SY = SIN(Y)
      CY = COS(Y)
      EZ = EXP(Z)
      S2Y = 1./CY/CY
      F1 = X + CY - .90710678
      F2 = X*X + SY/CY - EZ - .67212056
      F3 = Y + 5.*Z + 4.21460184
      D = 5.*S2Y + 10.*X*SY + EZ
      DX = (-F1*(5.*S2Y + EZ) - 5.*F2*SY -F3*EZ*SY)/D
      DY = (10.*X*F1 - 5.*F2 -EZ*F3)/D
      DZ = (-2.*X*F1 + F2 - F3*(S2Y + 2.*X*SY))/D
      X = X + DX
      Y = Y + DY
      Z = Z + DZ
      IF(ABS(DX) + ABS(DY) + ABS(DZ).GT..000003) GO TO 1
      PRINT 2, X,Y,Z
    2 FORMAT(1H ,1P3E15.7)
      STOP
      END
```

For results see Exercise 7.9

7.13

a) $\begin{bmatrix} \frac{8}{17} & \frac{31}{34} & \frac{29}{34} & -1 \\ -\frac{6}{17} & -\frac{18}{17} & -\frac{13}{17} & 1 \\ \frac{1}{17} & -\frac{14}{17} & -\frac{12}{17} & 1 \\ -\frac{1}{17} & \frac{11}{34} & \frac{7}{34} & 0 \end{bmatrix}$ **b) No solution possible**

c)
d) $\begin{bmatrix} \frac{3}{2} & -1 & 0 & \frac{1}{2} \\ -\frac{5}{2} & 1 & 1 & -\frac{1}{2} \\ \frac{3}{4} & 1 & -1 & -\frac{1}{4} \\ \frac{5}{4} & -1 & 0 & \frac{1}{4} \end{bmatrix}$ e) $\begin{bmatrix} -1 & 1 & 1 \\ 5 & -3 & -4 \\ 7 & -4 & -5 \end{bmatrix}$

7.14

```
      DIMENSION A(8,16),C(8),X(8)
    1 READ 2, N,((A(I,J),I=1,N),J=1,N)
    2 FORMAT(I3/(F10.0))
      READ 3, (C(I),I=1,N)
    3 FORMAT(F10.0)
      CALL INVERT(N,A)
      DO 4 I = 1,N
      DO 4 J = 1,N
      K = J + N
    4 A(I,J) = A(I,K)
      CALL VMULT(N,A,C,X)
      PRINT 5, (X(I), I=1,N)
    5 FORMAT(1H1/1H ,(1PE15.7))
      GO TO 1
      END
```

7.15 **See answers to Exercises 7.3 and 7.8**

7.16 Flowchart

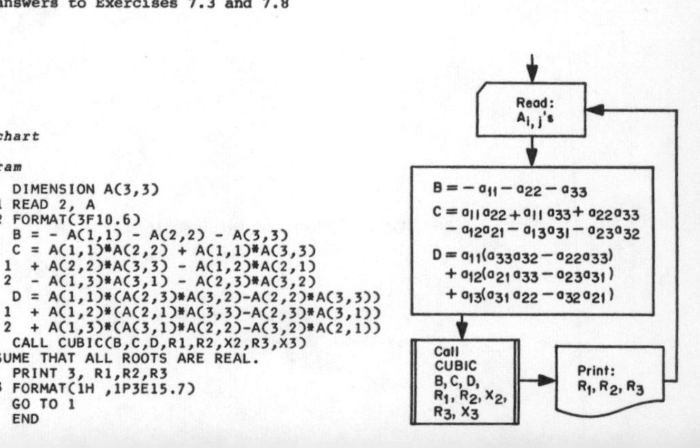

Program

```
      DIMENSION A(3,3)
    1 READ 2, A
    2 FORMAT(3F10.6)
      B = - A(1,1) - A(2,2) - A(3,3)
      C = A(1,1)*A(2,2) + A(1,1)*A(3,3)
     1 + A(2,2)*A(3,3) - A(1,2)*A(2,1)
     2 - A(1,3)*A(3,1) - A(2,3)*A(3,2)
      D = A(1,1)*(A(2,3)*A(3,2)-A(2,2)*A(3,3))
     1 + A(1,2)*(A(2,1)*A(3,3)-A(2,3)*A(3,1))
     2 + A(1,3)*(A(3,1)*A(2,2)-A(3,2)*A(2,1))
      CALL CUBIC(B,C,D,R1,R2,X2,R3,X3)
C ASSUME THAT ALL ROOTS ARE REAL.
      PRINT 3, R1,R2,R3
    3 FORMAT(1H ,1P3E15.7)
      GO TO 1
      END
```

7.11 Flowchart

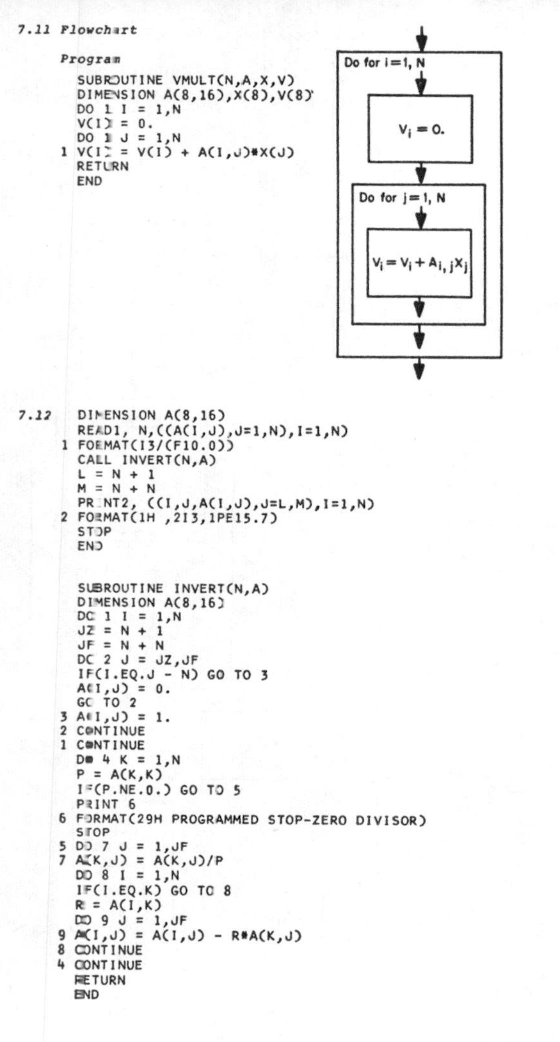

Program

```
   SUBROUTINE VMULT(N,A,X,V)
   DIMENSION A(8,16),X(8),V(8)
   DO 1 I = 1,N
   V(I) = 0.
   DO 1 J = 1,N
 1 V(I) = V(I) + A(I,J)*X(J)
   RETURN
   END
```

7.12

```
   DIMENSION A(8,16)
   READ1, N,((A(I,J),J=1,N),I=1,N)
 1 FORMAT(I3/(F10.0))
   CALL INVERT(N,A)
   L = N + 1
   M = N + N
   PRINT2, ((I,J,A(I,J),J=L,M),I=1,N)
 2 FORMAT(1H ,2I3,1PE15.7)
   STOP
   END

   SUBROUTINE INVERT(N,A)
   DIMENSION A(8,16)
   DO 1 I = 1,N
   JZ = N + 1
   JF = N + N
   DO 2 J = JZ,JF
   IF(I.EQ.J - N) GO TO 3
   A(I,J) = 0.
   GO TO 2
 3 A(I,J) = 1.
 2 CONTINUE
 1 CONTINUE
   DO 4 K = 1,N
   P = A(K,K)
   IF(P.NE.0.) GO TO 5
   PRINT 6
 6 FORMAT(29H PROGRAMMED STOP-ZERO DIVISOR)
   STOP
 5 DO 7 J = 1,JF
 7 A(K,J) = A(K,J)/P
   DO 8 I = 1,N
   IF(I.EQ.K) GO TO 8
   R = A(I,K)
   DO 9 J = 1,JF
 9 A(I,J) = A(I,J) - R*A(K,J)
 8 CONTINUE
 4 CONTINUE
   RETURN
   END
```

7.17 Flowchart

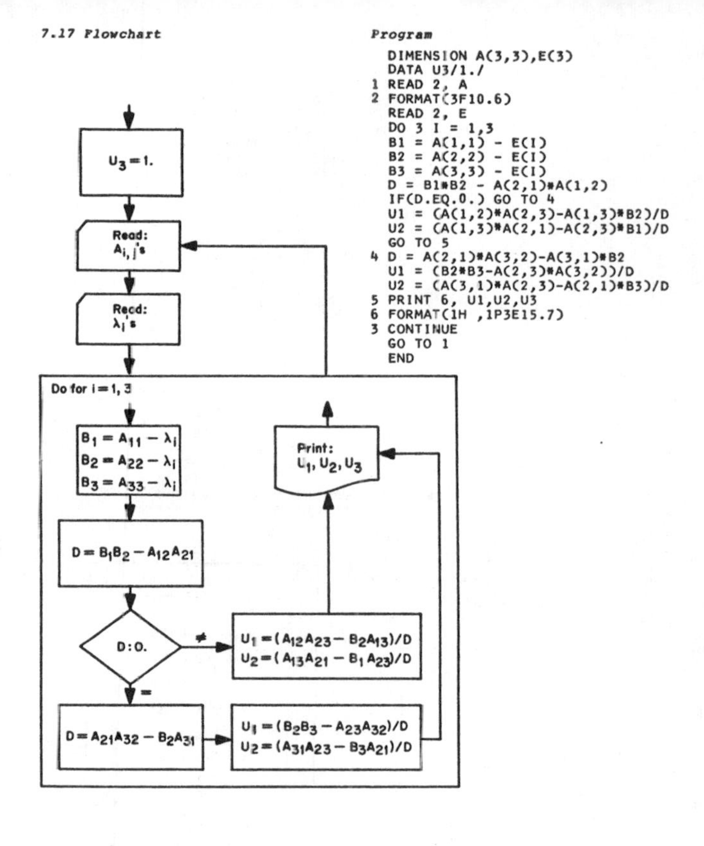

Program

```
   DIMENSION A(3,3),E(3)
   DATA U3/1./
 1 READ 2, A
 2 FORMAT(3F10.6)
   READ 2, E
   DO 3 I = 1,3
   B1 = A(1,1) - E(I)
   B2 = A(2,2) - E(I)
   B3 = A(3,3) - E(I)
   D = B1*B2 - A(2,1)*A(1,2)
   IF(D.EQ.0.) GO TO 4
   U1 = (A(1,2)*A(2,3)-A(1,3)*B2)/D
   U2 = (A(1,3)*A(2,1)-A(2,3)*B1)/D
   GO TO 5
 4 D = A(2,1)*A(3,2)-A(3,1)*B2
   U1 = (B2*B3-A(2,3)*A(3,2))/D
   U2 = (A(3,1)*A(2,3)-A(2,1)*B3)/D
 5 PRINT 6, U1,U2,U3
 6 FORMAT(1H ,1P3E15.7)
 3 CONTINUE
   GO TO 1
   END
```

7.18 $\lambda_1 = -5.$ $\quad\lambda_2 = 2.5$ $\quad\lambda_3 = 1.7$

$$\begin{bmatrix} \frac{-10}{11} \\ \frac{-26}{11} \\ 1 \end{bmatrix} \qquad \begin{bmatrix} \frac{5}{2} \\ \frac{1}{2} \\ 1 \end{bmatrix} \qquad \begin{bmatrix} \frac{-3}{10} \\ \frac{-5}{10} \\ 1 \end{bmatrix}$$

7.19 Flowchart

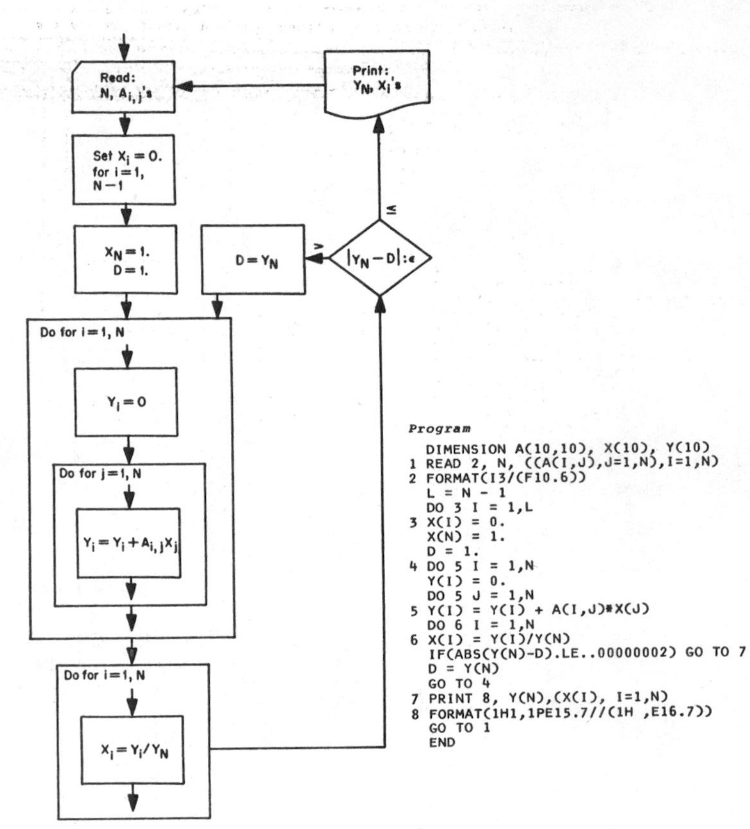

Program

```
  DIMENSION A(10,10), X(10), Y(10)
1 READ 2, N, ((A(I,J),J=1,N),I=1,N)
2 FORMAT(I3/(F10.6))
  L = N - 1
  DO 3 I = 1,L
3 X(I) = 0.
  X(N) = 1.
  D = 1.
4 DO 5 I = 1,N
  Y(I) = 0.
  DO 5 J = 1,N
5 Y(I) = Y(I) + A(I,J)*X(J)
  DO 6 I = 1,N
6 X(I) = Y(I)/Y(N)
  IF(ABS(Y(N)-D).LE..00000002) GO TO 7
  D = Y(N)
  GO TO 4
7 PRINT 8, Y(N),(X(I), I=1,N)
8 FORMAT(1H1,1PE15.7//(1H ,E16.7))
  GO TO 1
  END
```

7.20 Largest eigenvalue = 8.77500181

$$\text{Corresponding eigenvector} = \begin{bmatrix} 0.58924726 \\ 1.20145201 \\ 2.37994654 \\ 1.00000000 \end{bmatrix}$$

8.3 Flowchart

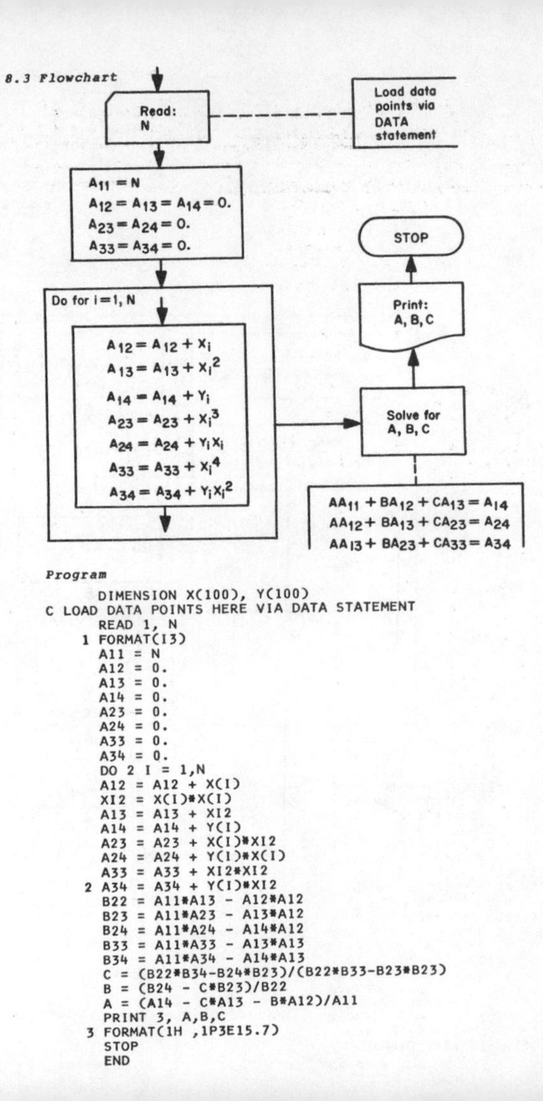

Program

```
      DIMENSION X(100), Y(100)
C LOAD DATA POINTS HERE VIA DATA STATEMENT
      READ 1, N
    1 FORMAT(I3)
      A11 = N
      A12 = 0.
      A13 = 0.
      A14 = 0.
      A23 = 0.
      A24 = 0.
      A33 = 0.
      A34 = 0.
      DO 2 I = 1,N
      A12 = A12 + X(I)
      XI2 = X(I)*X(I)
      A13 = A13 + XI2
      A14 = A14 + Y(I)
      A23 = A23 + X(I)*XI2
      A24 = A24 + Y(I)*X(I)
      A33 = A33 + XI2*XI2
    2 A34 = A34 + Y(I)*XI2
      B22 = A11*A13 - A12*A12
      B23 = A11*A23 - A13*A12
      B24 = A11*A24 - A14*A12
      B33 = A11*A33 - A13*A13
      B34 = A11*A34 - A14*A13
      C = (B22*B34-B24*B23)/(B22*B33-B23*B23)
      B = (B24 - C*B23)/B22
      A = (A14 - C*A13 - B*A12)/A11
      PRINT 3, A,B,C
    3 FORMAT(1H ,1P3E15.7)
      STOP
      END
```

8.1 Flowchart

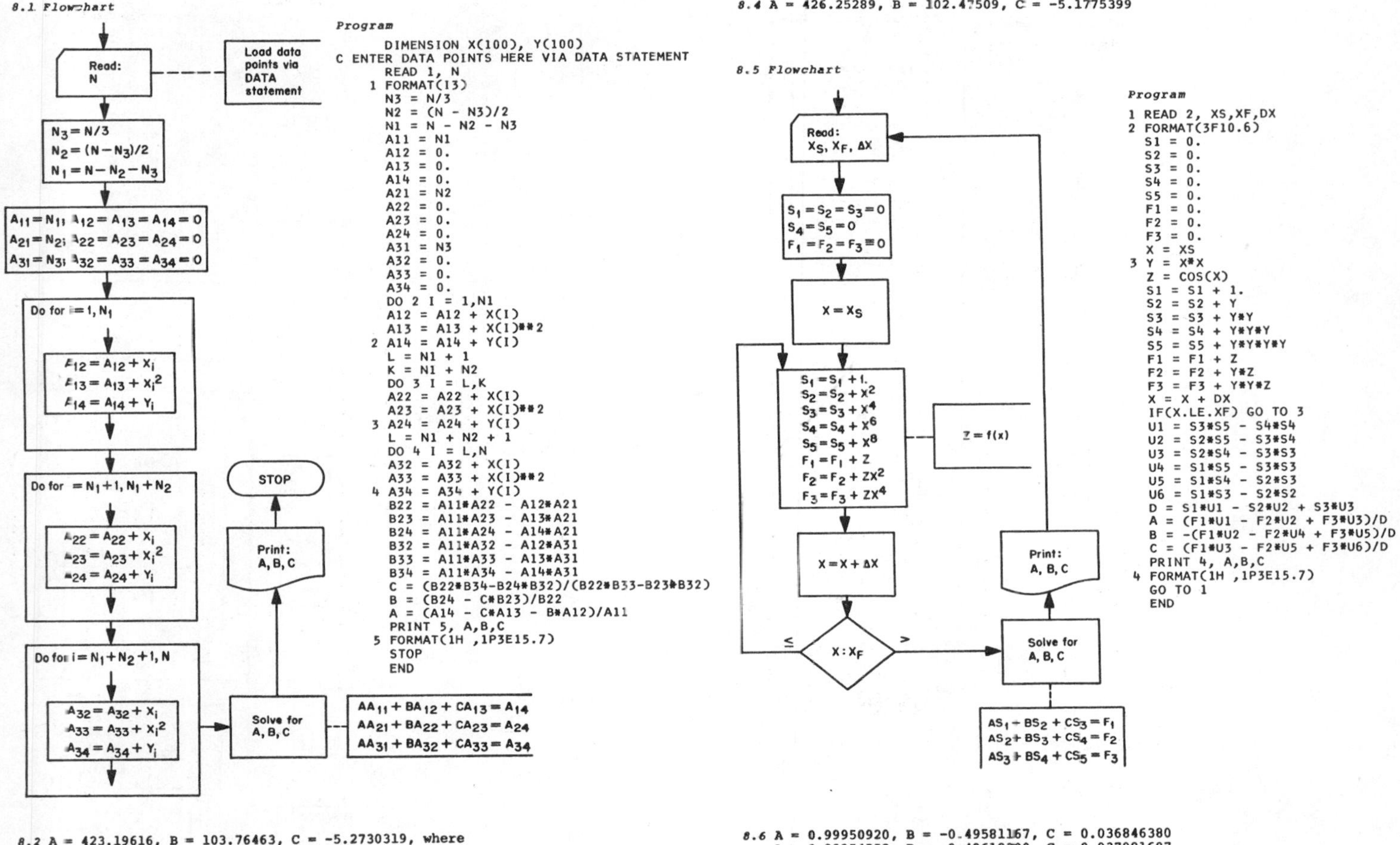

Program

```
      DIMENSION X(100), Y(100)
C ENTER DATA POINTS HERE VIA DATA STATEMENT
      READ 1, N
    1 FORMAT(I3)
      N3 = N/3
      N2 = (N - N3)/2
      N1 = N - N2 - N3
      A11 = N1
      A12 = 0.
      A13 = 0.
      A14 = 0.
      A21 = N2
      A22 = 0.
      A23 = 0.
      A24 = 0.
      A31 = N3
      A32 = 0.
      A33 = 0.
      A34 = 0.
      DO 2 I = 1,N1
      A12 = A12 + X(I)
      A13 = A13 + X(I)**2
    2 A14 = A14 + Y(I)
      L = N1 + 1
      K = N1 + N2
      DO 3 I = L,K
      A22 = A22 + X(I)
      A23 = A23 + X(I)**2
    3 A24 = A24 + Y(I)
      L = N1 + N2 + 1
      DO 4 I = L,N
      A32 = A32 + X(I)
      A33 = A33 + X(I)**2
    4 A34 = A34 + Y(I)
      B22 = A11*A22 - A12*A21
      B23 = A11*A23 - A13*A21
      B24 = A11*A24 - A14*A21
      B32 = A11*A32 - A12*A31
      B33 = A11*A33 - A13*A31
      B34 = A11*A34 - A14*A31
      C = (B22*B34-B24*B32)/(B22*B33-B23*B32)
      B = (B24 - C*B23)/B22
      A = (A14 - C*A13 - B*A12)/A11
      PRINT 5, A,B,C
    5 FORMAT(1H ,1P3E15.7)
      STOP
      END
```

8.2 A = 423.19616, B = 103.76463, C = -5.2730319, where
y = A + Bx + Cx2

8.4 A = 426.25289, B = 102.47509, C = -5.1775399

8.5 Flowchart

Program

```
1 READ 2, XS,XF,DX
2 FORMAT(3F10.6)
  S1 = 0.
  S2 = 0.
  S3 = 0.
  S4 = 0.
  S5 = 0.
  F1 = 0.
  F2 = 0.
  F3 = 0.
  X = XS
3 Y = X*X
  Z = COS(X)
  S1 = S1 + 1.
  S2 = S2 + Y
  S3 = S3 + Y*Y
  S4 = S4 + Y*Y*Y
  S5 = S5 + Y*Y*Y*Y
  F1 = F1 + Z
  F2 = F2 + Y*Z
  F3 = F3 + Y*Y*Z
  X = X + DX
  IF(X.LE.XF) GO TO 3
  U1 = S3*S5 - S4*S4
  U2 = S2*S5 - S3*S4
  U3 = S2*S4 - S3*S3
  U4 = S1*S5 - S3*S3
  U5 = S1*S4 - S2*S3
  U6 = S1*S3 - S2*S2
  D = S1*U1 - S2*U2 + S3*U3
  A = (F1*U1 - F2*U2 + F3*U3)/D
  B = -(F1*U2 - F2*U4 + F3*U5)/D
  C = (F1*U3 - F2*U5 + F3*U6)/D
  PRINT 4, A,B,C
4 FORMAT(1H ,1P3E15.7)
  GO TO 1
  END
```

8.6 A = 0.99950920, B = -0.49581167, C = 0.036846380
A = 0.99954552, B = -0.49618300, C = 0.037081607

8.7 Flowchart

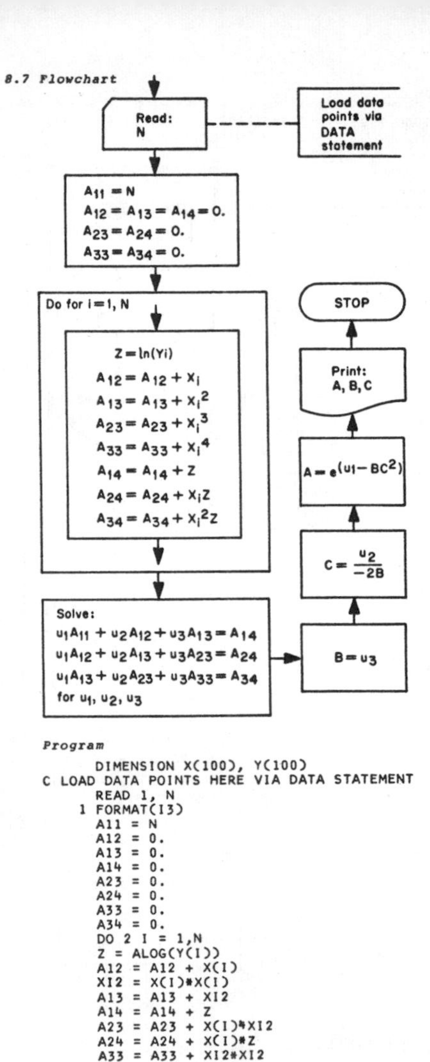

Program

```
      DIMENSION X(100), Y(100)
C LOAD DATA POINTS HERE VIA DATA STATEMENT
      READ 1, N
    1 FORMAT(I3)
      A11 = N
      A12 = 0.
      A13 = 0.
      A14 = 0.
      A23 = 0.
      A24 = 0.
      A33 = 0.
      A34 = 0.
      DO 2 I = 1,N
      Z = ALOG(Y(I))
      A12 = A12 + X(I)
      XI2 = X(I)*X(I)
      A13 = A13 + XI2
      A14 = A14 + Z
      A23 = A23 + X(I)*XI2
      A24 = A24 + X(I)*Z
      A33 = A33 + XI2*XI2
```

8.9 Program (continued)

```
      C = C + DC
      IF(ABS(DA/A) + ABS(DB/B) + ABS(DC/C) -.00000002)4,4,5
    4 PRINT 6, A,B,C
    6 FORMAT(1H ,1P3E14.7)
    7 STOP
    5 IF(K.LT.100) GO TO 8
      PRINT 9
    9 FORMAT(32H DOES NOT CONVERGE IN 100 TRIES.)
      GO TO 7
    8 K = K + 1
      GO TO 1
      END
```

Flowchart

```
    2 A34 = A34 + X12*Z
      B22 = A11*A13 - A12*A12
      B23 = A11*A23 - A13*A12
      B24 = A11*A24 - A14*A12
      B33 = A11*A33 - A13*A13
      B34 = A11*A34 - A14*A13
      D = B22*B33 - B23*B23
      U3 = (B22*B34 - B24*B23)/D
      U2 = (B24 - U3*B23)/B22
      U1 = (A14 - U3*A13 - U2*A12)/A11
      B = U3
      C = U2/(-2.*B)
      A = EXP(U1 - B*C*C)
      PRINT 3, A,B,C
    3 FORMAT(1H ,1P3E15.7)
      STOP
      END
```

8.8 A = 943.96137, B = -0.0085282872, C = 9.3231150

8.9 Program

```
      DIMENSION X(25), Y(25)
      DATA X/.0,.5,1.,1.5,2.,2.5,3.,3.5,4.,4.5,5.,5.5,6.,6.5,
     1       7.,7.5,8.,8.5,9.,9.5,10.,10.5,11.,11.5,12./
      DATA Y/439.6,477.5,515.5,563.9,603.8,
     1       647.5,686.4,718.7,754.7,785.1,
     2       810.8,836.4,855.4,874.4,889.6,
     3       902.8,916.1,924.7,930.4,932.3,
     4       934.2,931.3,927.5,918.0,908.5/
      READ 2,A,B,C
      K = 1
    2 FORMAT(3F10.0)
    1 S1 = 0.
      S2 = 0.
      S3 = 0.
      S4 = 0.
      S5 = 0.
      S6 = 0.
      F1 = 0.
      F2 = 0.
      F3 = 0.
      DO 3 I =1,25
      XC = X(I) - C
      YA = EXP(B*XC*XC)
      DY = Y(I) - A*YA
      YB = A*YA*XC*XC
      YC = -2.*B*XC*Z*YA
      S1 = S1 + YA*YA
      S2 = S2 + YA*YB
      S3 = S3 + YA*YC
      S4 = S4 + YB*YB
      S5 = S5 + YB*YC
      S6 = S6 + YC*YC
      F1 = F1 + YA*DY
      F2 = F2 + YB*DY
    3 F3 = F3 + YC*DY
      U1 = S4*S6 - S5*S5
      U2 = S2*S6 - S3*S5
      U3 = S2*S5 - S3*S4
      U4 = S1*S6 - S3*S3
      U5 = S1*S5 - S2*S3
      U6 = S1*S4 - S2*S2
      D = S1*U1 - S2*U2 + S3*U3
      DA = (F1*U1 - F2*U2 + F3*U3)/D
      DB = -(F1*U2 - F2*U4 + F3*U5)/D
      DC = (F1*U3 - F2*U5 + F3*U6)/D
      A = A + DA
      B = B + DB
```

This program can be used for fitting any function having three unknown parameters to a set of data points. This can be accomplished by replacing the statements marked with braces with other appropriate statements. The statements shown are for the solution of Exercise 8.10.

8.10 A = 941.43047, B = -0.0080812417, C = 9.4504930

8.11 A = 0.99957946, B = -0.49639049, C = 0.03720923

8.12 $A = 4/\pi$, $B = A/9$, $C = A/25$

8.13 $a_0 = 1$, $a_1 = 0$, $a_2 = -4\sqrt{2}/\pi^2$, $a_3 = 0$, $a_4 = 0$, $a_5 = 0$
$a_6 = 4\sqrt{2}/(9\pi^2)$, $a_7 = 0$, $a_8 = 0$
Therefore, $f(x) = \frac{1}{2} - \frac{4\sqrt{2}}{\pi^2}\left(\cos 2x - \frac{1}{9}\cos 6x\right)$

8.14

1.0
0
0
π

8.15 A = 0.99939829, B = -0.49556969, C = 0.036790487

9.1 Flowchart

Program

```
      DIMENSION Y(150), W(4), YB(150)
      DATA N/150/
      READ 1, Y
    1 FORMAT(10F5.0)
    2 READ 3, WZ, W
    3 FORMAT(5F12.8)
      DO 4 I = 1,4
    4 YB(I) = Y(I)
      L = N - 3
      DO 5 I = L,N
    5 YB(I) = Y(I)
      L = N - 4
      DO 6 I = 5,L
      YB(I) = WZ*Y(I)
      DO 6 J = 1,4
      IMJ = I - J
      IPJ = I + J
    6 YB(I) = YB(I) + W(J)*(Y(IMJ) + Y(IPJ))
      PRINT 7, YB
    7 FORMAT(1H1/(1H ,10F6.0))
      GO TO 2
      END
```

Read: w_0, w_1, w_2, w_3, w_4

Print: $\bar{Y}$ array

Move $Y_i \rightarrow \bar{Y}_i$ for $i = 1, 4$

Move $Y_i \rightarrow \bar{Y}_i$ for $i = N-3, N$

Do for $i = 5, N-4$

$\bar{Y}_i = w_0 Y_i$

Do for $j = 1, 4$

$\bar{Y}_i = \bar{Y}_i + w_j(Y_{i-j} + Y_{i+j})$

9.2 $w_0 = 0.25541126$, $w_1 = 0.23376623$, $w_2 = 0.16883117$, $w_3 = 0.060606061$, $w_4 = -0.090909091$

9.3

```
 287.  311.  335.  354.  366.  374.  377.  376.  373.  368.
 363.  356.  348.  341.  341.  351.  371.  396.  420.  438.
 445.  445.  442.  439.  439.  443.  449.  459.  473.  492.
 518.  553.  594.  639.  682.  719.  747.  766.  777.  780.
 780.  774.  765.  748.  725.  699.  671.  651.  640.  649.
 679.  730.  795.  865.  926.  971.  992.  995.  984.  967.
 948.  932.  918.  906.  899.  895.  895.  899.  907.  921.
 942.  969. 1000. 1032. 1063. 1090. 1109. 1122. 1128. 1128.
1123. 1116. 1107. 1096. 1087. 1079. 1075. 1074. 1077. 1085.
1034. 1106. 1117. 1129. 1143. 1159. 1175. 1188. 1199. 1210.
1221. 1232. 1246. 1257. 1265. 1271. 1280. 1297. 1320. 1347.
1373. 1390. 1400. 1403. 1402. 1401. 1408. 1424. 1445. 1465.
1481. 1490. 1490. 1489. 1494. 1506. 1523. 1542. 1561. 1575.
1584. 1590. 1591. 1591. 1589. 1589. 1588. 1588. 1587. 1584.
1583. 1587. 1601. 1621. 1647. 1674. 1696. 1714. 1715. 1714.
```

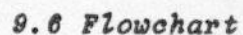

9.6 Flowchart

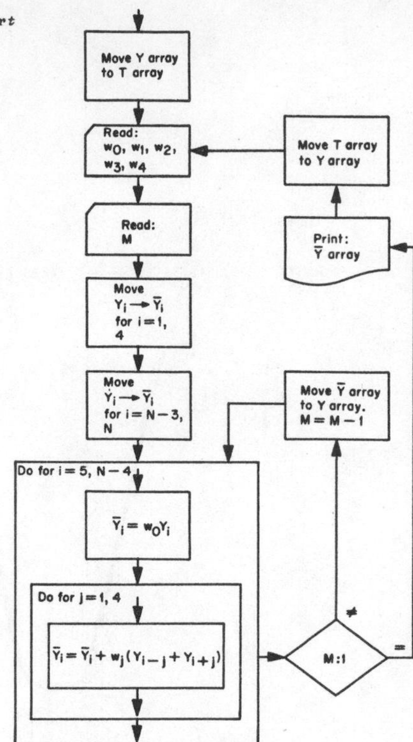

9.7

```
 287.  311.  335.  354.  366.  374.  377.  376.  374.  369.
 362.  354.  347.  343.  345.  355.  373.  395.  417.  433.
 442.  445.  443.  441.  440.  442.  448.  458.  473.  493.
 520.  555.  596.  639.  680.  717.  745.  765.  777.  781.
 781.  775.  763.  746.  723.  697.  671.  651.  643.  654.
 685.  734.  797.  862.  921.  964.  988.  994.  985.  969.
 951.  933.  918.  907.  899.  895.  895.  899.  908.  923.
 943.  970. 1000. 1032. 1062. 1088. 1108. 1121. 1127. 1128.
1123. 1116. 1106. 1096. 1087. 1080. 1075. 1075. 1078. 1085.
1094. 1105. 1117. 1130. 1144. 1159. 1173. 1187. 1199. 1211.
1222. 1234. 1245. 1254. 1262. 1271. 1283. 1300. 1322. 1346.
1369. 1387. 1398. 1401. 1402. 1405. 1413. 1426. 1444. 1463.
1477. 1485. 1489. 1492. 1498. 1508. 1523. 1542. 1559. 1574.
1584. 1589. 1591. 1591. 1590. 1589. 1588. 1587. 1585. 1583.
1584. 1590. 1603. 1623. 1648. 1673. 1696. 1714. 1715. 1714.
```

9.4 $w_0 = 0.41724942$, $w_1 = 0.31468531$, $w_2 = 0.069930070$, $w_3 = -0.12820513$, $w_4 = 0.034965035$

9.5

```
 287.  311.  335.  354.  368.  375.  377.  375.  372.  368.
 364.  358.  351.  342.  337.  343.  366.  398.  427.  443.
 447.  443.  440.  438.  439.  443.  449.  459.  473.  492.
 517.  550.  591.  639.  685.  723.  749.  765.  775.  780.
 780.  774.  763.  750.  730.  702.  671.  646.  638.  646.
 674.  723.  792.  867.  936.  979.  997.  994.  982.  964.
 948.  932.  917.  906.  899.  896.  896.  899.  908.  920.
 940.  967.  999. 1033. 1065. 1090. 1111. 1123. 1128. 1128.
1123. 1116. 1107. 1097. 1087. 1078. 1073. 1073. 1078. 1085.
1094. 1105. 1118. 1131. 1141. 1155. 1174. 1193. 1204. 1209.
1215. 1229. 1248. 1263. 1268. 1269. 1274. 1290. 1319. 1350.
1376. 1394. 1401. 1403. 1403. 1402. 1404. 1416. 1441. 1471.
1491. 1493. 1487. 1485. 1490. 1504. 1523. 1544. 1562. 1576.
1586. 1590. 1591. 1590. 1589. 1588. 1588. 1588. 1588. 1587.
1584. 1584. 1594. 1618. 1648. 1679. 1696. 1714. 1715. 1714.
```

9.6

Program

```
      DIMENSION Y(150), W(4), YB(150), T(150)
      DATA N/150/
      READ 1, Y
    1 FORMAT(10F5.0)
      DO 2 I = 1,N
    2 T(I) = Y(I)
    3 READ 4, WZ, W
    4 FORMAT(5F12.8)
      READ 5, M
    5 FORMAT(I3)
      DO 6 I = 1,4
    6 YB(I) = Y(I)
      L = N - 3
      DO 7 I = L,N
    7 YB(I) = Y(I)
    8 L = N - 4
      DO 9 I = 5,L
      YB(I) = WZ*Y(I)
      DO 9 J = 1,4
      IMJ = I - J
      IPJ = I + J
    9 YB(I) = YB(I) + W(J)*(Y(IMJ) + Y(IPJ))
      IF(M.EQ.1) GO TO 10
      DO 11 I = 1,N
   11 Y(I) = YB(I)
      M = M - 1
      GO TO 8
   10 PRINT 12, YB
   12 FORMAT(1H1/(1H ,10F6.0))
      DO 13 I = 1,N
   13 Y(I) = T(I)
      GO TO 3
      END
```

9.8

```
 287.  311.  335.  354.  368.  375.  377.  375.  372.  368.
 364.  358.  350.  341.  337.  345.  367.  398.  426.  442.
 447.  444.  440.  438.  439.  443.  449.  459.  473.  491.
 517.  550.  592.  639.  685.  723.  749.  766.  775.  780.
 779.  774.  765.  750.  729.  701.  671.  647.  638.  646.
 674.  724.  792.  868.  934.  979.  997.  995.  982.  965.
 948.  931.  917.  906.  899.  895.  896.  899.  907.  920.
 940.  967.  999. 1033. 1064. 1091. 1111. 1123. 1128. 1128.
1123. 1116. 1107. 1097. 1087. 1078. 1073. 1073. 1078. 1085.
1094. 1106. 1118. 1130. 1142. 1156. 1174. 1192. 1203. 1209.
1216. 1230. 1248. 1262. 1268. 1269. 1275. 1291. 1318. 1349.
1376. 1393. 1401. 1404. 1403. 1401. 1404. 1417. 1442. 1470.
1489. 1493. 1488. 1485. 1490. 1504. 1523. 1543. 1562. 1577.
1586. 1590. 1591. 1590. 1589. 1588. 1588. 1588. 1589. 1587.
1584. 1585. 1595. 1618. 1648. 1677. 1696. 1714. 1715. 1714.
```

9.9

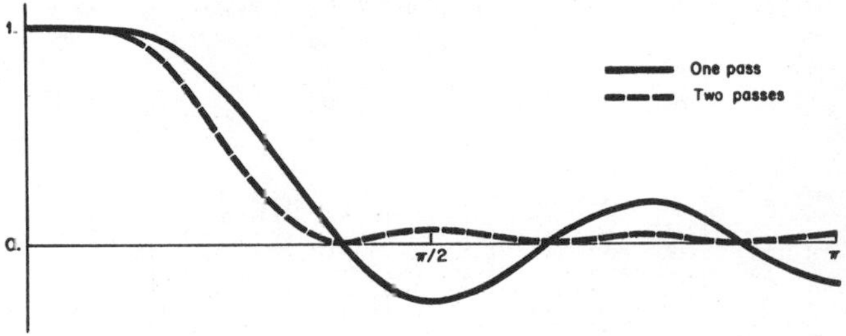

1.
0.
π/2
π
One pass
Two passes

9.10

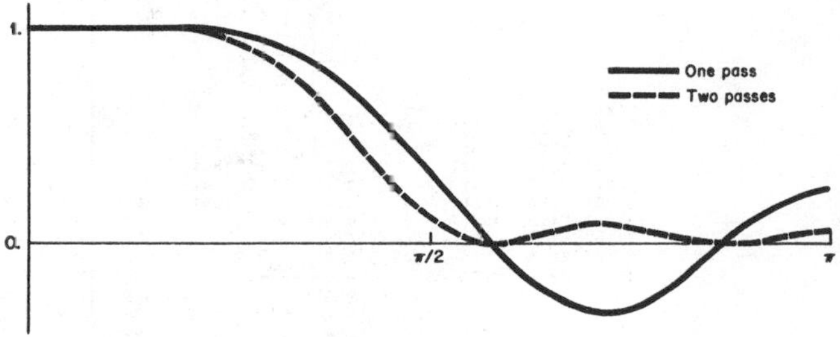

1.
0.
π/2
π
One pass
Two passes

9.11

	Un-Normalized	Normalized
W_0 =	0.16666667	0.18413213
W_1 =	0.15915494	0.17583322
W_2 =	0.13783222	0.15227604
W_3 =	0.10610329	0.11722214
W_4 =	0.06891611	0.07613802
W_5 =	0.03183099	0.03516665
W_6 =	0.00000000	0.00000000
W_7 =	-0.02273642	-0.02511903
W_8 =	-0.03445806	-0.03806901
W_9 =	-0.03536776	-0.03907404
W_{10} =	-0.02756644	-0.03045520
W_{11} =	-0.01446863	-0.01598484
W_{12} =	0.00000000	0.00000000

9.12 Flowchart

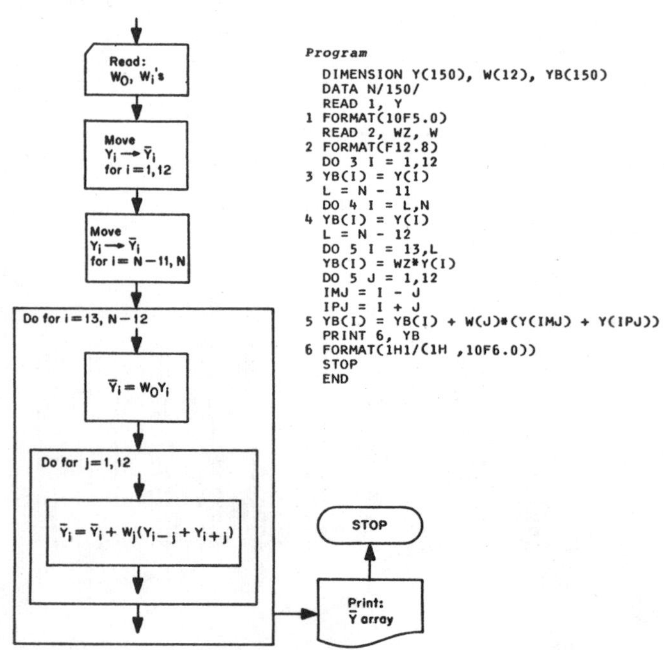

Program

```
  DIMENSION Y(150), W(12), YB(150)
  DATA N/150/
  READ 1, Y
1 FORMAT(10F5.0)
  READ 2, WZ, W
2 FORMAT(F12.8)
  DO 3 I = 1,12
3 YB(I) = Y(I)
  L = N - 11
  DO 4 I = L,N
4 YB(I) = Y(I)
  L = N - 12
  DO 5 I = 13,L
  YB(I) = WZ*Y(I)
  DO 5 J = 1,12
  IMJ = I - J
  IPJ = I + J
5 YB(I) = YB(I) + W(J)*(Y(IMJ) + Y(IPJ))
  PRINT 6, YB
6 FORMAT(1H1/(1H ,10F6.0))
  STOP
  END
```

9.16

```
 287.  294.  306.  321.  335.  347.  356.  362.  365.  366.
 365.  363.  360.  355.  349.  347.  353.  366.  385.  402.
 415.  424.  429.  431.  434.  436.  440.  446.  454.  465.
 481.  501.  528.  562.  599.  636.  670.  699.  721.  739.
 752.  758.  760.  757.  749.  736.  715.  695.  678.  668.
 671.  685.  718.  762.  815.  864.  905.  931.  946.  952.
 951.  945.  937.  927.  919.  912.  907.  905.  905.  910.
 920.  933.  953.  977. 1003. 1029. 1054. 1075. 1091. 1102.
1108. 1111. 1109. 1106. 1100. 1093. 1087. 1083. 1081. 1083.
1086. 1092. 1100. 1109. 1119. 1130. 1142. 1158. 1173. 1183.
1192. 1204. 1217. 1230. 1243. 1251. 1257. 1268. 1282. 1303.
1325. 1346. 1363. 1374. 1383. 1389. 1394. 1399. 1411. 1431.
1449. 1462. 1469. 1474. 1479. 1486. 1497. 1511. 1527. 1541.
1555. 1565. 1573. 1578. 1581. 1583. 1585. 1586. 1587. 1587.
1586. 1586. 1588. 1596. 1612. 1633. 1652. 1671. 1684. 1693.
```

9.17

```
 287.  304.  326.  345.  361.  372.  375.  375.  373.  369.
 366.  360.  354.  346.  340.  341.  358.  387.  416.  435.
 443.  443.  441.  438.  440.  441.  447.  455.  468.  484.
 508.  537.  574.  620.  666.  706.  736.  757.  769.  777.
 780.  775.  767.  755.  738.  715.  682.  658.  644.  645.
 668.  702.  767.  836.  908.  956.  987.  990.  983.  971.
 955.  938.  924.  911.  903.  897.  897.  898.  904.  916.
 934.  954.  987. 1020. 1050. 1079. 1101. 1116. 1125. 1127.
1124. 1118. 1110. 1101. 1091. 1082. 1076. 1074. 1077. 1083.
1090. 1101. 1113. 1125. 1138. 1150. 1164. 1186. 1201. 1204.
1210. 1225. 1241. 1256. 1266. 1268. 1270. 1286. 1308. 1337.
1366. 1385. 1397. 1400. 1402. 1403. 1405. 1410. 1430. 1462.
1482. 1490. 1486. 1486. 1489. 1499. 1516. 1535. 1555. 1570.
1581. 1587. 1590. 1589. 1590. 1589. 1587. 1588. 1589. 1587.
1586. 1586. 1589. 1609. 1638. 1668. 1688. 1706. 1712. 1713.
```

CHAPTER 10

10.1

x	Derivative	Error	x	Derivative	Error
0.00	0.99958338	0.00041662	1.00	0.54007721	0.00022510
0.05	0.99833417	0.00041609	1.05	0.49736375	0.00020730
0.10	0.99458963	0.00041454	1.10	0.45340714	0.00018898
0.15	0.98835914	0.00041194	1.15	0.40831726	0.00017018
0.20	0.97965827	0.00040831	1.20	0.36220679	0.00015096
0.25	0.96850876	0.00040366	1.25	0.31519099	0.00013137
0.30	0.95493848	0.00039801	1.30	0.26738739	0.00011144
0.35	0.93898135	0.00039136	1.35	0.21891545	0.00009124
0.40	0.92067727	0.00038372	1.40	0.16989633	0.00007081
0.45	0.90007197	0.00037513	1.45	0.12045257	0.00005020
0.50	0.87721695	0.00036561	1.50	0.07070773	0.00002947
0.55	0.85216934	0.00035518	1.55	0.02078616	0.00000867
0.60	0.82499177	0.00034385	1.60	-0.02918736	-0.00001216
0.65	0.79575214	0.00033166	1.65	-0.07908793	-0.00003296
0.70	0.76452354	0.00031865	1.70	-0.12879081	-0.00005368
0.75	0.73138404	0.00030483	1.75	-0.17817179	-0.00007427
0.80	0.69641645	0.00029026	1.80	-0.22710744	-0.00009466
0.85	0.65970819	0.00027496	1.85	-0.27547543	-0.00011482
0.90	0.62135100	0.00025897	1.90	-0.32315488	-0.00013469
0.95	0.58144075	0.00024234	1.95	-0.37002661	-0.00015422
			2.00	-0.41597346	-0.00017338

10.2 $\varepsilon = 0.00041667 \cos \xi$

9.13

287.	311.	335.	354.	367.	377.	376.	375.	372.	368.
364.	357.	343.	343.	348.	359.	373.	390.	406.	420.
431.	437.	439.	437.	433.	430.	432.	440.	456.	482.
515.	555.	599.	645.	689.	731.	766.	792.	809.	815.
810.	794.	768.	735.	699.	666.	639.	625.	627.	647.
683.	734.	794.	856.	914.	963.	997.	1014.	1016.	1004.
982.	955.	927.	903.	884.	873.	870.	876.	889.	909.
935.	966.	1000.	1035.	1068.	1097.	1120.	1136.	1144.	1145.
1139.	1128.	1114.	1100.	1087.	1076.	1069.	1065.	1067.	1072.
1081.	1093.	1108.	1123.	1139.	1155.	1171.	1186.	1199.	1211.
1221.	1229.	1238.	1248.	1260.	1274.	1291.	1310.	1328.	1346.
1363.	1377.	1390.	1401.	1411.	1420.	1428.	1436.	1445.	1454.
1464.	1475.	1485.	1497.	1508.	1521.	1534.	1547.	1561.	1573.
1585.	1593.	1598.	1599.	1596.	1590.	1583.	1577.	1590.	1586.
1585.	1586.	1591.	1617.	1650.	1681.	1696.	1714.	1715.	1714.

9.14

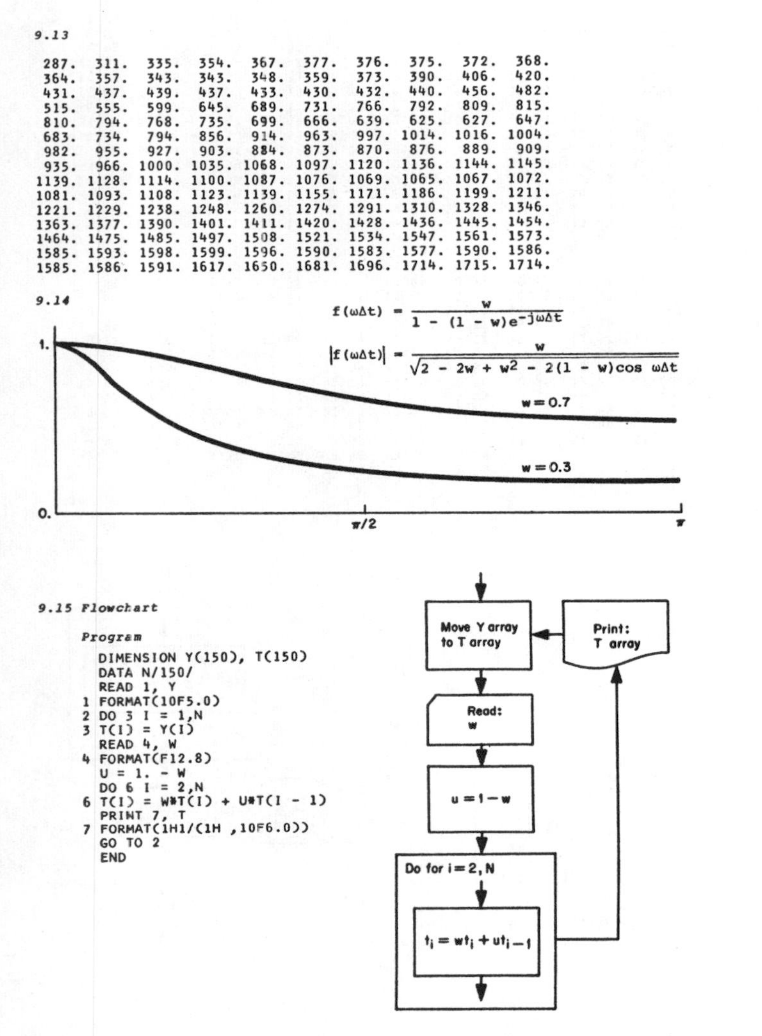

9.15 Flowchart

Program

```
  DIMENSION Y(150), T(150)
  DATA N/150/
  READ 1, Y
1 FORMAT(10F5.0)
2 DO 3 I = 1,N
3 T(I) = Y(I)
  READ 4, W
4 FORMAT(F12.8)
  U = 1. - W
  DO 6 I = 2,N
6 T(I) = W*T(I) + U*T(I - 1)
  PRINT 7, T
7 FORMAT(1H1/(1H ,10F6.0))
  GO TO 2
  END
```

10.3

V	2.0	2.4	2.8	3.2	3.6	4.0	4.4	4.8	5.2	5.6	6.0	6.4
P'_{comp}	-8.54	-5.25	-4.50	-2.63	-3.25	-1.79	-0.79	-2.04	-1.04	-0.21	-0.96	-0.50
Error	-1.38	-1.12	0.13	-0.53	0.88	-0.05	-0.66	0.86	0.07	-0.60	0.28	-0.09

10.4

V	2.4	2.8	3.2	3.6	4.0	4.4	4.8	5.2	5.6	6.0	6.4	6.8
P'_{comp}	-5.67	-4.25	-3.13	-2.75	-1.88	-1.29	-1.54	-1.04	-0.54	-0.71	-0.58	-0.50
Error	-0.70	-0.12	-0.03	0.38	0.04	-0.16	0.36	0.07	-0.27	0.03	-0.01	0.00

10.5 $f'_k = (f_{k-2} - 8f_{k-1} + 8f_{k+1} - f_{k+2})/(12h)$

$\epsilon = h^4 f^V(\xi)/30$

10.6 Same as answers to Exercise 10.5

10.7 Flowchart

Program

```
  DIMENSION V(15), P(15), PP(15), PPC(15), E(15)
  READ 1, (V(I),P(I),PP(I), I = 1,15)
1 FORMAT(3F12.8)
2 READ 1, C1, C2
  DO 3 I = 3,13
  PPC(I) = C1*(P(I + 1) - P(I - 1))
 1        + C2*(P(I - 2) - P(I - 2))
3 E(I) = PP(I) - PPC(I)
  PRINT 4, (V(I), I = 3,13)
4 FORMAT(1H ,11F6.1)
  PRINT 5, (PPC(I), I = 3,13)
5 FORMAT(1H ,11F6.2)
  PRINT 5, (E(I), I = 3,13)
  GO TO 2
  END
```

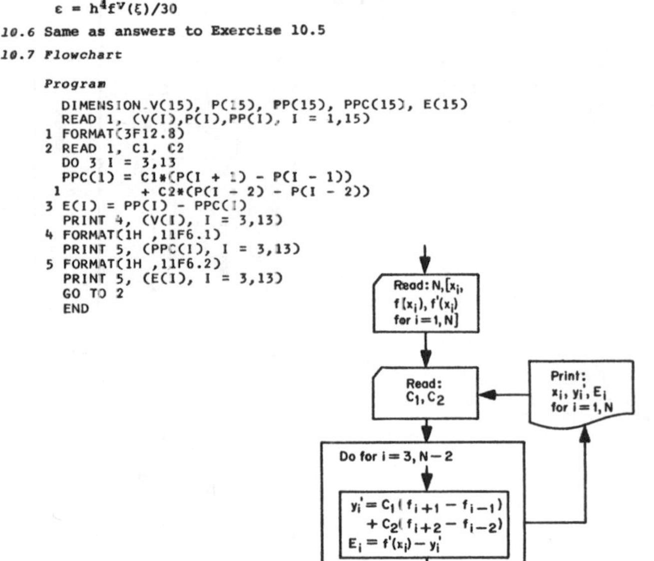

10.8

V	2.8	3.2	3.6	4.0	4.4	4.8	5.2	5.6	6.0	6.4	6.8
P'_{comp}	-4.15	-3.19	-2.62	-2.00	-1.27	-1.42	-1.17	-0.54	-0.63	-0.65	-0.48
Error	-0.22	0.03	0.25	0.16	-0.18	0.24	0.20	-0.17	-0.05	0.06	-0.02

10.9

V	2.8	3.2	3.6	4.0	4.4	4.8	5.2	5.6	6.0	6.4	6.8
P'_{comp}	-4.50	-3.40	-2.68	-2.00	-1.63	-1.28	-1.03	-0.83	-0.63	-0.58	-0.55
Error	0.13	0.24	0.31	0.16	0.18	0.10	0.06	0.02	-0.05	-0.01	0.05

10.10 Answers are the same as those of Exercise 10.8

10.11 Flowchart

Read: N, Y array

Read: C_1, C_2, C_3

Print: Y_i' for $i = 4, N-3$

Do for $i = 4, N-3$

$Y_i' = C_1(Y_{i+1} - Y_{i-1}) + C_2(Y_{i+2} - Y_{i-2}) + C_3(Y_{i+3} - Y_{i-3})$

Program

```
  DIMENSION Y(150), D(150)
  READ 1, Y
1 FORMAT(10F5.0)
2 READ 3, C1, C2, C3
3 FORMAT(3F12.8)
  DO 4 I = 4,147
4 D(I) = C1*(Y(I + 1) - Y(I - 1))
 1        + C2*(Y(I + 2) - Y(I - 2))
 2        + C3*(Y(I + 3) - Y(I - 3))
  PRINT 5, (D(I), I = 4,147)
5 FORMAT(1H1,20X,7F7.2/(1H ,10F7.2))
  GO TO 2
  END
```

10.12

```
                     15.39  10.61   5.79   1.79  -1.11  -3.25  -4.14
 -5.25  -5.96  -5.32  -1.46   5.43  13.18  19.29  21.18  18.21  11.71
  4.93   0.75  -0.79   0.32   2.57   5.43   8.71  12.64  17.57  23.04
 29.82  35.93  40.14  40.64  37.61  31.00  22.86  15.29   8.00   2.32
 -2.71  -7.50 -12.54 -18.21 -21.86 -23.07 -19.79 -10.96   1.89  20.00
 37.11  52.11  59.46  58.32  48.11  31.07  14.29  -0.54  -9.57 -14.29
-15.00 -14.11 -12.04  -8.71  -5.29  -1.79   2.36   6.79  11.29  17.00
 22.75  27.00  29.39  29.46  27.04  21.68  15.86   9.68   3.71  -1.21
 -4.79  -7.21  -8.61  -8.75  -7.64  -5.21  -2.00   1.39   4.79   7.75
  9.71  11.04  11.86  12.68  14.29  15.11  14.18  12.46  11.57  11.07
 10.89  12.11  12.00   9.75   8.54   9.61  13.57  18.89  22.96  23.36
 19.29  13.68   7.82   3.79   2.39   4.71  10.79  15.82  17.89  16.14
 11.89   6.07   2.57   4.43   8.75  13.64  16.29  16.89  14.93  11.36
  7.43   4.00   1.54  -0.07  -0.54  -0.29  -0.32  -0.57  -0.43   0.11
  3.18   8.86  15.93  20.79  23.50  21.86  16.21
```

10.13

```
                     17.14  10.61   4.42  -0.35  -3.05  -3.06  -4.53
 -5.06  -6.35  -8.04  -7.88  -0.40  14.93  28.81  31.87  22.49   9.38
 -0.13  -4.11  -2.54  -0.65   2.18   5.04   7.94  11.87  16.02  22.06
 28.65  37.68  45.00  48.23  42.66  32.17  21.11  12.17   7.81   2.13
 -3.10  -8.28 -12.34 -15.69 -24.97 -30.65 -28.73 -17.96   1.12  16.50
 39.05  59.11  74.83  74.07  56.27  31.65   5.54  -8.31 -15.79 -16.42
-16.94 -16.05 -12.81  -9.49  -5.09  -1.01   1.38   5.23  11.09  16.22
 23.14  29.92  33.67  33.55  28.40  23.62  16.44   8.51   2.16  -2.58
 -6.34  -8.58  -9.58 -10.11  -9.39  -7.16  -2.58   2.75   5.95   7.94
 10.30  12.01  13.02  12.10  11.37  16.08  19.82  15.96   7.10   3.88
 10.89  17.36  17.06  10.33   2.31   2.22  10.07  22.78  30.74  29.19
 22.98  11.93   3.74   0.48  -0.13   0.44   4.95  18.74  29.56  26.84
  9.95  -3.26  -4.23   0.93   9.92  16.17  20.76  20.00  16.48  11.94
  6.65   2.25  -0.21  -0.85  -1.12  -0.87   0.26   0.60  -0.82  -2.03
 -2.27   4.38  16.12  29.15  31.47  25.55  16.41
```

11.5 Flowchart

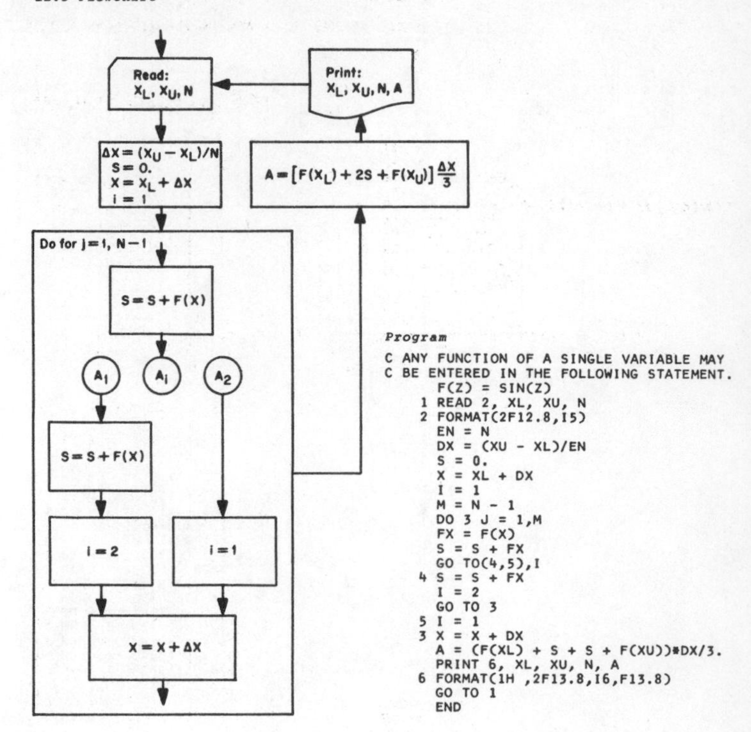

Program

```
C ANY FUNCTION OF A SINGLE VARIABLE MAY
C BE ENTERED IN THE FOLLOWING STATEMENT.
      F(Z) = SIN(Z)
    1 READ 2, XL, XU, N
    2 FORMAT(2F12.8,I5)
      EN = N
      DX = (XU - XL)/EN
      S = 0.
      X = XL + DX
      I = 1
      M = N - 1
      DO 3 J = 1,M
      FX = F(X)
      S = S + FX
      GO TO(4,5),I
    4 S = S + FX
      I = 2
      GO TO 3
    5 I = 1
    3 X = X + DX
      A = (F(XL) + S + S + F(XU))*DX/3.
      PRINT 6, XL, XU, N, A
    6 FORMAT(1H ,2F13.8,I6,F13.8)
      GO TO 1
      END
```

11.6 **N = 10, A = 1.0000034; N = 100, A = 0.99999978; N = 1000, A = 0.99999854 Exact answer is 1.00000000.**

11.7 **N = 10, A = 0.80436593; N = 100, A = 0.80436577; N = 1000, A = 0.80436283 Exact answer is 0.80436610**

10.14 $\varepsilon = -7h^2y'''(\xi)/6$

10.15 $\varepsilon = 131h^4y^{v}(\xi)/630$

10.16

V	P'_{comp}	*Error*	*V*	P'_{comp}	*Error*	*V*	P'_{comp}	*Error*
2.0	-9.38	-0.54	4.0	-1.79	-0.05	6.0	-0.68	0.00
2.4	-6.18	-0.19	4.4	-1.43	-0.02	6.4	-0.57	-0.02
2.8	-4.35	-0.02	4.8	-1.17	-0.01	6.8	-0.49	-0.01
3.2	-3.15	-0.01	5.2	-0.92	-0.05	7.2	-0.43	-0.01
3.6	-2.37	0.00	5.6	-0.78	-0.03	7.6	-0.37	-0.02

CHAPTER 11

11.1 Flowchart

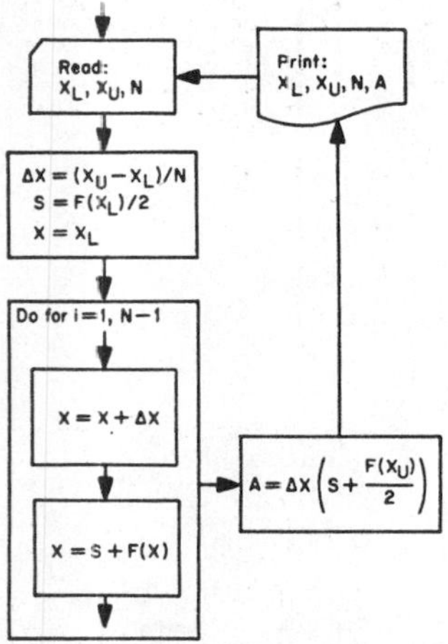

Read: X_L, X_U, N

$\Delta X = (X_U - X_L)/N$
$S = F(X_L)/2$
$X = X_L$

Do for i = 1, N − 1

$X = X + \Delta X$

$X = S + F(X)$

$A = \Delta X\left(S + \frac{F(X_U)}{2}\right)$

Print: X_L, X_U, N, A

Program

```
C ANY FUNCTION OF A SINGLE VARIABLE MAY
C BE ENTERED IN THE FOLLOWING STATEMENT.
      F(Z) = SIN(Z)
    1 READ 2, XL, XU, N
    2 FORMAT(2F12.8,I5)
      EN = N
      DX = (XU - XL)/EN
      S = F(XL)/2.
      X = XL
      M = N - 1
      DO 3 I = 1,M
      X = X + DX
    3 S = S + F(X)
      A = DX*(S + F(XU)/2.)
      PRINT 4, XL, XU, N, A
    4 FORMAT(1H ,2F13.8, I6, F13.8)
      GO TO 1
      END
```

11.2 N = 10, A = 1.9835235; N = 100, A = 1.9998355;
N = 1000, A = 1.9999987
Exact answer is 2.0000000

11.3 N = 2, A = 1.6857749; N = 5, A = 1.6857504;
N = 10, A = 1.6857503
Exact answer is 1.6857503548

11.4 N = 10, A = 0.80444460; N = 100, A = 0.80436666;
N = 1000, A = 0.80436412
Exact answer is 0.80436610
The errors in the integral have nearly odd symmetry
($\varepsilon_i = \varepsilon_{-i}$) about $x = \pi/4$; therefore, in Exercise 11.3,
the errors almost cancel each other.

11.8 Flowchart

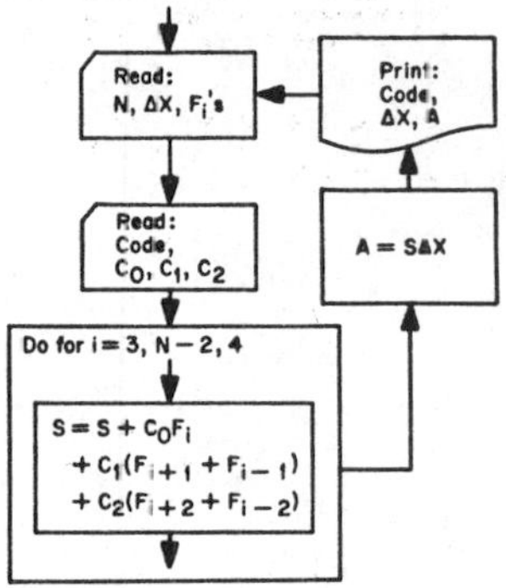

Read: N, ΔX, F_i's

Read: Code, C_0, C_1, C_2

Do for i = 3, N − 2, 4

$S = S + C_0F_i + C_1(F_{i+1} + F_{i-1}) + C_2(F_{i+2} + F_{i-2})$

A = SΔX

Print: Code, ΔX, A

Program

```
      DIMENSION F(25)
    1 READ 2, N, DX, (F(I), I = 1,N)
    2 FORMAT(I4,F12.8/(5F12.8))
      READ 3, CODE, CZ, C1, C2
    3 FORMAT(A6,3F12.8)
      S = 0.
      M = N - 2
      DO 4 I = 3,M,4
    4 S = S + CZ*F(I) + C1*(F(I + 1) + F(I - 1))
     1       + C2*(F(I + 2) + F(I - 2))
      A = S*DX
      PRINT 5, CODE, DX, A
    5 FORMAT(1H ,A6,2F13.8)
      GO TO 1
      END
```

11.9 A = 9508.02

11.10 A = 9505.56

11.11 A = 9499.72

11.12 A = 9474.64

11.13 A = 9507.34

11.14 A = 9506.56

11.15 Flowchart

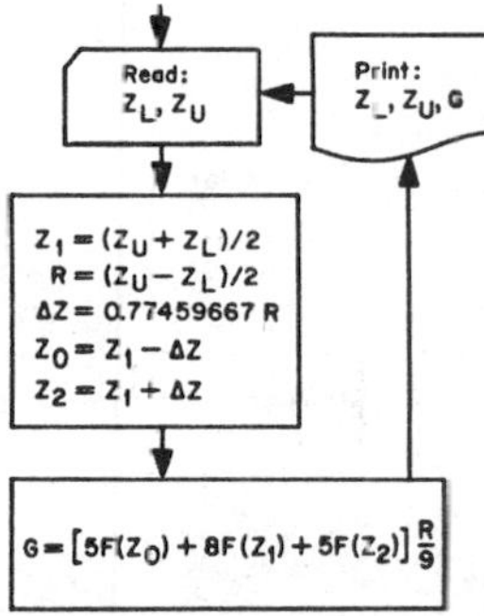

Read: Z_L, Z_U

$Z_1 = (Z_U + Z_L)/2$
$R = (Z_U - Z_L)/2$
$\Delta Z = 0.77459667\,R$
$Z_0 = Z_1 - \Delta Z$
$Z_2 = Z_1 + \Delta Z$

$G = [5F(Z_0) + 8F(Z_1) + 5F(Z_2)]\frac{R}{9}$

Print: Z_L, Z_U, G

Program

```
C ANY FUNCTION OF A SINGLE VARIABLE MAY
C BE ENTERED IN THE FOLLOWING STATEMENT.
      F(U) = 1./SQRT(1. - .25*SIN(U)**2)
    1 READ 2, ZL, ZU
    2 FORMAT(2F12.8)
      Z1 = (ZU + ZL)/2.
      R = (ZU - ZL)/2.
      DZ = .77459667*R
      ZZ = Z1 - DZ
      Z2 = Z1 + DZ
      G = (5.*F(ZZ) + 8.*F(Z1) + 5.*F(Z2))*R/9.
      PRINT 3, ZL, ZU, G
    3 FORMAT(3F13.8)
      GO TO 1
      END
```

11.16 G = 1.68561681; exact answer is 1.68575036

11.17 G = 0.80436618; exact answer is 0.80436610

CHAPTER 12

12.1 $\theta(t) = 1. - 4.72167532\ t^2 + 2.38582544\ t^4 + 3.02665306\ t^6$

$\dot{\theta}(t) = -9.44335063\ t + 9.54330175\ t^3 + 18.1599184\ t^5$

12.2 Flowchart

Program

```
C ANY FUNCTION OF TWO VARIABLES MAY
C REPLACE THE FOLLOWING FUNCTION.
      F(U,V) = V + SQRT(V)
    1 READ 2, N, DX, XZ, Y, XF
    2 FORMAT(I5,4F12.8)
      PRINT 3, DX
    3 FORMAT(10H1DELTA X=,F7.4//)
      EN = N
      DXN = DX*EN
    4 PRINT 5, XZ, Y
    5 FORMAT(1H ,1P2E15.7)
      IF(XZ.GE.XF) GO TO 1
      X = XZ
      DO 6 I = 1,N
      Y = Y + F(X,Y)*DX
    6 X = X + DX
      XZ = XZ + DXN
      GO TO 4
      END
```

Read: N, ΔX, X_O, Y, X_F
Top of page. Print heading: ΔX
Print: X_O, Y
$X_O = X_O + N\Delta X$
$X_O : X_F$
≥
<
Do for i = 1, N
$Y = Y + Y'\Delta X$
$X = X + \Delta X$
$X = X_O$

12.3

x	$Y_{(\Delta x = 0.100)}$	$Y_{(\Delta x = 0.010)}$	$Y_{(\Delta x = 0.001)}$	Y_{exact}
0.0	1.0000000	1.0000000	1.0000000	1.0000000
0.2	1.4295445	1.4610747	1.4645384	1.4649274
0.4	1.9913481	2.0718072	2.0806898	2.0816877
0.6	2.7174535	2.8701942	2.8871356	2.8890400
0.8	3.6470468	3.9031250	3.9316562	3.9348649
1.0	4.8280017	5.2283947	5.2732005	5.2782422

12.5

x	$Y_{(\Delta x = 0.100)}$	$Y_{(\Delta x = 0.010)}$	$Y_{(\Delta x = 0.001)}$	Y_{exact}
0.0	1.0000000	1.0000000	1.0000000	1.0000000
0.2	1.4656789	1.4649348	1.4649272	1.4649274
0.4	2.0836011	2.0817068	2.0816874	2.0816877
0.6	2.8926652	2.8890760	2.8890392	2.8890400
0.8	3.9409340	3.9349252	3.9348637	3.9348649
1.0	5.2877220	5.2783364	5.2782400	5.2782422

12.6 Flowchart

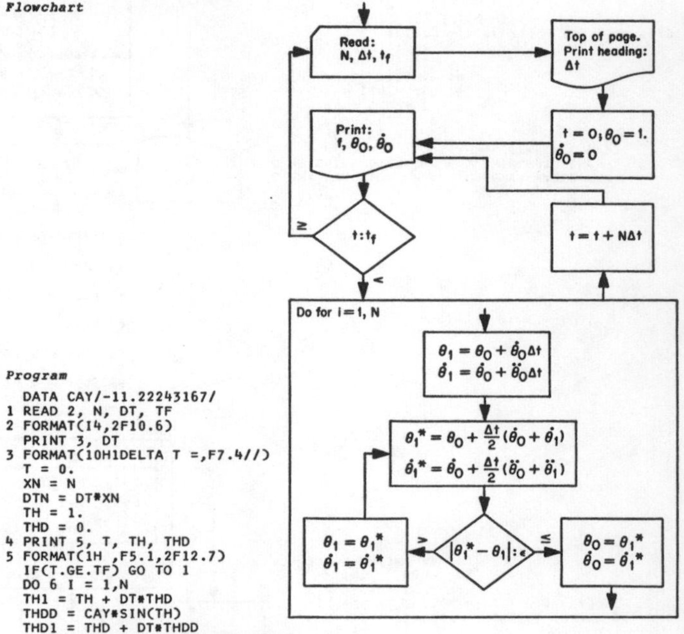

Program

```
  DATA CAY/-11.22243167/
1 READ 2, N, DT, TF
2 FORMAT(I4,2F10.6)
  PRINT 3, DT
3 FORMAT(10H1DELTA T =,F7.4//)
  T = 0.
  XN = N
  DTN = DT*XN
  TH = 1.
  THD = 0.
4 PRINT 5, T, TH, THD
5 FORMAT(1H ,F5.1,2F12.7)
  IF(T.GE.TF) GO TO 1
  DO 6 I = 1,N
  TH1 = TH + DT*THD
  THDD = CAY*SIN(TH)
  THD1 = THD + DT*THDD
7 TH1S = TH + DT*(THD + THD1)/2.
  THD1S = THD + DT*(THDD + CAY*SIN(TH1))/2.
  IF(ABS(TH1S - TH1).LE..0000001) GO TO 8
  TH1 = TH1S
  THD1 = THD1S
  GO TO 7
8 TH = TH1S
6 THD = THD1S
  T = T + DTN
  GO TO 4
  END
```

12.4 Flowchart

Program

```
C ANY FUNCTION OF TWO VARIABLES MAY
C REPLACE THE FOLLOWING FUNCTION.
      F(U,V) = V + SQRT(V)
    1 READ 2, N, DX, XZ, Y, XF
    2 FORMAT(I5,4F12.8)
      PRINT 3, DX
    3 FORMAT(10H1DELTA X =,F7.4//)
      EN = N
      DXN = DX*EN
    4 PRINT 5, XZ, Y
    5 FORMAT(1H ,1P2E15.7)
      IF(XZ.GE.XF) GO TO 1
      X = XZ
      DO 6 I = 1,N
      FXY = F(X,Y)
      Y1 = Y + FXY*DX
      X1 = X + DX
    7 Y1S = Y + (FXY + F(X1,Y1))*DX/2.
      IF(ABS(Y1S - Y1).LE..0000001) GO TO 8
      Y1 = Y1S
      GO TO 7
    8 Y = Y1S
    6 X = X1
      XZ = XZ + DXN
      GO TO 4
      END
```

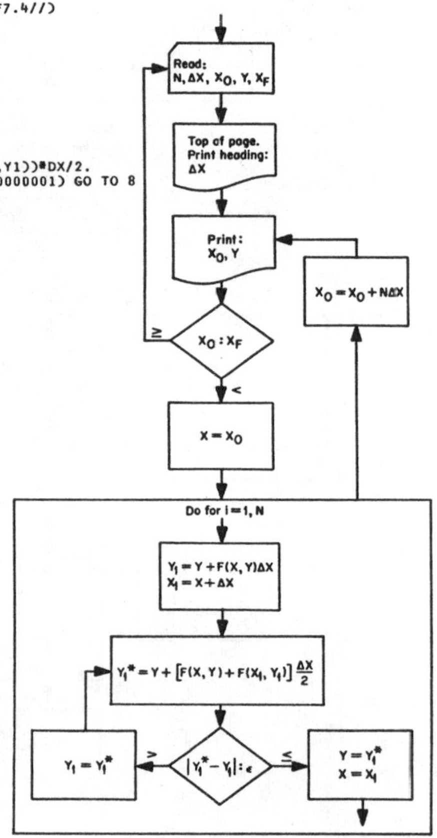

Read: N, ΔX, X_O, Y, X_F

Top of page. Print heading: ΔX

Print: X_O, Y

$X_O : X_F$ (≥, <)

$X_O = X_O + N\Delta X$

$X = X_O$

Do for i = 1, N

$Y_1 = Y + F(X,Y)\Delta X$
$X_1 = X + \Delta X$

$Y_1^* = Y + [F(X,Y) + F(X_1, Y_1)] \frac{\Delta X}{2}$

$|Y_1^* - Y_1| : \epsilon$ (>, ≤)

$Y_1 = Y_1^*$

$Y = Y_1^*$
$X = X_1$

12.7

t	Δt = 0.100 θ	Δt = 0.100 $\dot{\theta}$	Δt = 0.010 θ	Δt = 0.010 $\dot{\theta}$	Δt = 0.001 θ	Δt = 0.001 $\dot{\theta}$
0.0	1.0000000	0.0000000	1.0000000	0.0000000	1.0000000	0.0000000
0.1	0.9527832	-0.9294992	0.9530296	-0.9345638	0.9530249	-0.9346130
0.2	0.8165204	-1.7957581	0.8151541	-1.8068337	0.8151335	-1.8069392
0.3	0.6006417	-2.5218155	0.5963555	-2.5383900	0.5963063	-2.5385462
0.4	0.3236802	-3.0174148	0.3153587	-3.0349024	0.3152697	-3.0350622
0.5	0.0126601	-3.2029876	0.0001303	-3.2120720	0.0000003	-3.2121406
0.6	-0.2997103	-3.0444237	-0.3151209	-3.0351755	-0.3152692	-3.0350628
0.7	-0.5804826	-2.5710218	-0.5961567	-2.5388841	-0.5963059	-2.5385474
0.8	-0.8020322	-1.8599697	-0.8150126	-1.8074739	-0.8151332	-1.8069405
0.9	-0.9451219	-1.0018247	-0.9529565	-0.9352807	-0.9530247	-0.9346146
1.0	-0.9989702	-0.0751423	-1.0000001	-0.0007400	-1.0000000	-0.0000016

12.8 Flowchart

Program

```
      F1(U) = ((3.02665306*U*U+2.38582544)*U*U-4.72167532)*U*U+1.
      F2(U) = ((18.1599184*U*U+9.54330175)*U*U-9.44335063)*U
      F3(U) = -11.22243167*SIN(U)
      DIMENSION TH(5), THD(5), THDD(5)
    1 READ 2, N, DT, TF
    2 FORMAT(I3,2F10.6)
      EN = N
      DTN = DT*EN
      PRINT 3, DT
    3 FORMAT(10H1DELTA T =,F7.4//)
      T = -4.*DT
      DO 4 I = 1,4
      T = T+DT
      TH(I) = F1(T)
      THD(I) = F2(T)
    4 THDD(I) = F3(TH(I))
    5 PRINT 6, T, TH(4), THD(4)
    6 FORMAT (1H ,F5.1,2F12.7)
      IF(T.GE.TF) GO TO 1
      DO 7 I = 1,N
      TH(5) = TH(4)+(55.*THD(4)-59.*THD(3)+37.*THD(2)-9.*THD(1))*DT/24.
      THD(5) = THD(4)+(55.*THDD(4)-59.*THDD(3)+37.*THDD(2)-9.*THDD(1))
     1         *DT/24.
      Q1 = TH(4)+(19.*THD(4)-5.*THD(3)+THD(2))*DT/24.
      Q2 = THD(4)+(19.*THDD(4)-5.*THDD(3)+THDD(2))*DT/24.
    8 THS = Q1+.375*DT*THD(5)
      THDS = Q2+.375*DT*F3(TH(5))
      IF(ABS(THS-TH(5)).LE..0000001) GO TO 9
      TH(5) = THS
      THD(5) = THDS
      GO TO 8
    9 DO 10 J = 1,3
      TH(J) = TH(J+1)
      THD(J) = THD(J+1)
   10 THDD(J) = THDD(J+1)
      TH(4) = THS
      THD(4) = THDS
    7 THDD(4) = F3(THS)
      T = T+DTN
      GO TO 5
      END
```

12.8 Flowchart

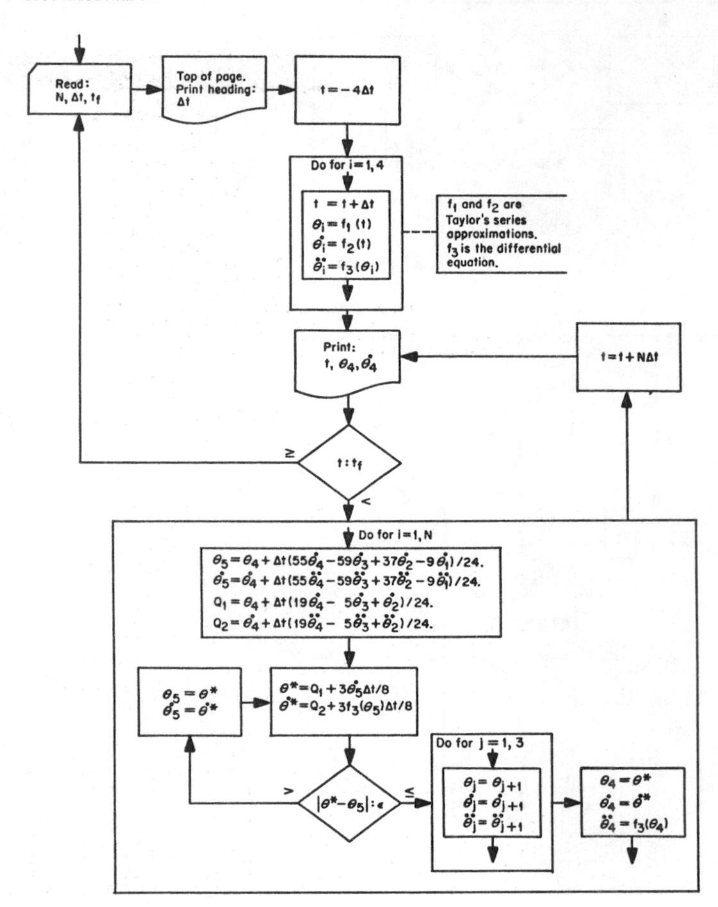

12.10 Flowchart

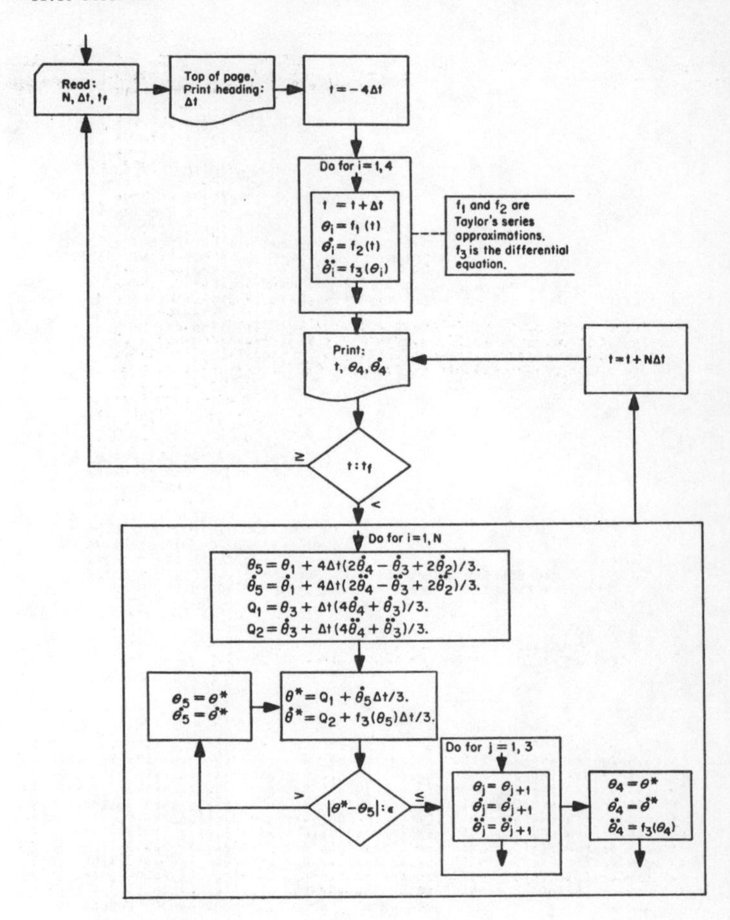

12.9

t	Δt = 0.100 θ	Δt = 0.100 $\dot{\theta}$	Δt = 0.010 θ	Δt = 0.010 $\dot{\theta}$	Δt = 0.001 θ	Δt = 0.001 $\dot{\theta}$
0.0	1.0000000	-0.0000000	1.0000000	0.0000000	1.0000000	-0.0000000
0.1	0.9530272	-0.9341024	0.9530248	-0.9346134	0.9530248	-0.9346135
0.2	0.8152449	-1.8060451	0.8151334	-1.8069401	0.8151333	-1.8069401
0.3	0.5965547	-2.5379228	0.5963058	-2.5385478	0.5963058	-2.5385477
0.4	0.3155354	-3.0355793	0.3152689	-3.0350641	0.3152688	-3.0350642
0.5	0.0000424	-3.2139917	-0.0000010	-3.2121416	-0.0000010	-3.2121418
0.6	-0.3156237	-3.0371116	-0.3152707	-3.0350620	-0.3152708	-3.0350624
0.7	-0.5969741	-2.5392589	-0.5963074	-2.5385440	-0.5963074	-2.5385444
0.8	-0.8158386	-1.8059047	-0.8151345	-1.8069352	-0.8151346	-1.8069357
0.9	-0.9535315	-0.9325850	-0.9530254	-0.9346079	-0.9530255	-0.9346084
1.0	-1.0002397	0.0020973	-1.0000000	0.0000058	-1.0000002	0.0000054

12.10 Flowchart

Program

```
      F1(U) = ((3.02665306*U*U+2.38582544)*U*U-4.72167532)*U*U+1.
      F2(U) = ((18.1599184*U*U+9.54330175)*U*U-9.44335063)*U
      F3(U) = -11.22243167*SIN(U)
      DIMENSION TH(5), THD(5), THDD(5)
    1 READ 2, N, DT, TF
    2 FORMAT(I3,2F10.6)
      EN = N
      DTN = DT*EN
      PRINT 3, DT
    3 FORMAT(10H1DELTA T =,F7.4//)
      T = -4.*DT
      DO 4 I = 1,4
      T = T+DT
      TH(I) = F1(T)
      THD(I) = F2(T)
    4 THDD(I) = F3(TH(I))
    5 PRINT 6, T, TH(4), THD(4)
    6 FORMAT(1H ,F5.1,2F12.7)
      IF(T.GE.TF) GO TO 1
      DO 7 I = 1,N
      TH(5) = TH(1)+(8.*THD(4)-4.*THD(3)+8.*THD(2))*DT/3.
      THD(5) = THD(1)+(8.*THDD(4)-4.*THDD(3)+8.*THDD(2))*DT/3
      Q1 = TH(3)+(4.*THD(4)+THD(3))*DT/3.
      Q2 = THD(3)+(4.*THDD(4)+THDD(3))*DT/3.
    8 THS = Q1+THD(5)*DT/3.
      THDS = Q2+F3(TH(5))*DT/3.
      IF(ABS(THS-TH(5)).LE..0000001) GO TO 9
      TH(5) = THS
      THD(5) = THDS
      GO TO 8
    9 DO 10 J = 1,3
      TH(J) = TH(J+1)
      THD(J) = THD(J+1)
   10 THDD(J) = THDD(J+1)
      TH(4) = THS
      THD(4) = THDS
    7 THDD(4) = F3(THS)
      T = T+DTN
      GO TO 5
      END
```

12.11

t	Δt = 0.100 θ	Δt = 0.100 $\dot{\theta}$	Δt = 0.020 θ	Δt = 0.020 $\dot{\theta}$	Δt = 0.001 θ	Δt = 0.001 $\dot{\theta}$
0.0	1.0000000	-0.0000000	1.0000000	0.0000000	1.0000000	-0.0000000
0.1	0.9530323	-0.9343852	0.9530248	-0.9346135	0.9530248	-0.9346135
0.2	0.8151879	-1.8068220	0.8151333	-1.8069401	0.8151333	-1.8069401
0.3	0.5963589	-2.5385296	0.5963058	-2.5385478	0.5963058	-2.5385478
0.4	0.3153069	-3.0354901	0.3152689	-3.0350640	0.3152688	-3.0350640
0.5	-0.0000780	-3.2126194	-0.0000009	-3.2121414	-0.0000010	-3.2121415
0.6	-0.3154033	-3.0353391	-0.3152707	-3.0350619	-0.3152707	-3.0350620
0.7	-0.5964870	-2.5382944	-0.5963073	-2.5385441	-0.5963073	-2.5385440
0.8	-0.8152361	-1.8064657	-0.8151344	-1.8069354	-0.8151344	-1.8069355
0.9	-0.9530939	-0.9340475	-0.9530253	-0.9346081	-0.9530254	-0.9346082
1.0	-0.9999774	0.0004174	-1.0000000	0.0000055	-1.0000000	0.0000055

12.12 Flowchart

Program

```
      F1(U) = ((3.02665306*U*U+2.38582544)*U*U-4.72167532)*U*U+1.
      F2(U) = ((18.1599184*U*U+9.54330175)*U*U-9.44335063)*U
      F3(U) = -11.22243167*SIN(U)
      DIMENSION TH(5), THD(5), THDD(5)
    1 READ 2, N, DT, TF
    2 FORMAT(I3,2F10.6)
      EN = N
      DTN = DT*EN
      PRINT 3, DT
    3 FORMAT(10H1DELTA T =,F7.4//)
      T = -4.*DT
      DO 4 I = 1,4
      T = T+DT
      TH(I) = F1(T)
      THD(I) = F2(T)
    4 THDD(I) = F3(TH(I))
    5 PRINT 6, T, TH(4), THD(4)
    6 FORMAT(1H ,F5.1,2F12.7)
      IF(T.GE.TF) GO TO 1
      DO 7 I = 1,N
      TH(5) = TH(3)+(8.*THD(4)-5.*THD(3)+4.*THD(2)-THD(1))*DT/3.
      THD(5) = THD(3)+(8.*THDD(4)-5.*THDD(3)+4.*THDD(2)-THDD(1))*DT/3.
      Q1 = (TH(4)+7.*TH(3))/8.+(243.*THD(4)+51.*THD(3)+THD(2))*DT/192.
      Q2 = (THD(4)+7.*THD(3))/8.+(243.*THDD(4)+51.*THDD(3)+THDD(2))
     1     *DT/192.
    8 THS = Q1+65.*DT*THD(5)/192.
      THDS = Q2+65.*DT*F3(TH(5))/192.
      IF(ABS(THS-TH(5)).LE..0000001) GO TO 9
      TH(5) = THS
      THD(5) = THDS
      GO TO 8
    9 DO 10 J = 1,3
      TH(J) = TH(J+1)
      THD(J) = THD(J+1)
   10 THDD(J) = THDD(J+1)
      TH(4) = THS
      THD(4) = THDS
    7 THDD(4) = F3(THS)
      T = T+DTN
      GO TO 5
      END
```

12.12 Flowchart

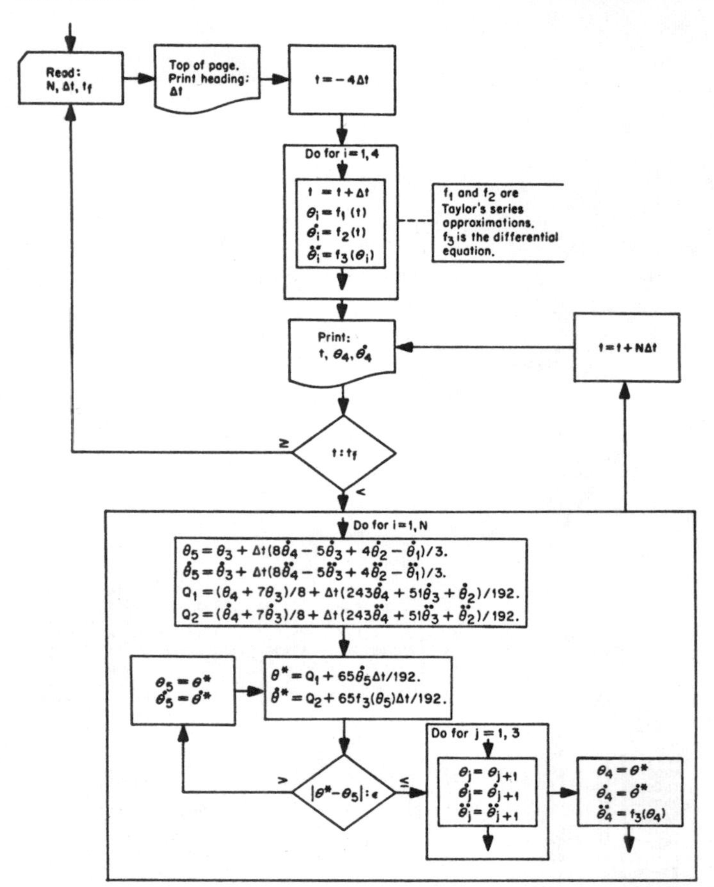

12.14 Flowchart

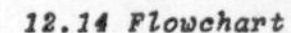

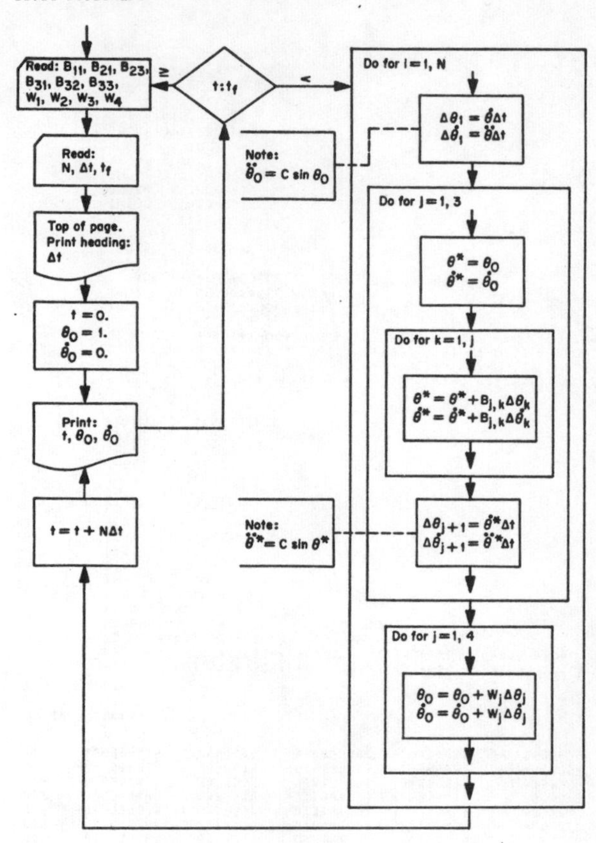

12.13

t	Δt = 0.100 θ	Δt = 0.100 $\dot{\theta}$	Δt = 0.010 θ	Δt = 0.010 $\dot{\theta}$	Δt = 0.001 θ	Δt = 0.001 $\dot{\theta}$
0.0	1.0000000	-0.0000000	1.0000000	0.0000000	1.0000000	-0.0000000
0.1	0.9530316	-0.9343498	0.9530248	-0.9346135	0.9530249	-0.9346135
0.2	0.8151959	-1.8067555	0.8151333	-1.8069402	0.8151334	-1.8069402
0.3	0.5963767	-2.5385119	0.5963058	-2.5385477	0.5963058	-2.5385480
0.4	0.3153196	-3.0355156	0.3152689	-3.0350639	0.3152689	-3.0350641
0.5	-0.0000735	-3.2127649	-0.0000009	-3.2121413	-0.0000010	-3.2121415
0.6	-0.3154351	-3.0354313	-0.3152707	-3.0350619	-0.3152707	-3.0350620
0.7	-0.5965185	-2.5383276	-0.5963073	-2.5385441	-0.5963074	-2.5385442
0.8	-0.8152824	-1.8063614	-0.8151344	-1.8069354	-0.8151345	-1.8069355
0.9	-0.9531048	-0.9339440	-0.9530254	-0.9346081	-0.9530254	-0.9346081
1.0	-0.9999976	0.0005456	-1.0000000	0.0000055	-1.0000001	0.0000056

12.14 Flowchart

Program

```
   DIMENSION B(3,3), W(4), DTH(4), DTHD(4)
 1 READ 2, B(1,1), B(2,1), B(2,2), B(3,1), B(3,2), B(3,3)
 2 FORMAT(6F12.0)
   READ 2, W
   READ 3, N, DT, TF
 3 FORMAT(I4,2F10.6)
   PRINT 4, DT
 4 FORMAT(10H1DELTA T =,F7.4)
   T = 0.
   TH = 1.
   THD = 0.
   XN = N
 5 PRINT 6, T, TH, THD
 6 FORMAT(1H ,F5.1,2F12.7)
   IF(T.GE.TF) GO TO 1
   DO 7 I = 1,N
   DTH(1) = DT*THD
   DTHD(1) = DT*(-11.22243167*SIN(TH))
   DO 8 J = 1,3
   THS = TH
   THDS = THD
   DO 9 K = 1,J
   THS = THS+B(J,K)*DTH(K)
 9 THDS = THDS+B(J,K)*DTHD(K)
   DTH(J+1) = DT*THDS
 8 DTHD(J+1) = DT*(-11.22243167*SIN(THS))
   DO 10 J = 1,4
   TH = TH+W(J)*DTH(J)
10 THD = THD+W(J)*DTHD(J)
 7 CONTINUE
   T = T+XN*DT
   GO TO 5
   END
```

12.15

t	Δt = 0.100 θ	Δt = 0.100 $\dot{\theta}$	Δt = 0.020 θ	Δt = 0.020 $\dot{\theta}$	Δt = 0.002 θ	Δt = 0.002 $\dot{\theta}$
0.	1.0000000	0.0000000	1.0000000	0.0000000	1.0000000	0.0000000
0.1	0.9530262	-0.9345308	0.9530248	-0.9346133	0.9530248	-0.9346135
0.2	0.8151450	-1.8067878	0.8151334	-1.8069399	0.8151333	-1.8069401
0.3	0.5963403	-2.5383630	0.5963059	-2.5385475	0.5963058	-2.5385476
0.4	0.3153410	-3.0349028	0.3152690	-3.0350637	0.3152688	-3.0350640
0.5	0.0001179	-3.2120561	-0.0000007	-3.2121414	-0.0000010	-3.2121415
0.6	-0.3151121	-3.0350869	-0.3152704	-3.0350621	-0.3152708	-3.0350618
0.7	-0.5961325	-2.5387047	-0.5963071	-2.5385447	-0.5963074	-2.5385439
0.8	-0.8149716	-1.8072491	-0.8151342	-1.8069362	-0.8151345	-1.8069353
0.9	-0.9528979	-0.9350764	-0.9530252	-0.9346091	-0.9530254	-0.9346080
1.0	-0.9999244	-0.0006025	-1.0000000	0.0000043	-1.0000000	0.0000057

12.16

t	Δt = 0.100 θ	Δt = 0.100 $\dot{\theta}$	Δt = 0.020 θ	Δt = 0.020 $\dot{\theta}$	Δt = 0.002 θ	Δt = 0.002 $\dot{\theta}$
0.0	1.0000000	0.0000000	1.0000000	0.0000000	1.0000000	0.0000000
0.1	0.9530276	-0.9345594	0.9530248	-0.9346134	0.9530248	-0.9346135
0.2	0.8151500	-1.8068319	0.8151334	-1.8069400	0.8151334	-1.8069403
0.3	0.5963485	-2.5384004	0.5963059	-2.5385475	0.5963059	-2.5385480
0.4	0.3153495	-3.0349171	0.3152691	-3.0350637	0.3152690	-3.0350640
0.5	0.0001234	-3.2120561	-0.0000007	-3.2121414	-0.0000009	-3.2121415
0.6	-0.3151098	-3.0350978	-0.3152704	-3.0350622	-0.3152706	-3.0350619
0.7	-0.5961299	-2.5387348	-0.5963070	-2.5385448	-0.5963072	-2.5385440
0.8	-0.8149653	-1.8072838	-0.8151342	-1.8069364	-0.8151343	-1.8069354
0.9	-0.9528875	-0.9350975	-0.9530252	-0.9346093	-0.9530252	-0.9346082
1.0	-0.9999123	-0.0006007	-1.0000000	0.0000042	-0.9999998	0.0000054

12.17

t	Δt = 0.100 θ	Δt = 0.100 $\dot{\theta}$	Δt = 0.020 θ	Δt = 0.020 $\dot{\theta}$	Δt = 0.002 θ	Δt = 0.002 $\dot{\theta}$
0.0	1.0000000	0.0000000	1.0000000	0.0000000	1.0000000	0.0000000
0.1	0.9530244	-0.9345669	0.9530248	-0.9346134	0.9530248	-0.9346135
0.2	0.8151381	-1.8068529	0.8151334	-1.8069400	0.8151333	-1.8069403
0.3	0.5963262	-2.5384407	0.5963059	-2.5385476	0.5963058	-2.5385477
0.4	0.3153195	-3.0349713	0.3152690	-3.0350639	0.3152689	-3.0350640
0.5	0.0000909	-3.2120984	-0.0000008	-3.2121415	-0.0000010	-3.2121417
0.6	-0.3151418	-3.0350977	-0.3152705	-3.0350622	-0.3152707	-3.0350620
0.7	-0.5961617	-2.5386830	-0.5963072	-2.5385446	-0.5963074	-2.5385439
0.8	-0.8149967	-1.8071913	-0.8151343	-1.8069360	-0.8151345	-1.8069355
0.9	-0.9529151	-0.9349779	-0.9530253	-0.9346089	-0.9530255	-0.9346082
1.0	-0.9999297	-0.0004626	-1.0000000	0.0000047	-1.0000001	0.0000055

12.18

t	Δt = 0.100 θ	Δt = 0.100 $\dot{\theta}$	Δt = 0.020 θ	Δt = 0.020 $\dot{\theta}$	Δt = 0.002 θ	Δt = 0.002 $\dot{\theta}$
0.0	1.0000000	0.0000000	1.0000000	0.0000000	1.0000000	0.0000000
0.1	0.9530229	-0.9345738	0.9530248	-0.9346134	0.9530248	-0.9346135
0.2	0.8151334	-1.8068651	0.8151333	-1.8069401	0.8151334	-1.8069402
0.3	0.5963176	-2.5384521	0.5963058	-2.5385476	0.5963059	-2.5385478
0.4	0.3153086	-3.0349733	0.3152689	-3.0350638	0.3152689	-3.0350640
0.5	0.0000803	-3.2120872	-0.0000008	-3.2121413	-0.0000009	-3.2121415
0.6	-0.3151503	-3.0350779	-0.3152705	-3.0350620	-0.3152707	-3.0350621
0.7	-0.5961682	-2.5386614	-0.5963071	-2.5385444	-0.5963073	-2.5385442
0.8	-0.8150024	-1.8071691	-0.8151342	-1.8069359	-0.8151344	-1.8069356
0.9	-0.9529197	-0.9349520	-0.9530253	-0.9346088	-0.9530254	-0.9346084
1.0	-0.9999317	-0.0004300	-1.0000000	0.0000047	-1.0000001	0.0000053

Index

Page numbers in *italics* refer to flowcharts.